STATISTICAL PHYSICS

STATISTICAL PHYSICS

A Probabilistic Approach

BERNARD H. LAVENDA

Universita di Camerino
Camerino, Italy

A Wiley-Interscience Publication

John Wiley & Sons, Inc.

New York • Chichester • Brisbane • Toronto • Singapore

In recognition of the importance of preserving what has been written, it is a policy of John Wiley & Sons, Inc., to have books of enduring value published in the United States printed on acid-free paper, and we exert our best efforts to that end.

Library of Congress Cataloging in Publication Data:

Lavenda, Bernard H.
 Statistical physics: a probalistic approach/Bernard H. Lavenda.

  p.   cm.
 "A Wiley-Interscience publication."
 Includes bibliographical references.
 ISBN 0-471-54607-0 (acid-free paper): $59.95 (est.)
 1. Statistical physics.   I. Title.

QC174.8.L38   1992
530.1'3 — dc20

                          91-15687
                          CIP

Printed in the United States of America

10 9 8 7 6 5 4 3 2 1

To the memory of my mother, Selma

30 September 1921–10 August 1990

Preface

As the title implies, this is a book on statistical physics, although there will be those who will disagree. There is a single, central theme to this book: "the connection between entropy and probability is through a law of error for extensive thermodynamic variables and Boltzmann's principle is a consequence of it." This is the missing link that removes the ambiguity from Boltzmann's principle and serves as an extremely powerful tool in the analysis of physical phenomena which can be subjected to a statistical analysis. Moreover, it will allow statistical thermodynamics to be viewed from the perspective of the theory of mathematical statistics.

I believe it is a mistake to divorce physical theories from the historical contexts in which they were conceived. In order to be able to appreciate a scientific breakthrough, one has to understand how the discoverer wrestled with the problem with all his doubts and perplexities. Great discoveries do not come from extending existing theories but, rather, when an impasse has been reached in a theory. A theory without definite boundaries, establishing its domain of validity, is never complete. Only contradictory results leading to confusion and individual perseverance in searching for truth give way to discovery.

An exemplary case was Planck's battle with black radiation. Planck was no revolutionary, in the sense of Einstein. He would have been perfectly satisfied to complete his program of giving the second law of thermodynamics an electromagnetic foundation without shaking classical physics to the bone. Planck had no competitors and his only interrogators were experimentalists. Having followed Planck's line of reasoning, it seems incredible how modern texts on statistical mechanics can, so nonchalantly, dissect the final result from the logical structure in which it was derived.

Planck had two pillars to rest upon: indisputable experimental facts and thermodynamics, including Boltzmann's principle. In the study of the origins of the universe and black holes we have few, if any, experimental facts. Reliance must therefore be shifted to theory, especially thermodynamics. It is all the more a shame that thermodynamics should suffer such mutilation, even beyond recognition, by people searching for specific results and taking those results out of the context in which they were derived. Thermodynamics seems one of those lost treasures of the past whose immense potentialities only a handful of historians of science appear capable of appreciating, just as Boltzmann,

Planck, and Einstein did.

It was an exhilarating experience to follow the footsteps of Planck and Einstein and to see exactly where logic gave way to intuition. All this and more could be accomplished with the aid of a single unifying principle cast within the framework of the modern theory of mathematical statistics. Certainly working in an area outside of the mainstream of present-day vogues does lead to problems, especially those with referees. I had one referee asking me "What is this 'Gauss' principle' which forms such an important role in the author's considerations yet appears in none of the texts on my shelf and which the author finds only in a Dover reprint on celestial mechanics?" Does it matter where it comes from, whether it be a Dover reprint or an out-of-print classic like *A Treatise on Probability* by the English economist John Maynard Keynes? This just points to the vacuum in which physicists work and the high degree of patronization that there is in science today.

The question is not "what is this Gauss' principle" but what does it do to clear up paradoxes in statistical physics that have been around for almost a century and new problems which have arisen out of a lack of understanding and confusion about the principles of thermodynamics and statistics. Heavy formalism is not required to get across fundamental ideas. If it can't be said simply, it's not worth being said at all.

The reader should not expect an encyclopedic treatment of the subject material. This approach is valid for encyclopedias but not when it comes to gaining an understanding and appreciation of the basic unity underlying diverse physical phenomena. Einstein used the same tools over and over again to analyze unrelated phenomena. One of his best known "pets" was Brownian motion, which he used so brillantly in the determination of molecular dimensions as well as in the calculation of the motion of molecules in a radiation field.

Each chapter is related to the previous one, and if I have repeated myself on occasion, I ask forgiveness from the reader. Nevertheless, it is always gratifying and reassuring to see an underlying unity in the presentation inasmuch as a variety of different approaches all lead to the same results. Much, but not all, of the work has already appeared in the literature.

Finally, I would like to thank my wife, Fanny, and my daughter, Marlene, for help in proofreading.

Bernard H. Lavenda

Camerino, Italy
February, 1991

Contents

Prologue

Although mathematical and physical statistics have employed many of the same concepts, they have developed largely independently of one another. In mathematical statistics, it is rather easy to obtain good approximate results, for large sample sizes, based on certain limit theorems in the theory of probability. These theorems, like the law of large numbers and the central limit theorem, are extremely elegant and thermodynamics—without realizing it— has captured their elegance.

The concept of probability in statistical mechanics is inherently related to entropy. According to Ludwig Boltzmann, the entropy is proportional to the logarithm of the number of microscopic complexions which are compatible with a given macroscopic state. The greater the number of microscopic complexions, the larger the entropy. And since thermodynamic equilibrium is characterized by maximum entropy, subject to imposed constraints, it is the most molecularly disordered state. Boltzmann argued that "an actual number of objects is a more perspicuous concept than a mere probability" and one "cannot speak of the permutation number of a fraction," which would be a true probability. Thus, Boltzmann introduced the notion of what Max Planck later termed a "thermodynamic" probability, being an extremely large number rather than a proper fraction.

Any attempt to reduce the thermodynamic probability, or the number of complexions, to a genuine probability as the ratio of the state with the greatest number of complexions to the total number of complexions is thwarted by the fact that the maximum of the thermodynamic probability is always of the order of the total number of complexions. Consequently, the logarithm of this ratio would vanish and so too would the entropy. Moreover, the connection between the additivity of the entropy of two subsystems and their statistical independence, which supposedly would be manifested by the thermodynamic probability being a product of the individual thermodynamic probabilities of the subsystems, is not self-evident. The additivity property of the entropy hinges on the fact that the individual subsystems are initially at the same temperature, and hence it cannot be claimed that they are *a priori* statistically independent. It would therefore appear that any reference to probability theory should be abandoned and that we simply *define* the entropy as the logarithm of the number of complexions. Justification of such a definition would then be relegated to *a posteriori* comparison with classical thermodynamic results and

1

not by *a priori* arguments that are open to criticism. However, there would be nothing *absolute* about such a definition of entropy for we could never be sure that our counting procedure really exhausts all the possibilities.

On the basis of the foregoing criticisms, we would conclude that Boltzmann's principle, relating the entropy to the logarithm of the number of microscopic complexions, which are consistent with a given macroscopic state, introduces insurmountable obstacles to providing a probabilistic basis for thermodynamics. Yet, provided we adopt the correct counting procedure, it does give correct thermodynamic results, so there must be something more to it than merely good guessing.

It is not always a simple matter to justify the expression of the thermodynamic probability. Depending on how the counting is done, the particles can appear either as distinguishable or indistinguishable objects. In what circumstances are they to be treated as distinguishable and in what other circumstances as indistinguishable? Almost 100 years have passed and the question has not been laid to rest.

The apparent success of Boltzmann's principle—provided we can justify our choice of the expression of the thermodynamic probability—lends support to the belief that there must be something more profound behind it. But where can we look to find that "something"? Surely, it cannot come from the causal dynamics of the molecules themselves because no matter how well we refine potential interactions and molecular dynamics, it is always necessary to introduce the concept of temperature, which is a purely statistical notion. Once we introduce a statistical notion, we sever our links with the time reversible behavior of the world of atoms. With no direct connection in sight between the reversible, microscopic world of atoms and the apparent irreversible behavior of macroscopic systems, why not focus our attention on the statistical aspects of large systems for which there is a well-developed mathematical theory? What protects thermodynamics from becoming still another exercise in mathematical statistics is the notion of entropy and, with it, the notion of an unnormalizable, or *improper*, probability density.

In mathematical statistics, as the sample size increases without bound, probability gives way to certainty. In thermodynamics, the "sample size" is replaced by the number of subdivisions of our original system. Suppose we are interested in estimating the temperature of the system from observations made on the energies of the subsystems. If we could imagine a subdivision so fine that we could tell how the individual molecules share their energy among their internal degrees-of-freedom, it would be tantamount to having a Maxwell demon at our disposal that could do better than the macroscopic estimate of the temperature. If such fine subdivisions could give a better estimate of the absolute temperature than the macroscopic estimate, it would at some stage violate the additivity, or the statistical independence, of the energies of the subsystems. Moreover, we know that the probability of the energy of any subsystem, given the total energy of all the subsystems, is independent of their common temperature. In mathematical statistics, this property is referred to

as a *sufficient* statistic, and where additivity holds, so too will sufficiency. It is
this concept, coupled to the uncertainty that exists between thermodynamic
conjugate variables, which stands in the defense of a purely phenomenological
approach: nothing can beat the macroscopic estimate of temperature, not even
a Maxwell demon.

Although this was discovered by Leo Szilard as far back as 1921, it was
completely ignored by the scientific community up until the mid-1950s.[1] Only
Albert Einstein and Szilard's dissertation advisor, Max von Laue, could have
appreciated its significance inasmuch as it led to a phenomenological theory of
thermodynamics that could include the role of fluctuations and yet *make no
reference to any atomic model.* Until that time, the phenomenological theory
regarded thermodynamic equilibrium as the final resting place for all macroscopic systems which have been detached from the outside world and left alone
for a sufficiently long period of time. In effect what Szilard showed was that
thermodynamic equilibrium could be associated with a distribution of inces-
santly fluctuating configurations so that one could talk about the *stability* of
such a state in terms of the moments (notably, the second) of the distribution.
Sadly enough, the inherent connection of thermodynamic stability criteria with
the concavity condition of the entropy and the relation of the latter to the error
law went unnoticed.

Yet, Szilard's approach appeared to provide a way for justifying why thermodynamics gave results that could be judged as "best." The situation appears
akin to an earlier one in the development in the theory of statistics. Uncertainty in astronomical measurements was recognized at a much earlier stage
than uncertainty in thermodynamic measurements. It was for this reason that
the method of least squares was introduced, at the turn of the nineteenth century, because it produced the best results that minimized the sum of squared
errors leading to a state of mechanical equilibrium. But without having any
way of determining what best was, it made the adjective best somewhat superfluous. The synthesis of "error" and "probability" was achieved by Carl
Freidrich Gauss, who "elevated" the negative of the sum of squares to a law
of error and thereby determined just how good best really was. Szilard could
have been the Gauss of statistical thermodynamics were it not for his over
pessimistic opinion of what he had wrought and his lack of appreciation that
the exponential family of distributions are, in fact, laws of error for extensive thermodynamic variables that are determined uniquely in terms of the
concavity of the entropy. The synthesis of thermodynamics with probability
theory had still to wait another 30 years before any definite progress could be
registered.

The appearance of an exponential family of distributions for the random
variable of energy can be traced back even farther to the pioneering work of
Josiah Willard Gibbs during the second half of the nineteenth century. Gibbs
introduced a prior probability measure on the space of extensive fluctuating

---

[1] Perhaps this explains the four year delay in publication of Szilard's paper.

thermodynamic variables which completely determines the mechanical features of an *isolated* system. Like Boltzmann's thermodynamic probability, Gibbs' prior probability measure is not normalizable since it represents the number of systems whose energies lie below a given value. Because this probability measure increases as a finite fixed power of the energy, it is an *improper* distribution to which the statistical theory of conjugate, or associate, distributions, developed by Harald Cramér in the late 1930s, could be applied.

The method of conjugate distributions introduces an exponential factor, dependent not only upon the random variable but also on some quantity that can be estimated through observation, so that the product of this exponential factor and the prior probability measure forms such a sharp peak that there can be no distinction between the means and mode of the distribution. This is accomplished by varying the parameter in the exponential factor so that the peak occurs at the experimentally observed value of the extensive chance variable. Alternatively, we may consider the optimal value of the parameter to be estimated in terms of the most probable value of the quantity observed. This is tantamount to the method of *maximum likelihood* in mathematical statistics, where the maximum likelihood estimate of the unknown parameter, determined from the extremum condition of the "likelihood" function, is expressed as a function of the average of its thermodynamically conjugate extensive variable.

Through the process of observation, we convert our originally isolated system, which is fully characterized by the defective prior probability measure, into a closed, or possibly open, system that is characterized by a *proper*, posterior probability measure. Prior to measurement, the defective prior measure summarizes all the information we have about the mechanical structure of the isolated system. In order to perform a measurement, we must place the system in contact with a reservoir that will allow fluctuations in the appropriate extensive variable, the most primitive reservoir being a thermostat. Without the possibility of fluctuations, no experiment would be informative. Placing the system in contact with a reservoir has the effect of converting a defective, prior measure into a proper, posterior probability measure. For ideal systems the latter turns out to belong to the exponential family of distributions.

Solomon Kullback called the average of the logarithm of the ratio of the two probability measures the mean "discrimination information," while A. Ya. Khinchin called the negative of it the entropy. If it were true that the prior probability measure was normalizable, then Kullback's interpretation of the minimum discrimination information would be applicable. The discrimination information would then be a measure of how closely the sample "resembles" the population that is characterized by the prior probability measure. But because the prior probablity distribution is improper, the discrimination information interpretation is not appropriate; consequently, entropy and discrimination information do not differ merely by a change in sign. The differences between the two are much more profound and are related to the fact that even if an experiment could be repeated an innumerable number of times, a finite entropy would still prevent us from achieving certainty. Lurking in the background is still Boltzmann's connection between entropy and an improper prior, or "thermodynamic," probability density.

The ratio of the posterior to the prior measure defines a likelihood function which can be used to estimate the parameter that characterizes the reservoir. It turns out that the maximum likelihood estimate coincides with the thermodynamic definition of the conjugate intensive variable. In the case where the system is placed in contact with a thermostat, the energy is allowed to fluctuate and the maximum likelihood value of the conjugate parameter is the inverse temperature. The fact that the maximum of the logarithm of the likelihood function is equivalent to an entropy minimum principle, as a function of the conjugate intensive variable, displays the intimate connection between the mathematical theory of statistical optimization and the general extremum principles of thermodynamics.

The estimators of the true value of the intensive conjugate parameter are functions of the observations made on the conjugate extensive quantity and so too must fluctuate. When a system is placed in thermal contact with a thermostat, the energy of the system ceases to be a thermodynamic function since it is no longer uniquely determined in terms of the external parameters required to specify the state of the system. Sample values of the energy can be used to estimate the temperature. For a vanishing thermostat, the energy is fixed and a definite temperature cannot be assigned to the system, while for an infinite thermostat, the temperature becomes precise at the expense of the energy. Conjugate thermodynamic variables thus satisfy *uncertainty relations* in which the precision in the measurement of one variable varies inversely with that of the other. At thermodynamic equilibrium, the dispersion in the energy reaches its lower bound, given in terms of the Fisher "information." The Fisher information agrees with our intuition that the more information we have, the smaller will the uncertainty, or the more *efficient* the estimator, will be. Moreover, grouping or transforming our observations will, in general, result in a loss of information. It is only when the statistic is *sufficient* that grouping observations causes no loss in information. We have already seen that the mathematical property of sufficiency is essential to thermodynamics.

Beginning in the mid-1950s and lasting for almost a decade, Benoit B. Mandelbrot made a series of attempts to resuscitate Szilard's phenomenological approach to thermodynamics by further developing the concept of sufficiency. But it met with no better reception than the original formulation. The exponential family of distributions are the only ones to give sufficient statistics for any sample size or any number of observations—including a single observation! Viewed from the perspective of almost a century later, it appears truly remarkable how the forefathers of statistical thermodynamics hit upon precisely those statistical properties which are characteristic of thermodynamics.

The exponential family of probability distributions not only gives sufficient statistics, for any sample size, but is also the one for which the uncertainties in conjugate thermodynamic variables is a minimum. We can appreciate here an analogy with the probabilistic interpretation of quantum theory which is defended by the Heisenberg uncertainty relations. If there would be some way to violate the uncertainties in simultaneous measurements of conjugate variables, like momentum and position or energy and time, it would be possible to replace the probabilistic interpretation by a causal, or deterministic, description of quantum phenomena. It makes no difference whether the conjugate variables are Fourier or Laplace duals. The very fact that uncertainty relations exist in thermodynamics between conjugate thermodynamic quantities, like energy and inverse temperature, makes it all but impossible that a probabilistic interpretation of thermodynamics would ever be superseded by deterministic one, rooted in the dynamics of large assemblies of molecules. Something is missing and that something is a hypothesis of "randomness" which always creeps in during the process of measurement. This is not a technological point which can be resolved with time and money. Rather, it appears that nature has limited our capability to understand and describe physical phenomena in anything but a statistical framework.

The facts that sufficiency is a characteristic of the exponential family of distributions and that maximum probability corresponds to maximum entropy are alone insufficient to identify the relevant distributions without having an expression for the entropy. This can only be obtained by considering the physical processes which admit a stationary probability distribution. To say that the entropy is $-\sum_i p_i \ln p_i$, where the p_i are the individual probabilities, means little or nothing if we don't know what these probabilities are. In fact, the more successful information approach of Edwin T. Jaynes, measured in terms of the number of adherents, turns the argument around. Using this expression for the entropy, which was brought into the realm of communication theory by Claude Shannon in the late 1940s, Jaynes assigns the least prejudiced, or least biased, probabilities to those which maximize the entropy "subject to the given information." Not unsurprisingly, Jaynes is led to the exponential family of distributions. However, in going from the discrete expression to the continuous, integral expression for the entropy, Jaynes realized that the later expression is not invariant under coordinate changes unless one introduces a prior probability measure. Although he faults Shannon for not having appreciated this, he subsequently dodged the problem by adopting the uniform measure by "regarding classical statistical mechanics merely as a limiting form of the (presumably more fundamental) discrete quantum statistical mechanics." Since the prior probability measure is tantamount to the knowledge of Boltzmann's thermodynamic probability, the maximum entropy formalism tells us nothing that we don't already know or can't obtain by other means. The problem is to obtain a unique and unambiguous expression for the thermodynamic probability, and this can only be achieved by relating Boltzmann's principle to the probability distribution governing the physical processes under consideration.

In fact, representing knowledge as average values which are used as constraints on the probability distribution is inconsistent with the initial premise that all *a priori* probabilities are equal. Distributing a given number of particles over a given number of disjoint cells in phase space, Boltzmann assumed the thermodynamic probability to be given by the multinomial coefficient. If connection is to be made with probability theory, then one could argue that the multinomial coefficient is actually the multinomial distribution with all *a priori* probabilities equal to the inverse of the number of cells in the phase space since this will have the innocuous effect of introducing a constant in the expression for the entropy. If we now introduce the constraint of constant total energy, which is entirely foreign to probability theory, we find that the cells in phase space are not all *a priori* equal. Consider the case where all but one particle has been distributed; the remaining particle must go into that cell whose energy is such that the total energy is conserved. The *a posteriori* distributions or observed frequencies which maximize the thermodynamic probability, or equivalently its logarithm or the entropy, turn out to be exponential functions of the energies of the different cells in phase space so that they cannot be equated with the *a priori* probabilities which are assumed to be equal to one another.

Moreover, Jaynes' information approach does not discriminate between observed and predicted results so that it can only be expected to hold in the asymptotic limit where the number of observations increases without bound. And it is precisely in this limit that one derives the exponential family of distributions which maximize the entropy. A rather peculiar characteristic of the exponential family of distributions is that the most likely value of the sample mean is equal to the expected value. But this is precisely the claim made by statistical mechanics which identifies the average values of extensive quantities with their thermodynamic counterparts, and thus these average values must correspond to the most probable values of the quantity measured. The individual values of the extensive quantity measured will therefore be governed by a law of error leading to the average value as the most probable value of the quantity measured. This error law can be expressed in terms of the concavity property of the entropy; the deviations from mean values, or fluctuations, are responsible for the monotonic increase in the entropy as a function of the extensive variables. In other words, the entropy will show a tendency to increase only when there are deviations from the mean or most probable values of the extensive variables.

The identification of the entropy in terms of a Gaussian error law for which the average value of an extensive quantity is the most probable value of the quantity measured eliminates the arbitrariness of Boltzmann's principle in which the entropy is defined as the logarithm of the number of complexions. It is indeed ironic that Maxwell's derivation of his distribution of molecular speeds, which is based on Gauss' error law, actually precedes chronologically Boltzmann's principle. In Boltzmann's own words, it was Maxwell who proved that the various speeds have the same distribution as the errors of observation that always creep in when the same quantity is repeatedly determined by measurement under the same conditions. That these two laws agree cannot, of course, be taken as accidental, since both are determined by the same laws of probability.

The procedure we follow is to consider a physical process which admits an invariant, or stationary, probability distribution belonging to the exponential family. Comparing this distribution with Gauss' error law, given in terms of the concavity criterion of the entropy, identifies not only the entropy in terms of the number of complexions but also the type of ensemble being considered since the normalizing factor, or "partition function," is obtained in terms of the Massieu transform of the entropy. The thermodynamic estimate of the intensive parameter, whose extensive conjugate is undergoing fluctuations, places the exponential distribution at its maximum value, just as if we had replaced the constant *a priori* probabilities by their maximum likelihood values in the original probability distributions. The maximum likelihood estimate is expressed in terms of the average value, which by hypothesis, coincides with the most probable value of the quantity measured.

Hence, we shall show that Boltzmann's statistical interpretation of the entropy and the thermodynamic concavity property of the entropy can be united in terms of Gauss' law of error leading to the average value as the most probable value of the quantity measured. The principal theme of this book is that one should *work directly with probability distributions*, and not just with their moments.

The thermodynamic stability criteria are couched in terms of the concavity of the entropy and so, too, is the existence of a stationary probability distribution belonging to the exponential family or the law of error giving the probability of a deviation from the most probable value. Thus, *Boltzmann's principle is a consequence of a probabilistic line of reasoning, rather than its point of departure*. Once an explicit expression for Gauss' principle has been derived, the physicist's definition of "statistics" is obtained by appealing to the second law of thermodynamics which introduces the notion of temperature. However, the statistics is determined by the probability distribution rather than its first moment. Therefore, by dealing directly with the probability distributions themselves, rather than their vestiges, we can fulfill the dream of combining the rigor of modern mathematical statistics with the intuitive vigor of thermodynamics.

In this way we hope to clear up a great deal of the misconceptions about statistical thermodynamics that have recently made their way into the literature. Statistical considerations highlight the fundamental properties of thermodynamics. This is a very powerful tool and, if used properly, it can tell us what is physically realizable and what is not. Based on statistical considerations, we will appreciate that the defining property of the entropy is *concavity*. Hence, any putative expression for the entropy which sets it proportional to the square of an extensive quantity is complete nonsense. This wipes out any credibility that black hole thermodynamics could have claimed.

The entropy of an adiabatically isolated system cannot change. It is a real miracle then that an increase in the entropy is found[2] in a universe which obeys Einstein's *adiabatic* equations of general relativity. Specifically, it is claimed that the entropy density is the same before and after "inflation" has occurred in the early universe. But when the entropy density is multiplied by the volume, it leads to an enormous increase in entropy after inflation has occurred. However, in an adiabatically isolated system, the internal energy just compensates the work done by expansion leaving the entropy unchanged.

Another fallacy that has recently surfaced is the claim that "the consistent and consensual definition of temperature admits no fluctuation." Common sense tells us that if the energy fluctuates, that which measures it should also fluctuate. The argument used to support such a claim is that in the canonical ensemble the temperature is fixed and the energy is allowed to fluctuate. According to the uncertainty relation, between energy and inverse temperature, the limit where the standard deviation in energy vanishes would mean an unlimited large value of the standard deviation in the temperature.

Anyone who makes such an accusation has undoubtedly never heard of statistical inference or how degrees-of-belief are altered by data. We take a sample of a given size from a population where the set of independent random variables are identically distributed. Their common probability density may depend upon a parameter which is fixed but may be unknown. Any distribution that we can attach to the unknown parameter must be considered in the sense of degree-of-belief, as opposed to a probability distribution in the frequency sense. The density of beliefs about the parameter will be changed by sampling in accordance with Bayes' theorem, or the principle of inverse probability, where "cause" and "effect" are interchanged. The idea of inverse inference, reasoning probabilistically from the effect to the cause, was a conceptual liberation to eighteenth-century statisticians which was later formalized into a "fiducial" argument by Fisher. There is absolutely no conceptual difficulty in combining frequency and degree-of-belief concepts of probability into a single uncertainty relation, where fluctuations in the energy are to be interpreted in the frequency sense, while those in temperature are intended in the sense of degree-of-belief. Statisticians have been using it for over two centuries!

The most remarkable feature of the thermodynamic and probabilistic synthesis is their complete compatibility. Without realizing it, the forefathers of statistical thermodynamics constructed a theory connecting seemingly unrelated quantities which echoed a probabilistic structure that was unknown to them. Statistical ensembles or collections of macroscopically similar systems were constructed in such a way that they fit naturally into a probabilistic framework. In thermodynamics one begins with a fundamental relation where, in the entropy representation, the entropy is a function of the set of extensive

---

[2]It is claimed to be of the order 10^{87}!

variables comprising the energy, volume, and number of particles. In order that the additivity of these extensive variables be transferred to the entropy, which is really a reflection of their statistical independence, a relation is required among the three intensive variables, temperature, pressure, and chemical potential. The intensive variables make their entrance by placing our previously isolated system in contact with a characteristic reservoir. In this way, the isolated system can be transformed into a *closed* system (heat reservoir) or *open* system (heat + particle reservoir). However, thermodynamics tells us that this cannot be done in an arbitrary fashion, and it is here that we can find a probabilistic justification for the ordering.

The Gibbs–Duhem equation is a differential relation that either expresses the pressure as a function of the temperature and chemical potential, leading to the grand-canonical ensemble, or the chemical potential as a function of the temperature and pressure, which is often referred to as the isothermal–isobaric ensemble because the system is in contact with reservoirs that permit the temperature and pressure to be held constant. Since the three intensive variables stand on equal footing in the Gibbs–Duhem relation, it would seem possible to create an additional ensemble in which the temperature is taken to be a function of the pressure and chemical potential. It is here that thermodynamics intervenes by demanding that there be a hierarchy in the establishment of the different types of equilibria with maximum priority given to thermal equilibrium. Once thermal equilibrium has been secured, it is immaterial whether we then secure mechanical equilibrium or equilibrium with respect to the transport of matter. Thermal equilibrium must precede all other forms of equilibria.

We begin with a microcanonical ensemble which has a purely mechanical structure, devoid of all notions of heat and temperature. Placing the system in contact with a heat reservoir converts it into a closed system and introduces a parameter whose "most likely" value turns out to be the inverse temperature. Equilibrium is achieved when both system and reservoir reach a common temperature. The partition between the two consists of a diathermal wall. We can then make the partition either movable or permeable to matter, but not both simultaneously. In the former, the energy and volume fluctuate at a constant number of particles while in the latter the energy and particle number fluctuate at constant volume. Since the entropy cannot be considered to be a concave function of the three independent extensive variables simultaneously,[3] these two possibilities are open to us together with the possibility of having the particle number and volume fluctuate at constant energy. But this would correspond to making the temperature a function of the pressure and chemical potential, and it is precisely this situation which is forbidden by the fact that thermal equilibrium must precede all other forms of equilibrium. How is this justified probabilistically?

Probabilistic considerations show that there is a *statistical equivalence* be-

---

[3]This is ensured by the fact that the so-called generating function of the ensemble would vanish identically on account of the Gibbs–Duhem relation.

tween fluctuations in the volume and the number of particles. They correspond to two types of experiments which although physically distinct are nonetheless statistically equivalent. For instance, we may count the number of pollen grains dispersed homogeneously in a liquid using a square grid, containing r squares, under a microscope. We find an average number of particles per grid, $\bar{n}$. This is a problem involving the placement of $r\bar{n}$ particles in r squares each of which can go into any of the squares with probability $1/r$, independently of each other. The probability of finding a given number of particles in a given square is given by the binomial distribution which, in the limit of large r, transforms into the Poisson law for the spatial distribution of particles. This is representative of "homogeneous chaos."

However, we can perform another experiment in which the size of the squares in the grid can be varied until they contain a given number of pollen grains. In place of a random number of particles being found in a given square, it is now the size of the square which is the random variable for a fixed number of particles. The probability distribution that the size of a square of the grid has more than the given number of particles is the gamma density. This gamma density offers a statistical equivalent description to the Poisson distribution; the selection of either one depends upon whether we are considering the size of a square of the grid, for a fixed number of pollen grains, or the number of pollen grains, for a given size of a square of the grid, as the random variable. Therefore, simultaneous fluctuations in the particle number and volume, at constant energy, are meaningless. Fluctuations in the volume or particle number necessitate fluctuations in the energy, and this is the statistical statement that thermal equilibrium must be secured prior to mechanical equilibrium or equilibrium with respect to the transport of matter.

This is just one instance where probabilistic considerations have profound repercussions on the foundations of thermodynamics. As we have mentioned earlier, the sheer presence of uncertainty relations and sufficient statistics warn us that any attempt to fathom the world of atoms and come away with an explanation of macroscopic behavior compatible with thermodynamics is doomed to failure. Mandelbrot summed up the situation in a rather nice way:

> Since, therefore, the kinetic foundations of thermodynamics are not sufficient in the absence of further hypotheses of randomness, are they still quite necessary in the presence of such hypotheses? Or else, could not one "short-circuit" the atoms, by centering upon any elements of randomness....

Chapter 1

Entropy and Probability

1.1   The Predecessors of Boltzmann

There are two basic categories of thermodynamic theories: phenomenological and probabilistic. Chronologically, the former precedes the latter because thermodynamics evolved from the observations made on steam engines by engineers, like Sadi Carnot, at the beginning of the nineteenth century. These observations were formalized into principles by physicists, like Rudolf Clausius, during the middle part of the nineteenth century. According to these principles, "heat is energy" (first law) and "heat flows spontaneously from hot to cold" (second law). According to Clausius, the first principle can be phrased as "the energy of the universe is constant" while the second law introduces the abstract concept of "entropy" in which the entropy of the universe tends to a maximum. It is precisely this property of the entropy that had apocalyptic consequences since it predicted that the universe would end in a *heat death* caused by thermal interactions that lead to an unending increase in entropy.

But in what sense is a system ever "left to itself" or completely isolated? For if it were completely isolated, there would be no way for the energy, or for that matter any other thermodynamic variable, to change and, consequently, the entropy could not increase. If there would be some means by which we could alter the energy or the other thermodynamic parameters necessary to specify the state of the system, the entropy could be made to vary at will and it would therefore violate the second law of thermodynamics, as formulated by Clausius.

By the turn of the century it became evident that there was something incomplete about Clausius' formulation. There began a search for an alternative approach that would avoid such cataclysmic consequences. This alternative approach became known as the "statistical" or "probabilistic" formulation. It asserts that "heat is a form of random molecular motion" and "entropy is a measure of disorder or the probability of realizing a given macroscopic state in terms of the number of microscopic 'complexions' that are compatible with it." Equilibrium would be that state with maximum probability or, equivalently, one with maximum entropy. In contrast to the phenomenological formulation, it would not be a static final resting state of the system but, rather, thermal equilibrium would be characterized by a distribution of constantly fluctuating configurations.

How can this be brought about in a truly isolated system? Planck's response was that

> no system is ever truly isolated; there is always an exchange of energy with the outside world, no matter how small or irregular. This exchange allows us to measure the temperature of the system. The "loose coupling" between system and environment allows the system to change its state—albeit on a rather irregular basis.

The consequence of this was to lower the second law, from the status of an absolute truth to one of high probability. However, it still left open the question of how to determine the entropy from molecular considerations which necessarily had to agree with the phenomenological expression for the entropy. Since the molecular variables are necessarily random, the resulting entropy which is a function of those variables would also be a random quantity. It thus appeared that one abstract definition of entropy was being substituted for another.

Ideally, one would like to begin with the dynamics of large assemblies of molecules and show how the system evolves to its final state. However, as Boltzmann realized, these reversible laws must always be completed by some hypothesis of randomness. More recently, it has been shown that randomness can be introduced in a deterministic framework by going to an appropriate asymptotic limit, such as the Brownian motion limit. Since nature never goes to such limits,[1] our real interest lies in the random hypothesis. However, once the random hypothesis is introduced, the link with the reversible world of molecular motions has been severed. With no direct link-up with the kinetic foundations, we are led to focus our attention on the elements of randomness. By concentrating on the nature of the randomness itself, we are able to avoid entering the reversible microscopic world of atoms and treat the aspect of irreversibility which appears at the macroscopic stage. This is the inception of a probabilistic formulation of thermodynamics, devoid of any particular dynamical feature that may filter through to the macroscopic world from the microscopic one. The stage has now been set for Boltzmann's contribution to this probabilistic formulation.

1.2   Boltzmann's Principle

The following inscription (in our notation),

$$S = k \ln \Omega, \tag{1.1}$$

---

[1]Brownian motion is a mathematical idealization since no physical system can have a constant spectral density over the entire real axis. Although such processes would exhibit an infinite variance, they do prove useful in the description of rapidly fluctuating processes which are virtually uncorrelated at different instants in time.

relating the entropy S to the logarithm of the so-called thermodynamic probability Ω, is engraved on Boltzmann's tombstone. The constant of proportionality in Boltzmann's principle is k.[2] Planck did not at all appreciate k being referred to as Boltzmann's constant since it was he who discovered it. According to Planck, "Boltzmann never calculated with molecules but only with moles" and it therefore never occurred to him to introduce such a factor. Planck was preoccupied with the *absolute* nature of the entropy, and without the factor of proportionality between the entropy S and thermodynamic probability Ω, there would necessarily appear an undetermined additive constant in (1.1). Whereas Boltzmann considered the enumeration of the microscopic complexions belonging to a macroscopic state to be an "arithmetical device of a certain arbitrary character," Planck was to learn that it certainly was not.

This universal constant, Planck contended, "is the same for a terrestrial as for a cosmical system, and when it is known for one, it is known for the other; when k is known for radiant phenomena, it is also known and is the same for molecular motions." The importance of Boltzmann's principle cannot be overestimated since it led the way toward the theory now known as statistical mechanics.

Early in Boltzmann's career, he thought that a theory of heat could be reduced to a purely mechanical interpretation. The years between 1869 and 1872 saw a large infusion of probabilistic notions in Boltzmann's predominantly mechanistic view. The culmination came in 1872 with the first enunciation of what will later be known as the "H-theorem." Slowly, he became converted to the idea that the full content of the second law will only be grasped when its roots are sought in the theory of probability, abandoning his earlier belief that thermodynamics can be reduced to mechanics. Boltzmann asserted that "if the initial distribution amongst the bodies did not correspond to the laws of probability, it will tend increasingly to become so." The entropy will be a measure of the tendency of the system to become allied with these probabilistic laws. He argued that it cannot be fortuitous that, in its most probable state, the velocities of a very large aggregate of gas molecules possess the same distribution as the "errors of observation that always creep in when the same quantity is repeatedly determined by measurement under the same conditions."

The very sharpest definition of a macroscopic state we have is the number of its microscopic complexions that are compatible with it. The most probable state of the system is one in which the number of microscopic complexions is the greatest; this corresponds to the state of maximum entropy or the most "disorderly" state and coincides with the thermodynamic notion of equilibrium. It may be said that Clausius elevated the second law to an absolute truth for which the entropy of a composite system, obtained by bringing two isolated systems into thermal contact, cannot be smaller than the sum of the

---

[2]In most of the subsequent formulas, Boltzmann's constant will be omitted implying that the temperature will be measured in energy units. This introduces a greater symmetry in the formulas, especially in regard to dual Legendre functions. To convert back to a temperature measured in degrees, $T \rightarrow kT$ and $S \rightarrow S/k$.

entropies of the individual systems, while Boltzmann lowered it to one of high probability. To quote Gibbs, "The impossibility of an uncompensated decrease of entropy seems to be reduced to an improbability."

Boltzmann transferred his attention from the fatalistic implications of the second law, where energy is continually being degraded into "purely thermal vibrations [which] slip through our hands and escape our senses and which for us is synonymous with rest," to the amazing uniformity of the final state for which the law of large numbers had to be responsible. Boltzmann drew the analogy with a probability that a large number of people should be found in complete agreement. Such a circumstance is not *impossible*, he claimed, although it is highly *improbable*: it simply would not conform to the laws of probability. Every dissension would amount to a degradation of energy, and in the state of complete disagreement, the "degraded energy forms will be none but the most probable forms; or better, it will be the energy that is distributed amongst the molecules in the most probable way." The final distribution conforms to the laws of probability where deviations from the most probable values will be governed by laws of error, the statistical independence of the observations being completely harmonious with the thermodynamic property of additivity.

It took Planck a period of years to abandon Clausius' interpretation and to reconcile his ideas with those of Boltzmann. Planck's conversion was due to the simple fact that it was only Boltzmann's approach which provided a theoretical basis for the expression of the entropy of black radiation that Planck had so luckily guessed and which fitted the data remarkably well—perhaps too well for it to be a lucky guess! It was one of those macabre twists of fate that Boltzmann did not live to see the fruit of his efforts, for which he struggled so laboriously in his lifetime to gain universal acceptance.

However, the unity of thermodynamics and probability theory afforded by Boltzmann's relation (1.1) brought with it a certain opaqueness to the concept of probability. For one thing, the "thermodynamic" probability is an integer— and a large integer at that—so that it cannot be considered a probability (which, of course, must be a proper fraction) at all. Boltzmann's argument that the actual number of objects is a more perspicuous concept than a *mere* probability is not convincing. If Ω represents the number of microscopic complexions corresponding to a macroscopic state, then we might try to divide Ω by the total number of complexions in order to get a proper fraction which can represent a probability. Apart from the fact that it does not lead to the correct thermodynamic result, not any value of Ω can be related to the thermodynamic entropy but rather its maximum value, which has been determined by imposing the constraints that the total number of particles and total energy are constant. Since the thermodynamic entropy coincides with the state of thermodynamic equilibrium, for which there is an overwhelmingly large number of indistinguishable microstates, the ratio of the maximum of Ω to its total value is effectively unity, giving a trivial result for the entropy.

We may, however, ask for the probability of any given state. This probability may be represented as the fraction representing the ratio of the statistical weight of this state, Ω, to the sum of the statistical weights of all the macroscopic states that are compatible with the given constraints. Since the ratio of the maximum of Ω to its total value is effectively unity, we may replace the sum in the denominator by the maximum of Ω, which we shall denote by Ω_{max}. The probability of any state whose statistical weight is Ω is thus

$$\Omega/\Omega_{max} = \exp\left(S - S_{max}\right),$$

where S_{max} is the maximum value of the entropy. It is this form of Boltzmann's principle that was used so successfully by Einstein in his study of thermodynamic fluctuations.

In a sense, Einstein "inverted" Boltzmann's principle (1.1) to obtain information regarding fluctuations about the state of equilibrium. According to Planck in his *Theory of Heat*, the entropy difference,

$$\Delta S = S - S_{max},$$

is negative, and introducing this into the above expression, we get

$$\Omega/\Omega_{max} = \exp\left(\Delta S\right). \tag{1.2}$$

Since ΔS is negative, this is a proper fraction, as it should be. If we reinstate Boltzmann's constant, it becomes apparent that due to the smallness of k, moderate values of ΔS will lead to vanishing small probabilities so large deviations from equilibrium will be extremely rare. It is only when the deviation from equilibrium, $|\Delta S|$, is of the same order of magnitude as Boltzmann's constant will there be a finite probability for fluctuations to occur. These fluctuations will necessarily have an exceedingly small amplitude unless conditions for magnifying their effect are present.

In the atmosphere, for example, small adjacent regions of widely different densities act as scattering centers for light. Scattering will be appreciable only when the wavelength of the light is of the same order as the dimension of the region. Since large fluctuations are more probable in small regions, blue light is more likely to be scattered than red light.

Another situation in which fluctuations will be observable is the irregular motion of small particles caused by the chaotic thermal motions of the lighter surrounding molecules. This is the phenomenon known as Brownian motion. Also, one should expect nonnegligible fluctuations in first-order phase equilibria due to large differences in the densities of the two phases. In the phenol–water system, a curious phenomenon is observed as the critical temperature is approached from above. Immediately before the single homogeneous phase splits up into two phases, a milky iridescence appears. This *critical opalescence* is caused by the scattering of light from neighboring regions having slightly different densities. It was this phenomenon which was analyzed by Einstein in 1910 on the basis of his formula (1.2) for a spontaneous fluctuation [cf. §1.11].

Einstein's formula, (1.2), allow us to determine the relative frequency that deviations from equilibrium occur. To implement it, we could attempt a Taylor series expansion of S about equilibrium, and if the fluctuations are small, we could neglect higher than second-order terms. Since the entropy is a maximum at equilibrium, the coefficient of the first-order term would vanish and we would be left with a negative quadratic form in the exponent. The negativity of the (straight) coefficients of the quadratic terms comprise thermodynamic stability criteria.

The problem with the above proposal is that the entropy change is not that of the system but rather the total entropy change of system + reservoir. The usual form of the entropy maximum principle is

> The entropy tends to a maximum for a given value of the total energy.

Intended in this statement is that the entropy be a maximum with respect to some internal parameter for which the first derivative vanishes. But we are dealing with the extensive variables consisting of internal energy E, volume V, and number of particles N. And the derivative of the entropy with respect to any one of these variables does not vanish at equilibrium; rather, it defines the conjugate intensive variable. So a naïve Taylor series expansion will certainly not do. And certainly, Planck and his contemporaries did not make such a banal error.

Rather, they considered a composite system in which two subsystems are separated by a thin piston, occupying no volume, which is a good conductor of heat, is impermeable to the passage of matter, and moves without friction. Let the suffixes 1 and 2 denote the subsystems. For fixed subvolumes, we can ask for the probability that the energy of subsystem 1 is between E_1 and $E_1 + dE_1$ at the same time the energy of subsystem 2 is between E_2 and $E_2 + dE_2$. The "probability" for the individual subsystems will be given by $\Omega_1(E_1, V_1)\, dE_1$ and $\Omega_2(E_2, V_2)\, dE_2$ so that the combined probability should be given by

$$\Omega_1(E_1, V_1)\, \Omega_2(E_2, V_2)\, dE_1\, dE_2. \qquad (1.3)$$

This would be the case if we were considering independent events and a realization of the energy in subsystem 1 had no effect upon a realization of the energy in 2. But the composite system is isolated from the rest of the world and this implies energy conservation, namely, $E_1 + E_2 = E = $ const. So E_1 and E_2 are not independent random variables, and the joint probability will not reduce to the product of the individual probabilities as (1.3) implies.

Thus, we are only allowed to vary E_1 and V_1 since there is also a closure condition on the volume, namely, $V_1 + V_2 = V = $ const. We now want to determine the most probable state of the system for which

$$\Omega(E_1, V_1)\, \Omega_2(E_2, V_2)$$

is a maximum with respect to E_1 and V_1. In other words, we realize that the extensive variables of both subsystems cannot be simultaneously independent variables but still insist that their joint probability reduces to a product of their individual probabilities because the entropy of the composite system will be the sum of their individual entropies,

$$S = S_1 + S_2. \tag{1.4}$$

The argument proceeds by considering small deviations in the energy, ΔE_1, and volume, ΔV_1, such that

$$
\begin{aligned}
\Delta E_1 + \Delta E_2 &= 0, \\
\Delta V_1 + \Delta V_2 &= 0.
\end{aligned} \tag{1.5}
$$

Since the deviations are small in the sense that $|\Delta E_1|/E \ll 1$ and $|\Delta V_1|/V \ll 1$, the corresponding change in the entropy of the composite system is

$$
\begin{aligned}
\Delta S \;=\;& \Delta S_1 + \Delta S_2 \tag{1.6} \\
=\;& \sum_i \left\{ \left(\frac{\partial S}{\partial E} \right)_i \Delta E_i + \left(\frac{\partial S}{\partial V} \right)_i \Delta V_i \right. \\
& + \frac{1}{2} \left[\left(\frac{\partial^2 S}{\partial E^2} \right)_i (\Delta E_i)^2 + 2 \left(\frac{\partial^2 S}{\partial E\, \partial V} \right)_i (\Delta E_i)(\Delta V_i) \right. \\
& + \left. \left. \left(\frac{\partial^2 S}{\partial V^2} \right)_i (\Delta V_i)^2 \right] \right\}.
\end{aligned}
$$

This is to say that for small deviations we can safely neglect higher than quadratic terms in the Taylor series expansion of the entropy about the state of equilibrium.

If we now impose the closure conditions (1.5), the linear terms will cancel provided

$$\left(\frac{\partial S}{\partial E} \right)_1 = \left(\frac{\partial S}{\partial E} \right)_2$$

and

$$\left(\frac{\partial S}{\partial V} \right)_1 = \left(\frac{\partial S}{\partial V} \right)_2.$$

Now from thermodynamics we know that

$$\frac{\partial S}{\partial E} = \frac{1}{T}$$

and

$$\frac{\partial S}{\partial V} = \frac{P}{T},$$

where T is the absolute temperature and P is the pressure. Consequently, the first condition implies the equality of the temperatures of the two subsystems, $T_1 = T_2$, while the second condition implies the equality of their pressures, $P_1 = P_2$. Hence, the entropy difference reduces to the quadratic form

$$\Delta S = \tfrac{1}{2} \sum_i \left\{ \left(\frac{\partial^2 S}{\partial E^2} \right)_i (\Delta E_i)^2 + 2 \left(\frac{\partial^2 S}{\partial E\, \partial V} \right)_i (\Delta E_i)(\Delta V_i) + \left(\frac{\partial^2 S}{\partial V^2} \right)_i (\Delta V_i)^2 \right\}.$$

In order that the change in entropy be strictly negative, we require

$$\left(\frac{\partial^2 S}{\partial E^2}\right)_i < 0, \qquad \left(\frac{\partial^2 S}{\partial E^2}\right)_i \left(\frac{\partial^2 S}{\partial V^2}\right)_i - \left(\frac{\partial^2 S}{\partial E \, \partial V}\right)_i^2 > 0, \qquad (1.7)$$

for each of the subsystems. We have thus reduced the entropy difference to a negative quadratric form for small fluctuations.

This is the desired result, but is it physically meaningful? According to the second law, only in the event where the temperatures of the two subsystems are equal will the entropy of the composite system be the sum of the entropies of the two subsystems. Hence, the two subsystems cannot be statistically independent of one another—although this is what is implied when the joint probability is expressed as a product of the individual probabilities. If we bring two completely arbitrary subsystems into thermal contact, the probability that their temperatures will be equal is negligible, and the entropy of the composite system will be greater than the sum of the entropies of the subsystems.

Although we can now invoke statistical independence, the lack of additivity of the entropies destroys the argument leading to the reduction of the joint probability into a product of the individual probabilities. Furthermore, the entropies of each of the subsystem are not at their maximum values since the linear term is obviously different from zero and the entropy change of either system can be negative so long as the total entropy change is positive. The statistical weight, Ω, belongs to the composite system and not to the system under study. If we write out the first line of (1.6) in detail, we have

$$S(E') - S(E) = S_1(E'_1) - S_1(E_1) + S_2(E'_2) - S_2(E_2).$$

But because the composite system is isolated, not only do we have $E = E_1 + E_2$ but, in addition,

$$E' = E'_1 + E'_2 = E.$$

Hence, the total entropy cannot change and $\Delta S = 0$. The use of composite systems therefore seems to be a contrivance rather than an aid to the understanding of Boltzmann's principle.

All this points to the fact that the relation between entropy and probability is not so straightforward as one is usually made to believe. It is clear, however, that Boltzmann's principle, which identifies the equilibrium state as the state of maximum entropy corresponding to that state having the greatest number of microscopic complexions, is meaningless since it does not involve the specification of the temperature. The question remains: maximum with respect to what? Boltzmann's principle should be the end result of a statistical formulation rather than a postulate which muddles the concept of probability.

However, the method of composite systems has provided some hints as to how we should proceed. First, inequalities (1.7) define a concave function in two variables so it is this property of the entropy which is the characterizing one [cf. §1.8]. Second, in view of the reduction of the entropy difference to a negative quadratic form, Boltzmann's principle implies that the fluctuations in energy and volume obey a normal law. But in order to do so, we had to introduce the notion of temperature, meaning that our original system was not isolated but could sustain fluctuations in the energy. There is then no reason to insist upon the entropy difference to be negative.

To the second power of the error, the normal law is equivalent to any law of error which is a function of the error only and for which positive and negative errors are equally probable. These two facts lead us to believe that there is a relation between the concavity property of the entropy and a Gaussian law of error. Since the connection between thermodynamics and statistics is through average values, the Gaussian law we are looking for should lead to the average value as the most probable value of the quantity measured. We shall develop this theme in the sequel.

1.3   Thermodynamic Probability

Since the thermodynamic probability is the number of ways a given number of molecules can be arranged in a fixed number of "cells," or states, it can be expressed in terms of a binomial or multinomial coefficient.

Suppose a system of N members has r possible states, or "cells," with a *priori* probabilities p_i. The probability of the set is given by the distribution

$$f(\{n_i\}) = \frac{N!}{n_1! \cdots n_r!} p_1^{n_1} \cdots p_r^{n_r}, \tag{1.8}$$

which is known as the multinomial distribution. We want to maximize (1.8) (or, equivalently, its logarithm), subject to the constraint

$$\sum_{i=1}^{r} n_i = N. \tag{1.9}$$

Since the n_i are usually sufficiently large numbers, we can use Stirling's approximation,

$$\frac{d}{dn} \ln n! \approx \ln n,$$

to evaluate the logarithms of the factorial in the expression for $\ln f$. We thus obtain

$$\ln f = N \ln N - N - \sum_{i=1}^{r} \left[n_i \ln \left(\frac{n_i}{p_i} \right) - n_i \right]. \tag{1.10}$$

We want the maximum value of (1.10) subject to the number constraint (1.9). This is most easily accomplished by introducing the constraint directly into the variational equation

$$\delta(\ln f - \alpha N) = 0, \tag{1.11}$$

by the method of Lagrange multipliers. Equation (1.11) determines the stationary value where α is the Lagrange undetermined multiplier. The variational equations are

$$\sum_{i=1}^{r} \left\{ \ln N - \ln \left(\frac{n_i}{p_i} \right) - \alpha \right\} \delta n_i = 0.$$

Since the introduction of the constraint has freed all the n_i, the only way that the variational equations can be satisfied is to set each coefficient of the δn_i equal to zero. And since the second variation of (1.11) is negative, this gives us the most likely values,

$$\hat{n}_i = p_i N e^{-\alpha}, \tag{1.12}$$

of the occupation numbers. The Lagrange multiplier is now determined by summing both sides. Since the p_i are normalized to unity,

$$\sum_{i=1}^{r} p_i = 1, \tag{1.13}$$

$\alpha = 0$ and the most probable value (1.12) is the average number,

$$\hat{n}_i = p_i N = \bar{n}_i.$$

But how are we to determine the *a priori* probabilities p_i? We have no reason to assume that any state is preferential to any other state. So, it seems entirely reasonable to set all the *a priori* probabilities equal to one another, $p_i = 1/r$, in the multinomial distribution (1.8), and this is precisely what Boltzmann did. Hence, the statistical weight Ω is proportional to the multinomial distribution

$$\Omega = \frac{N!}{n_1! \cdots n_r!} \left(\frac{1}{r} \right)^N, \tag{1.14}$$

and by maximizing this statistical weight, we should be able to derive the "distribution" that maximizes the entropy through Boltzmann's principle (1.1). However, if we maximize it subject to only the number constraint, (1.9), we would get the uniform distribution, $\bar{n}_i = N/r$, which is certainly not the distribution Boltzmann sought. According to Hendrik A. Lorentz, this "will certainly not correspond to reality. There is thus manifestly some error, or rather incompleteness in our reasoning."

In order to remedy the situation, we must also specify the additional condition that the total energy of all molecules shall have a completely determined value. When a molecule is found in cell 1, it will have a kinetic energy E_1; similarly, when it is found in cell 2, it will have a kinetic energy E_2, and so on. Obviously, we have a finite number of molecules in the container and since it is isolated from the outside world, the energy of the assembly must be fixed:

$$\sum_{i=1}^{r} n_i E_i = E. \tag{1.15}$$

Now, we must realize that all our drawings, $n_1, \ldots, n_r$, will not satisfy the energy constraint (1.15), and those configurations which do not agree with it must be rejected. Suppose that we have found two sets of values

$$n_1, \ldots, n_r \quad \text{and} \quad n_1', \ldots, n_r'$$

that satisfy the energy constraint. The probabilities of these two systems will be as

$$\Omega = \frac{N!}{n_1! \cdots n_r!} \tag{1.16}$$

is to

$$\Omega' = \frac{N!}{n_1'! \cdots n_r'!},$$

so that we are really interested in maximizing the statistical weight (1.16) for an arbitrary configuration with respect to the number and energy constraints, (1.9) and (1.15), respectively. In place of the variational equation (1.11) we now have

$$\delta(\ln \Omega - \alpha N - \beta E) = 0, \tag{1.17}$$

where β is the Lagrange multiplier for the energy constraint. The variational set of equations is given explicitly as

$$\sum_{i=1}^{r} \left\{ \ln \left(\frac{n_i}{N} \right) + \alpha + \beta E_i \right\} \delta n_i = 0,$$

which on account of the independence of all the δn_i can only be satisfied if each and every coefficient vanishes. This results in the most probable values being given by

$$\hat{n}_i = N e^{-\alpha} e^{-\beta E_i}. \tag{1.18}$$

The Lagrange multiplier α is again determined by summing (1.18) over all cells. We then obtain

$$e^{\alpha} = \sum_{i=1}^{r} e^{-\beta E_i}.$$

Introducing this expression in (1.18) shows that the most probable value

$$\hat{n}_i = N e^{-\beta E_i} \Big/ \sum_{i=1}^{r} e^{-\beta E_i} \tag{1.19}$$

is equal to the average value

$$\bar{n}_i = N \frac{e^{-\beta E_i}}{\mathcal{Z}(\beta)}, \tag{1.20}$$

where

$$\mathcal{Z}(\beta) = \sum_{i=1}^{r} e^{-\beta E_i} \tag{1.21}$$

is referred to as the "sum-over-states" or "partition function," coming from the German *Zustandssumme*.

Since $\bar{n}_i/N$ is nothing but the probability of finding a member of the set in the ith state with energy E_i, we run into an incongruity: we started with the assumption that all states were *a priori* equally probable and ended up with the result that the probability of any state is an exponentially decreasing function of its energy, (1.20). It is not difficult to see that the energy constraint invalidates the assumption that the states are equally probable. This can be seen *a fortiori* by comparing our variational principles with the number constraint alone, (1.11), and the number and energy constraints taken together, (1.17).

In the former case, the Lagrange multiplier for the number constraint was equal to the logarithm of the sum of the *a priori* probabilities, which vanishes on account (1.13). In the second case, it turns out to be equal to the logarithm of (1.21), where each term in the sum can be interpreted as an unnormalized *a priori* probability. Seen in this light, it becomes clear that the energy constraint has destroyed the assumption of equal *a priori* probabilities. Not only is Boltzmann's argument *nonsequitur*, it actually contradicts the fact that the statistical weight (1.16) is directly related to the probability. But before we attempt to correct this state of affairs, let us follow through the conventional argument to the end. In so doing we will uncover more problems than we solve.

1.4   When Entropy Is Not Entropy

The only parameter remaining to be determined in Boltzmann's distribution (1.20) is β. The energy constraint (1.15) holds true for the most probable occupation numbers (1.20),

$$E = \sum_{i=1}^{r} \bar{n}_i E_i = N \sum_{i=1}^{r} p_i E_i, \tag{1.22}$$

where the probabilities $p_i = \bar{n}_i/N$. The increment in the total energy, (1.22), is

$$
\begin{aligned}
dE &= N \sum_{i=1}^{r} E_i \, dp_i + N \sum_{i=1}^{r} p_i \, dE_i \\
&= -\frac{N}{\beta} \sum_{i=1}^{r} (\ln p_i + \ln \mathcal{Z}) \, dp_i + N \sum_{i=1}^{r} p_i \, dE_i, \tag{1.23}
\end{aligned}
$$

where we have introduced (1.20).

When we compare two neighboring conditions of equilibrium, the infinitesimal differences dE_i in the energies of the states of the system are due to small displacements of certain macroscopic bodies that are described by the generalized coordinates. The energy levels as well as the total energy are functions of these coordinates. For instance, if the system is a fluid that is in a cylinder fitted with a movable piston, the piston is such a macroscopic body and the coordinate is the variable volume of the cylinder. Thus, the last term in (1.23) can be written as

$$N \sum_{i=1}^{r} p_i \left(\frac{\partial E_i}{\partial V}\right) dV,$$

where V is the volume of the system.

Noting that

$$\sum_{i=1}^{r} \ln p_i \, dp_i = d\left(\sum_{i=1}^{r} p_i \ln p_i\right),$$

which follows from the fact that (1.13) holds and consequently $\sum_{i=1}^{r} dp_i = 0$, the increment in the energy can be written as

$$\begin{aligned}
dE &= -\frac{N}{\beta} d\left(\sum_{i=1}^{r} p_i \ln p_i\right) - PV \\
&= T\,dS - P\,dV,
\end{aligned}$$

where

$$P = -N \sum_{i=1}^{r} p_i \left(\frac{\partial E_i}{\partial V}\right)$$

is the pressure and the second line is a combination of the first, $dE = \delta Q - \delta W$, and second, $\delta Q = T\,dS$, laws. Then, by association,

$$T\,dS = -\frac{N}{\beta} d\left(\sum_{i=1}^{r} p_i \ln p_i\right).$$

From this it appears that the quantity $-N d\left(\sum_{i=1}^{r} p_i \ln p_i\right)/\beta$ is the average heat added to the system in the change between two neighboring conditions of equilibrium. It therefore follows that β is an integrating factor for the heat added. From the second law, $\delta Q/T = dS$, it is known that $1/T$ is such a factor. Furthermore, it can be shown that $1/\beta$, like T, is an intensive variable. If T is measured in degrees,

$$1/\beta T = k,$$

where k is a constant. This constant is independent of such changes that may be caused by changing the external parameters and average energy. That it is a universal constant, independent of the structure as well as the condition of the system, follows from the fact that if any two systems are in thermal contact, they have the same β and T and hence the same k. Therefore, the constant k must be *universal*.

We have thus shown that the entropy is given by

$$S(\{\bar{n}_i\}) = -\sum_{i=1}^{r} \bar{n}_i \ln\left(\frac{\bar{n}_i}{N}\right). \tag{1.24}$$

This substantiates the fact that the entropy is related to the maximum value of the statistical weight (1.16) under particle number and energy constraints through Boltzmann's principle (1.1). But expression (1.16) has been criticized inasmuch as the number of states is overcounted exactly $N!$ times, where N is the number of *identical* particles. It is usually justified by the asymptotic equivalence of Ω for the two forms of quantum statistics. This so-called corrected Boltzmann counting divides the volume of classical phase space by $N!$. It would lead to an expression for the entropy of the form

$$S(\{\bar{n}_i\}) = -\sum_{i=1}^{r} \bar{n}_i \ln \bar{n}_i + N$$

$$= \sum_{i=1}^{r} \bar{n}_i[\alpha + \beta E_i + 1] - N \ln N. \qquad (1.25)$$

For an ideal gas [cf. §3.6.1],

$$\alpha = N \ln \left[\left(\frac{2\pi M T}{h^2} \right)^{3/2} V \right],$$

where M is the mass of the particle, so that the entropy becomes

$$S = \frac{5}{2}N - N \ln \left[\left(\frac{2\pi M T}{h^2} \right)^{3/2} \frac{V}{N} \right]. \qquad (1.26)$$

The division of Ω by $N!$ has made the entropy extensive, as it should be, since it amounts to dividing the volume in the argument of the logarithm by the number of particles and adding a factor N to the entropy.

Gibbs wrote this factor in by hand in order to obtain the correct expression for the entropy of an ideal gas, (1.26). It was later attributed to the quantum mechanical result that the particles are identical. This is known as the Gibbs paradox, which will be discussed in §1.13. Nevertheless, this leaves us with the uneasy feeling that our statistical approach is in some way incomplete. We would be more reassured if the statistics itself would yield the desired result rather than having to resort to additional considerations. What does the multinomial distribution (1.14) tell us about the distinguishability or indistinguishability of the particles?

The multinomial distribution (1.14) gives the probability for the occurrence of $n_1, \ldots, n_r$ occupancy numbers in which N *indistinguishable* particles are distributed among r *distinguishable* cells, under the assumption that all r^N possible arrangements are equally probable. Note that we are concerned with the number of occurrences and not with the particular particles that are involved. For instance, insurance statistics are relevant to the number of accidents in a given period of time but not to the individuals involved in those accidents. The multinomial distribution (1.14) actually takes into account the fact that the particles are *indistinguishable*. Now the big question: why must we divide by $N!$ which we previously argued took into consideration that the particles were originally considered to be distinguishable and must be corrected to make them indistinguishable? Moreover, the two entropy expressions (1.24) and (1.25) do not agree. Since the former was obtained from a thermodynamic identification, we have no reason to doubt its validity, although we have seen that it is in conflict with the expression for the entropy of an ideal gas.

The problems we have raised are due to an incomplete statistical specification of the problem. The fact that we have been treating large numbers rather than "mere" probabilities has allowed a certain arbitrariness to creep in. The only way in which this can be eliminated is by treating the probability distributions themselves rather than their moments.

Another point worth worrying about is that we have derived an expression for the canonical ensemble starting from an isolated system. If the system is isolated, as we have claimed, then the total differential of the energy (1.23) is zero. The fact that it is not zero means that we have implicitly placed the system in thermal contact with another system. In other words, the temperature cannot be defined for an *isolated* system! Placing the system in contact with a larger system allows fluctuations in the internal energy of our system which can be used to estimate the undetermined parameter β. The larger the thermostat, the more precise becomes our estimation of the temperature; alternatively for a vanishing thermostat, the temperature cannot be measured [cf. §4.8.2].

1.5 Maxwell's Error Law

It is rather ironic that the statistical approach which would lead to Boltzmann's principle was already around when Boltzmann introduced his relation in 1877. In fact, he was even familiar with it! In an 1868 paper, Boltzmann introduced a generalization of the Maxwell distribution law for the velocity of a particle which is acted upon by a force. Maxwell derived the law that bears his name in 1850. The then 19-year-old Maxwell recognized the fact that collisions between molecules of a gas install a statistical distribution of velocities rather than tend to equalize the velocities of all the molecules.

Until Maxwell, there was little interest in studying deviations from average behavior, and what Maxwell did was to wed probability theory to the kinetic theory of gases. Undoubtedly Maxwell was under the influence of Herschel's review of the works on probability theory of the Belgian statistician Adolphe Quetelet that appeared in the *Edinburgh Review* of July 1850. In particular, Herschel believed the law of errors, which identifies the average value as the most probable value, to be so important that he gave a mathematical proof that could be understood even by the well-informed layman.

Maxwell's derivation of his probability distribution for the velocities of gas molecules makes use of the same postulates that Herschel used:

- the joint probability of two statistically independent events is the product of the individual probabilities,

- there is a relation between the probability of committing an error and the amount of error committed, and

- negative and positive errors of the same absolute amounts are equally likely.

Maxwell's basic tenet was that the collisions between the molecules of the gas, instead of tending to equalize the molecular velocities, establish a statistical distribution. His first proof, appearing in 1860, was prefaced by the remark that in the elastic collision of two spheres all directions of rebound are equally likely. From this, Maxwell concluded that all directions in the motion of the molecules are equally likely and *each* component of the velocity is independent of the value of the other components. It is this second assumption that Maxwell later referred to as "precarious."

Suppose, said Maxwell, that there are N particles in a given container and let v_x, v_y, and v_z be the components of the velocity of each particle in an orthogonal coordinate system. If the number of particles whose velocity component v_x lies between v_x and $v_x + dv_x$ is $Nf(v_x)$ while, at the same time, $Nf(v_y)$ and $Nf(v_z)$ are the number of particles whose velocity components v_y and v_z lie, respectively, between v_y and $v_y + dv_y$ and v_z and $v_z + dv_z$, then the number of particles in the volume element $dv_x\,dv_y\,dv_z$ centering about the origin will be $Nf(v_x)f(v_y)f(v_z)$. This number must be a function solely of the distance from the origin. Thus,

$$Nf(v_x)f(v_y)f(v_z) = \varphi\left(v_x^2 + v_y^2 + v_z^2\right).$$

Maxwell found the solution to this functional equation to be

$$f(v_i) = Ce^{Av_i^2}, \qquad \varphi(v) = C^3 e^{Av^2}.$$

He then argued that the constant A had to be negative for, otherwise, the number of particles would increase with velocity, the former tending to infinity with the latter, which is surely unreasonable. The constant C is the normalization factor.

Since the speed $v = \sqrt{v_x^2 + v_y^2 + v_z^2}$ is fixed, only two of the velocity components can be statistically independent. Realizing that this assumption "may appear precarious," Maxwell offered an alternative derivation of his velocity distribution. Now, said Maxwell, suppose that the velocities of two molecules are v_a and v_b before they collide and v_a' and v_b' after collision. The number of such encounters will be proportional to the product of their densities, $f_1(v_a)f_2(v_b)$, while other factors remain invariant. If we reverse the direction of the collision, the number of encounters is proportional to $f_1(v_a')f_2(v_b')$. At equilibrium, the rates of these two processes must balance one another,[3] and so

$$f_1(v_a)f_2(v_b) = f_1(v_a')f_2(v_b'),$$

---

[3]This is probably the first application of the principle of "microscopic reversibility," or "detailed balancing," at equilibrium.

subject to the condition of conservation of kinetic energy,

$$\tfrac{1}{2}\left\{ M_a v_a^2 + M_b v_b^2 \right\} = \tfrac{1}{2}\left\{ M_a v_a'^2 + M_b v_b'^2 \right\},$$

where M_a and M_b are the masses of the two particles.

The two equations imply the functional relation

$$f_1(v_a) = C_1 e^{-v_a^2/\alpha^2}, \qquad f_2(v_b) = C_2 e^{-v_b^2/\beta^2},$$

where

$$M_1 \alpha^2 = M_2 \beta^2$$

and C_1 and C_2 are constants of normalization. Maxwell made the additional comment that this is the *only* stable distribution, which implied, of course, that it was unique. The existence of a unique distribution to which all others would tend in the course of time was not proved until much later and remained a subject of controversy for a long time.

We now take the view that the particle velocities are random quantities and apply the methods of probability theory to obtain their error law without having to employ the additional information that the total energy is fixed, since this would necessarily mean that the system is isolated and hence it would have no notion of temperature. We replace the constraint of total fixed energy by the second law, in which the system is related *to the outside world*.

1.6   Gauss' Principle

To set Gauss' principle in the correct historical perspective, let us return to the year 1809. Four years earlier, Adrien Marie Legendre had published his method of least squares, which gave results which could be considered the "best" when they minimized the sum of the square of the errors. Yet, without any knowledge of the probability of a deviation, how could one surmise what best was? The key to the solution was given by Carl Friedrich Gauss in his *Theoria Motus Corporum Coelestium in Sectionibus Conicis Solum Ambientium*, or "The Theory of the Motion of Heavenly Bodies Moving about the Sun in Conic Sections."

The book dealt primarily with the mathematics of the orbits of the planets, but it did contain a section on the combination of observations which had been attacked previously by Euler, Laplace, and Legendre among others. The new angle was the probabilistic flavor of the discussion. Gauss started from a set of expected values which were linear functions of the observables whose constants of proportionality were certain parameters. He then assumed that the errors Δ, Δ', etc., between these values and the observed values had probabilities $f(\Delta), f(\Delta')$, etc. Having more equations than unknowns, he followed Pierre Simon Laplace's lead and considered the unknowns to be *a priori* equally likely, which Laplace referred to as the "principle of insufficient reason." The unknown parameters could then be estimated by maximizing the product $f(\Delta)f(\Delta') \cdots$ with respect to them. The stationary condition would then determine the "most probable" values of these parameters, but Gauss first needed an explicit expression for his error curve, $f(\Delta)$. Here, we can see the origins of what would be later called the method of "maximum likelihood" emerging.

So far Gauss did little that was new. Rather than taking his cue from Laplace, who used certain principles, such as that of insufficient reason, to arrive at the error curve and then a method of combining observations, he chose to reverse the steps and assumed the conclusion. From the form of the error curve he could make some general statements about it: the curve would be insensitive to whether the errors where positive or negative. Thus, $f(\Delta)$ would be symmetric about its maximum, $\Delta = 0$. At this point Gauss jumped to the conclusion that the error curve would be a maximum when the arithmetic mean of the measurements is equal to the most probable value of the quantity measured. This could only occur when the error curve had the form

$$f(\Delta) = \frac{h}{\sqrt{\pi}} e^{-h^2 \Delta^2}$$

for some positive constant h. This constant could be used as a measure of the precision of the observation. That the errors had to be normally distributed brought him back to the method of least squares.

As Stephen M. Stigler concludes in his book, *The History of Statistics*, Gauss' reasoning was circuitous and the conclusion did not follow from his premises. It may very well have gone without notice if Laplace hadn't observed the connection between the error curve and his central limit theorem involving the same form of the probability distribution. Replacing Gauss' disputable proof of his error curve, Laplace showed that if the errors were aggregates, then the limit theorem demanded that they be approximately distributed according to the normal, or "Gaussian," curve. The fusion between probability and least squares had been accomplished: the unknowns were distributed according to $\exp(-h^2 W)$, where $W = \Delta^2 + \Delta'^2 + \cdots$. Minimizing the sum of quadratic terms gave the greatest posterior probability, and it provided a way of determining how good best was in the method of least squares. We now derive the law of error for which the arithmetic mean is the most probable value of the quantity measured in a more modern setting using Maxwell's distribution as an illustration.

Let the true value of the velocity be v. Suppose we make a set of measurements and obtain the values $v_1, v_2, v_3, \ldots, v_r$. From this set of r values we want to determine a value of the velocity that will be most representative of the measurements. The most obvious is the arithmetic mean

$$\bar{v} = \frac{1}{r} \sum_{i=1}^{r} v_i.$$

If there are no errors in measurement, then the same value of v will result from every observation. The differences,

$$\Delta_i = v_i - v, \qquad (1.27)$$

are the *errors* committed, and we want to determine the frequency or error curve in which the ordinate gives the number of cases in which the error lies in any given interval $d\Delta_i$. The probability that the error shall lie in this interval is $f(\Delta_i)d\Delta_i$, and the density $f(\Delta_i)$ is the error function we wish to determine. Here we are assuming that the error function is a function only of the deviation.

The probability that in a set of r measurements we will commit errors Δ_i is $\prod_{i=1}^{r} f(\Delta_i)$, since the errors themselves are independent of one another and are assumed to have a common distribution. The value of v which we will accept as the true or most representative value of the velocity will maximize the error function. Here we are inverting the dependence of the different measurements v_i and the true, but unknown, value of the velocity v. In other words, the v_i are now the parameters which are to be held constant and v is the variable. Since the integral of the error function with respect to v is not unity, we cannot talk about the different probabilities for v but rather compare their likelihoods.

The stationary condition is

$$\sum_{i=1}^{r} \frac{f'(\Delta_i)}{f(\Delta_i)} = 0, \qquad (1.28)$$

since differentiating with respect to Δ_i is equivalent to differentiating with respect to v apart from a change in sign.

Assume now that the most probable value of v is the arithmetic mean, $\bar{v}$, which coincides with the mean of the distribution.[4] By definition of the arithemetic mean we have

$$\sum_{i=1}^{r} \Delta_i = \sum_{i=1}^{r} (v_i - \bar{v}) = 0. \qquad (1.29)$$

Since $\bar{v}$ is the most probable value of the velocity, (1.28) must be equivalent to (1.29). Their equivalence may be established by setting the same terms in both sums proportional to one another,

$$(\ln f(\Delta_i))' = -\lambda \Delta_i, \qquad (1.30)$$

where λ is a constant of proportionality. It cannot depend upon the individual velocities v_i, for if it did depend on the velocity v_i, the jth equation would have a quantity on the left that depends only on the velocity v_j while the quantity on the right would depend upon both v_i and v_j, which would lead to a contradiction. The constant of proportionality may, however, depend upon the mean velocity $\bar{v}$.

---

[4]In general, this is rigorously so when the number of measurements increases without bound.

In regard to the difficulty raised by Maxwell's proof of his velocity distribution, we may look upon the definition of the arithmetic mean (1.29) as a constraint on the variational expression (1.28). The coefficient of proportionality, λ, then becomes a Lagrange multiplier, and introducing the constraint directly into the variational expression liberates the Δ_i so that the only way the variational equations can be satisfied identically is to require that (1.30) holds for each i.

Integrating (1.30), we get

$$\ln f(\Delta_i) = -\tfrac{1}{2}\lambda\Delta_i^2 + \text{const.} \tag{1.31}$$

The constant in the expression will determine the proper normalization. Our problem now is to determine the parameter λ.

In order for the error function to be a proper density, we must have

$$\frac{\partial^2 \ln f}{\partial \Delta_i^2} = -\lambda < 0$$

for all i. We may relate this to the property of the convexity of the total energy or, still better, to the concavity of the entropy since, in view of Boltzmann's principle (1.2), the latter seems more appropriate. For a body which is in macroscopic motion, the entropy S is a function of the internal energy U, which is the difference between its total energy E and its kinetic energy $M\bar{v}^2/2$ where M is its mass. Hence,

$$S = S(U) = S\left(E - \tfrac{1}{2}M\bar{v}^2\right)$$

and the second law can thus be expressed as

$$S' = \frac{\partial S}{\partial U}\frac{dU}{d\bar{v}} = \frac{1}{T} \cdot (-M\bar{v}), \tag{1.32}$$

where T is the absolute temperature.

Differentiating a second time, we find

$$S'' = -M/T < 0.$$

The concavity property of the entropy guarantees the existence of the error function since it identifies the undetermined multiplier as

$$\lambda = M/T.$$

Thus, the error law (1.31) is none other than Maxwell's distribution

$$f(|v_i - \bar{v}|) = \sqrt{\frac{M}{2\pi T}} \exp\left(-\frac{M(v_i - \bar{v})^2}{2T}\right). \tag{1.33}$$

Maxwell proceeded along a different line in order to relate the constant λ to the absolute temperature of the gas: he compared the theoretical pressure of the gas with the experimental relation between the pressure and temperature (Gay–Lussac's law). In particular, if the gas is at rest, $\bar{v} = 0$, formula (1.33) reduces to the usual form of Maxwell's law.

This is a particular case where the law of error leading to the average value as the most probable value determines the physical statistics. Even more important is the fact that it illustrates how the entropy is related to probability. By integrating the second law (1.32), we can determine the form of the entropy

$$S(\bar{v}) = \text{const.} - \frac{M\bar{v}^2}{2T}. \tag{1.34}$$

Expanding the quadratic function in Maxwell's distribution (1.33), we readily see that it can be written as

$$\begin{aligned} \ln f(v_i; \bar{v}) &= -\frac{M}{2T}(v_i - \bar{v})^2 + \text{const.} \\ &= \frac{M}{2T}\left\{\bar{v}^2 + 2(v_i - \bar{v})\bar{v} - v_i^2\right\} + \text{const.} \\ &= S(v_i) - S(\bar{v}) - S'(\bar{v})(v_i - \bar{v}) + \text{const.}, \end{aligned} \tag{1.35}$$

where the "stochastic" entropy $S(v_i)$ is the same function of the random variable v_i as the entropy $S(\bar{v})$ is of the average velocity [cf. Eq. (1.34)]. This may seem like a trivial statement, hardly worth being dignified as a principle. But as we shall see in §3.6, it is the cornerstone of statistical mechanics. In situations where the two expressions do not have the same functional form, or perhaps where the stochastic entropy even vanishes, one cannot apply statistical mechanical reasoning.

Expression (1.35) is the most general form of Gauss' error law for which the average and most probable values coincide. In its most general form, Gauss' law does not require that negative and positive errors of the same magnitude are equally likely. It is in this form that the error law is comparable to Boltzmann's principle (1.2). Whereas the first two terms are identical to Boltzmann's principle, the term containing the entropy derivative is missing. However, it is precisely this term that introduces the notion of temperature via the second law (1.32). Hence, are we to assume that Boltzmann's principle is valid *only* for isolated systems?

1.7   Boltzmann versus Gauss

The error law for the distribution in the number of particles can be derived in exactly the same way as the law for the velocities that we derived in the previous section. We want the arithmetic mean of the number of particles,

$$\sum_{i=1}^{r} (n_i - \bar{n}) = 0, \tag{1.36}$$

to be equivalent to the likelihood equation

$$\sum_{i=1}^{r} \frac{\partial}{\partial \bar{n}} \ln f(n_i; \bar{n}) = 0. \qquad (1.37)$$

This can be accomplished by setting the corresponding terms in the two sums proportional to one another,

$$\frac{\partial}{\partial \bar{n}} \ln f(n_i; \bar{n}) = -\varphi''(\bar{n}) (n_i - \bar{n}), \qquad (1.38)$$

where we have introduced a constant of proportionality as the second derivative of a continuous and twice differentiable function, $\varphi(\bar{n})$, for convenience. Again we note that it must be independent of the n_i, but it may depend upon the average, $\bar{n}$.

Integrating (1.38) gives

$$
\begin{aligned}
\ln f(n_i; \bar{n}) &= -\int (n_i - \bar{n}) \, d\varphi'(\bar{n}) \\
&= -(n_i - \bar{n}) \varphi'(\bar{n}) - \varphi(\bar{n}) + \psi(n_i), \qquad (1.39)
\end{aligned}
$$

where ψ must be independent of $\bar{n}$ but it can be, and usually is, a function of the variates n_i. Since f will be maximum when $n_i = \bar{n}$, we take $\psi(n_i)$ to be the same function of n_i that $\varphi(\bar{n})$ is of the average, $\bar{n}$. We will formalize this into a principle in §3.6. Here we argue that such a choice would guarantee that f be a proper probability distribution, based on the concavity of the function $\varphi(\bar{n})$ since a *strictly* concave function satisfies the inequality

$$\varphi(\bar{n}) - \varphi(n_i) - (\bar{n} - n_i)\varphi'(\bar{n}) > 0,$$

as we shall see in the next section.

Expression (1.39) is Gauss' law of error leading to the average number of particles as the most probable value of the quantity measured. Armed with this error law, we now return to our discussion of the correct expression for the entropy that we began in §1.4.

If we set

$$\varphi(\bar{n}) = -\bar{n} \ln \bar{n} + \bar{n} \qquad (1.40)$$

and write down an analogous expression for $\varphi(n_i)$ as a function of n_i, the error law (1.39) becomes

$$
\begin{aligned}
f(n_i; \bar{n}) &= \exp\{(n_i - \bar{n}) \ln \bar{n} + \bar{n} \ln \bar{n} - \bar{n} - n_i \ln n_i + n_i\} \\
&= \frac{\bar{n}^{n_i}}{n_i^{n_i}} \frac{e^{-\bar{n}}}{e^{-n_i}} = \frac{\bar{n}^{n_i}}{n_i!} e^{-\bar{n}},
\end{aligned}
$$

which will be easily recognized as the Poisson distribution under the condition that Stirling's approximation is valid.

We would like to build up the multinomial distribution from products of the Poisson distribution, and in order to do so, we have to change our perspective.

Instead of considering n_i as any observable, we consider it to be the number of particles in the ith cell whose average value is $\bar{n}_i$. All this necessitates is putting an index on $\bar{n}$. Then since n_i is distributed according to the Poisson distribution, the distribution of particles among the r cells will be given by

$$\prod_{i=1}^{r} f(n_i; \bar{n}_i) = \prod_{i=1}^{r} \frac{\bar{n}_i^{n_i}}{n_i!} e^{-\bar{n}_i}$$

$$= N! \prod_{i=1}^{r} \frac{1}{n_i!} \left(\frac{\bar{n}_i}{N}\right)^{n_i}. \qquad (1.41)$$

This is none other than the multinomial distribution where the *a priori* probabilities are given by their maximum likelihood values, $\hat{p}_i = \bar{n}_i/N$. The maximum of the multinomial distribution, (1.41), can be written as the error law

$$f(\{n_i\}; \{\bar{n}_i\}) = \exp\left\{\sum_{i=1}^{r} \left[(n_i - \bar{n}_i)\ln\bar{n}_i + \bar{n}_i\ln\bar{n}_i - n_i\ln n_i\right]\right\}, \qquad (1.42)$$

whose potential is seen to be given by (1.40) since adding n_i and subtracting $\bar{n}_i$ in the sum does not alter its value; both sums are equal to N, the total number of particles.

We would therefore like to identify the sum of (1.40) over all cells with the entropy. But we cannot since (1.40) is not *extensive*. It can be made extensive by introducing a new set of extensive scale parameters, C_i, such that

$$\varphi(\{\bar{n}_i\}) = \sum_{i=1}^{r} \bar{n}_i \left[\ln\left(\frac{C_i}{\bar{n}_i}\right) + 1\right] = S(\{\bar{n}_i\}). \qquad (1.43)$$

These scale parameters have absolutely no effect on the law of error (1.42). In this sense, the error law is scale invariant. The scale parameters, C_i, can be interpreted as the number of cells of type i which we will see to be equal to the number of states in a given energy interval.

The multinomial distribution (1.41) is also invariant to a change of scale. This would not have been so had we been working with the statistical weight where we would append the factors $C_i^{n_i}$ that would represent the number of ways n_i particles could be put into C_i cells. But then we would have to divide by $N!$ in order to come out with an entropy expression that is extensive since there would remain the term $N\ln N$ in addition to $-\sum_i^r \bar{n}\ln(\bar{n}_i/C_i)$. Rather, we have seen that according to Gauss' principle (1.42), the term $N\ln N$ cancels, and it is a matter of choice how we introduce the extensive scale factor. Thermodynamics demands that the scale factors should be the C_i and not N which render the entropy extensive.

Differentiating (1.43) with respect to $\bar{n}_i$, we have

$$\left(\frac{\partial S}{\partial \bar{n}_i}\right)_V = -\ln\left(\frac{\bar{n}_i}{C_i}\right)$$

$$= \frac{1}{T}\left(\frac{\partial E}{\partial \bar{n}_i}\right)_V + \left(\frac{\partial S}{\partial \bar{n}_i}\right)_{E,V}$$

$$= \frac{E_i - \mu}{T},$$

where μ is the chemical potential. For correct normalization we must set

$$e^{-\mu/T} = \frac{1}{N} \sum_{i=1}^{r} C_i e^{-E_i/T},$$

or equivalently,

$$\mu = T[\ln N - \ln \mathcal{Z}(\beta)]. \tag{1.44}$$

Instead of our old definition of the sum-over-states, (1.21), we now have

$$\mathcal{Z}(\beta) = \sum_{i=1}^{r} C_i e^{-\beta E_i} \tag{1.45}$$

as the definition of the partition function. Often, distinction is made between a sum over (quantum) states, where the C_i are present, and a sum over "cells," or (quantum) states of the molecules, where the C_i are absent. Expression (1.45) shows that the sum is over the cells and C_i is present. Since one always goes over to an integral, where C_i represents the number of states in the energy interval from E_i to $E_i + \Delta E_i$, the distinction has no physical relevance. Nevertheless, the correct expression for the partition function is (1.45) and not (1.21).

All this points to the fact that the multinomial distribution is just about the worst place to begin a derivation of the canonical ensemble. Historical prejudice has led to a confusion of concepts and terms, and this is the culprit that is responsible for the continuing debate over whether particles are really distinguishable or indistinguishable. Division of $N!$ has nothing to do with distinguishability or the lack of it but rather with trying to get out an expression for the entropy that coincides with the thermodynamic result. If the multinomial distribution is chosen as the correct distribution, the particles, by definition, are indistinguishable since all we want to know is how many particles there are in a cell and not which particles are in which cells. In other words, particles are always treated as being indistinguishable unless we are told otherwise (e.g., by discriminating them according to their energies, cf. §3.8).

1.8   The Essence of the Second Law

The relation between Gauss' principle (1.39) and the concavity of the function φ brings to light a very deep and intrinsic connection among fluctuations, thermodynamics, and probability.

Let $\bar{X}$ be any extensive variable; that is, we associate the average value with the corresponding extensive thermodynamic quantity. Fluctuations in measurements of this quantity will be governed by Gauss' principle,

$$\ln f(X) = S(X) - S(\bar{X}) - (X - \bar{X}) \left(\frac{\partial S}{\partial \bar{X}} \right)_{\bar{Y}} + \text{const.}, \tag{1.46}$$

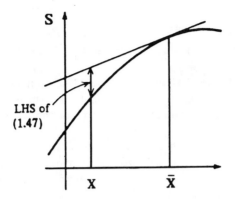

Figure 1.1: Strict concavity.

where $\bar{Y}$ stands for the remaining extensive variables which are held constant. As we have seen in the previous section, in order for f to be a *proper* probability distribution (i.e., a proper fraction), it is necessary that

$$S(\bar{X}) - S(X) - (\bar{X} - X)\left(\frac{\partial S}{\partial \bar{X}}\right)_{\bar{Y}} > 0, \qquad (1.47)$$

which is one definition of a *strictly concave* function. The left side of (1.47) can never be negative and vanishes only at $X = \bar{X}$,[5] as shown in Fig. 1.1. Geometrically speaking, the entropy function never rises above its tangent plane at any point.

By interchanging X and $\bar{X}$ in (1.47) and adding the two, we get

$$(\bar{X} - X)\left(\frac{\partial S}{\partial \bar{X}} - \frac{\partial S}{\partial X}\right) \leq 0, \qquad (1.48)$$

which says that the slope of a strictly concave function is monotonically decreasing. This inequality embodies the second law of thermodynamics: if $X = E$, inequality (1.48) guarantees that heat will not spontaneously flow from cold to hot. Alternatively, if $\bar{X} = \bar{V}$, the inequality negates the possibility of the pressure increasing with volume while if $\bar{X} = \bar{N}$, it states that the chemical potential is an increasing function of the number of particles. However, in this interpretation it is not clear how the average value $\bar{X}$ distinguishes itself from any (nonequilibrium) state X.

The value $\bar{X}$ is distinguished *probabilistically* among all other possible values of the energy since for $X = \bar{X}$, the probability density is a maximum; that is, the average value is the most probable value of the extensive quantity

---

[5] *Weakly concave* is defined by writing $\geq$ in (1.47) so that the equality can be valid for *some* distinct pairs of states.

measured. The concavity criterion (1.47) does not guarantee that each X is less than $\bar{X}$ but rather that the entropy of the equilibrium state $S(\bar{X})$ cannot be inferior to the average entropy, $\overline{S(X)} = \sum_i f_i S(X_i)/\sum_i f_i$. This can be seen by multiplying (1.47) by f_i and summing to give

$$\sum_i f_i \, S(\bar{X}) - \sum_i f_i S(X_i) = -\sum_i f_i \ln f_i \geq 0, \qquad (1.49)$$

where we have made use of Gauss' principle, (1.46). This inequality was first derived by Gibbs, and the quantity on the right is known as the Gibbs entropy. As we have shown, the inequality is due to the concavity property of the entropy.

We cannot assert that each $S(X)$ will be less than the equilibrium entropy $S(\bar{X})$. Nevertheless, this is precisely what Boltzmann's principle (1.2) implies! Then what is the meaning of the state of maximum entropy? We must first ask maximum with respect to what? If S is a function of any extensive variable, X, then a necessary condition for an extremum is that $\partial S/\partial X = 0$, but from thermodynamics we know that

$$S'(\bar{X}) = \frac{\alpha}{T}, \qquad (1.50)$$

where if $\bar{X} = \bar{E}$, then $\alpha = 1$, while if $\bar{X} = \bar{V}$, then $\alpha = P$ or if $\bar{X} = \bar{N}$, $\alpha = -\mu$. Since

$$S''(\bar{X}) < 0 \qquad (1.51)$$

is equivalent to (1.47) for a twice differentiable function, the entropy will not reach its maximum unless the temperature is practically infinite.

We can, however, consider $S(\bar{X})$ as *constrained* maximum. In thermodynamics, the intensive variables are defined according to (1.50), where S is the *thermodynamic* entropy. However, our S is a *statistical* entropy and (1.50) should be interpreted as a necessary condition for a constrained maximum, where α/T plays the role of a Lagrange multiplier for the appropriate constraint. We may thus consider the maximum of S at $\bar{X}$ as a constrained maximum. The equilibrium state is defined by fixing the value of the temperature as well as the external parameters needed to specify the state of the system.

Moreover, we have no way of guaranteeing that the deviation $X - \bar{X}$ is always negative. And since the entropy is a concave and monotonically increasing function of X, a positive deviation will lead to a state of greater entropy. The original interpretation of Boltzmann's principle in terms of (1.2) therefore falls apart. But how could a spontaneous fluctuation in an isolated system ever lead to an increase in entropy beyond its equilibrium value? The answer is that the system is not isolated for, according to (1.50), the temperature of the system must be defined—to ask for the temperature of an isolated system has no meaning. And since the system is in contact with a heat reservoir, there is nothing to prevent the deviation $X - \bar{X}$ from being positive.

Rearranging (1.49) gives

$$S(\bar{X}) \geq \sum_i f_i S(X_i) \Big/ \sum_i f_i = \overline{S(X)}, \qquad (1.52)$$

where $\bar{X}$ is the weighted average

$$\bar{X} = \sum_i f_i X_i \Big/ \sum_i f_i, \qquad (1.53)$$

or the mean of the distribution. It is not necessary to require that the sum of all the weights f_i be unity since the events are not mutually exclusive. Expressed in words, inequality (1.52) states that the "statistical" entropy, $S(\bar{X})$, which we assume always coincides with the thermodynamic entropy, cannot be inferior to the average of the "stochastic" entropies, $S(X_i)$, which are functions of the random events, X_i. The equality in (1.52) holds only when every X_i coincides with the average value, $\bar{X}$, in which case there are no fluctuations or deviations from the average behavior, and consequently, the entropy of the average is equal to the average of the entropies.

We shall assume that the arithmetic mean of the measurements,

$$\bar{X} = \frac{1}{r} \sum_i X_i, \qquad (1.54)$$

is equal to the mean of the distribution. In general, this is rigorously so when the number of measurements increases without bound. But for the particular family of distributions we will be dealing with (i.e., the so-called exponential family of distributions), it will be true for *any* size of the sample—even for a sample of size 1. Replacing the mean of the distribution, (1.53), by the arithmetic mean, (1.54), in (1.52) and observing that

$$S\left(\sum_i X_i\right) \geq S\left(\frac{1}{r} \sum_i X_i\right),$$

we have

$$\sum_i f_i S\left(\sum_i X_i\right) \geq \sum_i f_i S(X_i), \qquad (1.55)$$

which is a necessary and sufficient condition that $S(X)$ should increase (in the wide sense) for $X > 0$. Inequality (1.55) characterizes monotonic increasing functions just as (1.52) characterizes continuous concave functions.

Thus far, X has stood for E, V, or N. Can it stand for the entire set? Consider a composite system in which the individual subsystems are labeled by the subscripts 1 and 2. Since the entropy is a universal property of all systems, it suffices to consider a single experiment. Consider an ideal gas whose fundamental relation is

$$S = N \ln \left(\frac{E^a V^b}{N^{a+b}}\right) + N s_0, \qquad (1.56)$$

where s_0 is a constant. It is clear that the entropy is a first-order homogeneous function for if we increase the energy λ times, both the number of particles and volume are increased exactly λ times. The property of first-order homogeneity is stated as

$$S(\lambda \mathbf{X}) = \lambda S(\mathbf{X}), \tag{1.57}$$

where $\mathbf{X}$ is a shorthand notation to denote the entire set, (E, V, N), of extensive variables.[6] The condition that the entropy be concave in all three variables entails

$$(N_1 + N_2) \ln \left[\frac{(E_1 + E_2)^a (V_1 + V_2)^b}{(N_1 + N_2)^{a+b}} \right] \geq N_1 \ln \left(\frac{E_1^a V_1^b}{N_1^{a+b}} \right) + N_2 \ln \left(\frac{E_2^a V_2^b}{N_2^{a+b}} \right).$$
$$\tag{1.58}$$

Let $E = E_1 + E_2$, $N = N_1 + N_2$, and $V = V_1 + V_2$ be the total energy, number of particles, and volume of the composite system. If subsystem 1 has a fraction, λ, of the total energy, it will also have the same fraction of particles and volume on account of the first-order homogeneous property of the entropy. Then, with $E_1 = \lambda E$, $[E_2 = (1 - \lambda)E]$, $N_1 = \lambda N$, $[N_2 = (1 - \lambda)N]$, and $V_1 = \lambda V$, $[V_2 = (1 - \lambda)V]$, the concavity criterion becomes

$$N \ln \left(\frac{E^a V^b}{N^{a+b}} \right) \geq \lambda N \ln \left(\frac{E^a V^b}{N^{a+b}} \right) + (1 - \lambda) N \ln \left(\frac{E^a V^b}{N^{a+b}} \right).$$

This shows that on account of the first-order homogeneous property of the entropy, the equality holds and the entropy is *not a strictly concave function of all three extensive variables simultaneously*.[7]

At most, the entropy can be considered as a strictly monotonically increasing and concave function in any two of the three extensive variables which are indicated by X and Y in Fig. 1.2. Furthermore, in §6.4 we will argue that one of these variables must always be the energy since fluctuations in the volume and particle number cannot be achieved at a constant energy. In fact, we will show that such types of fluctuations cannot occur simultaneously and actually represent complementary descriptions of the same phenomenon!

---

[6]It would appear from the fundamental relation (1.56) that the entropy is a first-order homogeneous function of N only. However, by defining the intensive variables as $1/T = aN/E$, $P/T = bN/V$, and $-\mu/T = \ln \left(E^a V^b / N^{a+b} \right) - (a + b)$, the fundamental relation (1.56) can be written as

$$S = \frac{1}{T} E + \frac{P}{T} V - \frac{\mu}{T} N,$$

apart from an undetermined constant. This shows that the entropy is, indeed, a first-order homogeneous function of all three extensive variables.

[7]An even more compelling reason why the entropy cannot be concave in all the extensive variables, simultaneously, is that the Hessian vanishes. In the case of an ideal gas, the determinant of the matrix of second derivatives in the entropy vanishes,

$$\begin{vmatrix} -\frac{aN}{E^2} & 0 & \frac{a}{E} \\ 0 & -\frac{bN}{V^2} & \frac{b}{V} \\ \frac{a}{E} & \frac{b}{V} & -\frac{(a+b)}{N} \end{vmatrix} = 0,$$

indicating the presence of *degenerate* critical points.

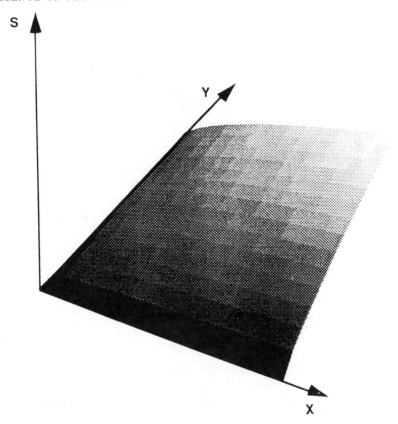

Figure 1.2: Entropy as a monotonically increasing function of two extensive variables.

The inequality in (1.58) is often referred to as the "superadditivity" property of the entropy, and it has been suggested that it is the characterizing property of the entropy rather than concavity. However, once the first-order homogeneous property of the entropy is introduced, the inequality reduces to an equality which demonstrates the illusiveness of the superadditivity property of the entropy.[8] *The characterizing property of the entropy is concavity.* In the next section we will borrow an example, devised by Lorentz, to show that the

---

[8]A case in point is the definition of the entropy of a black hole. Setting the entropy proportional to the area of a black hole gives it a quadratic dependence upon the (irreducible) mass M. Then if two black holes collide, the entropy will increase because $(M_1 + M_2)^2 > M_1^2 + M_2^2$. The argument is, however, fallacious simply because the putative expression for the entropy is a convex function. Among other things, the heat capacity turns out to be negative. Also the first-order homogeneity of the entropy is thrown to the winds. Moreover, it condradicts the rudiment that when two systems are placed in thermal contact which were originally at the same temperature, proportional to $1/M$, the entropy will be equal to the sum of the entropies of the individual subsystems. The expression for the "entropy" of a black hole is anything but an entropy.

entropy can actually be *subadditive* when it is taken to be a particular concave function of a single variable.

That the entropy cannot be a concave function of all three extensive variables is implicit in the definition of the different ensembles in statistical mechanics. The different ensembles are derived from the fact that the probability distribution f is normalized,

$$\exp\left\{S(\bar{E}) - \bar{E}S'(\bar{E})\right\} = \sum_i B \exp\left\{S(E_i) - E_i S'(\bar{E})\right\}, \qquad (1.59)$$

where B is a norming function which may be a function of the E_i but not of $\bar{E}$. The quantity in the exponent of the left side of (1.59) is the Massieu transform of the entropy with respect to the energy, which is $-A/T$, where A is the Helmholtz potential. This is the so-called canonical ensemble.

Expression (1.59) may be generalized to include other ensembles by introducing additional extensive variables. The normalization condition then becomes

$$\exp\left\{S(\bar{E}, \bar{X}) - \bar{E}\left(\frac{\partial S}{\partial \bar{E}}\right)_{\bar{X}} - \bar{X}\left(\frac{\partial S}{\partial \bar{X}}\right)_{\bar{E}}\right\}$$
$$= \sum_i B \exp\left\{S(E_i, X_i) - E_i\left(\frac{\partial S}{\partial \bar{E}}\right)_{\bar{X}} - X_i\left(\frac{\partial S}{\partial \bar{X}}\right)_{\bar{E}}\right\}.$$

If $\bar{X} = -\bar{N}$, then the Massieu function is the work potential, $W = -P\bar{V}/T$, of the "grand-canonical" ensemble. Alternatively, if $\bar{X} = \bar{V}$, the Massieu function is $-G/T$, corresponding to the temperature–pressure ensemble, where G is the Gibbs potential. However, if we were to consider the entropy as a function of the three extensive variables simultaneously, the Massieu transform would vanish identically on account of the first-order property of the entropy; this is guaranteed by the Gibbs–Duhem relation [cf. §3.7].

Hence, *it is the first-order homogeneous property of the entropy that prevents the entropy from being considered as a concave function of all the extensive quantities simultaneously.* The entropy can be considered as a concave function simultaneously of the energy and either the number of particles or the volume.[9] More will be said in §3.7 and §5.3 concerning the properties of

---

[9]There is nothing to prevent us from mathematically considering the entropy as a concave function of N and V. The concavity condition,

$$S\left(\frac{N_1 + N_2}{2}, \frac{V_1 + V_2}{2}\right) \geq \tfrac{1}{2}\left\{S(N_1, V_1) + S(N_2, V_2)\right\},$$

for an ideal gas is

$$(N_1 + N_2)\ln\left[\frac{2^a(V_1 + V_2)^b}{(N_1 + N_2)^{a+b}}\right] \geq N_1 \ln\left(\frac{V_1^b}{N_1^{a+b}}\right) + N_2 \ln\left(\frac{V_2^b}{N_2^{a+b}}\right).$$

Rearranging this inequality, we obtain

$$b\left\{(N_1 + N_2)\ln\left(\frac{N_1 + N_2}{V_1 + V_2}\right) - N_1 \ln\left(\frac{N_1}{V_1}\right) - N_2 \ln\left(\frac{N_2}{V_2}\right)\right\}$$

the different ensembles.

1.9 Concavity and Irreversibility

We have previously considered the entropy to be a concave function of the energy. The equilibrium state is fixed somewhere on the entropy curve by the temperature and external parameters that are needed to specify the state of the system. Deviations from the state of equilibrium can arise because the system is coupled to a heat reservoir which leaves the energy free to fluctuate. The points on the entropy curve consist of nonequilibrium states that can be realized by the system. An alternative view is also possible. We may imagine the entropy curve to be a locus of points representing equilibrium states which differ according to the external constraints. It is this latter interpretation that we will use in this section to develop the relationship between concavity and irreversibility.

An illuminating example of the interrelationship between concavity and irreversibility has been given by Lorentz, in his Leiden lectures, who considered the splitting of a pencil of rays of a certain frequency into reflected and transmitted rays by a thin plate. Such an entropy curve is shown in Fig. 1.3 as a function of the energy $\bar{E}$ at a given frequency.[10] The curve has a vertical tangent at $\bar{E} = 0$ (i.e., $T = 0$) and rises continuously, turning toward the $\bar{E}$ axis to which it is asymptotically horizontal. In the absence of radiation $\bar{E} = 0$ and so too will be the entropy.

Let the energies of the split pencils be denoted by $\bar{E}_1$ and $\bar{E}_2$, and by the conservation of energy, the original energy is their sum,

$$\bar{E} = \bar{E}_1 + \bar{E}_2.$$

The two partial beams will have entropies S_1 and S_2 and we want to show

$$\leq a \left\{ N_1 \ln N_1 + N_2 \ln N_2 - (N_1 + N_2) \ln \left(\frac{N_1 + N_2}{2} \right) \right\},$$

which, on account of the inequality

$$X \ln \frac{X}{A} + Y \ln \frac{Y}{A} \geq (X + Y) \ln \left(\frac{X + Y}{A + B} \right),$$

is satisfied identically since the left-hand side is always negative while the right-hand side is always positive for all values of a and b unless $N_1/V_1 = N_2/V_2$, in which case the equality sign applies. But how can the number of particles and volume vary at constant energy? The hierarchy set forth by Gibbs: microcanonical → canonical → grand-canonical ensemble or temperature–pressure ensemble is not superfluous: Energy can vary without matter flow or mechanical work being done. But matter flow or mechanical work cannot occur without an energy exchange!

[10] We don't know it yet, but the argument holds for radiation at relatively high frequencies only. This corresponds to the so-called Wien region of the spectrum as opposed to the Rayleigh–Jeans region of small frequencies where the following argument will not work. For greater detail see §2.3.2.

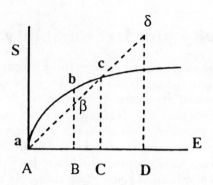

Figure 1.3: Entropy versus energy curve for radiation at a given frequency.

that

$$S < S_1 + S_2,$$

stating that *splitting of the two beams occurs with an increase in entropy and hence is an irreversible process.*

The tracts $AB = \bar{E}_1$, $AC = \bar{E}_2$, and $AD = \bar{E}_1 + \bar{E}_2 = \bar{E}$ so that $AB = CD$. The entropies of the split beams are $Bb = S_1$ and $Cc = S_2$ while the entropy of the unsplit beam is $Dd = S$. If the entropy were a linear function, then it would be given by the straight line $Ac\delta$, where $D\delta = B\beta + Cc$. But the entropy curve for radiation at a given frequency is represented by the concave curve so that

$$Dd < Cc + Bb.$$

Hence, the entropy of the single beam is less than the sum of the entropies of the split beams. We have thus established that the entropy of radiation is, in fact, *subadditive*. A sufficient condition for the subadditive property of the entropy,

$$S\left(\sum_i X_i\right) \le \sum_i S(X_i),$$

is that $X^{-1}S(X)$ be a strictly decreasing function. Combining this with the concavity inequality leads to the string of inequalities

$$rS\left(\frac{\sum_i^r X_i}{r}\right) \ge \sum_i^r S(X_i) \ge S\left(\sum_i^r X_i\right),$$

where the equality signs will hold only in the case that the entropy is a first-order homogeneous function of X.

In information theory, where $S = -X \ln X$, the entropy is subadditive. However, it would be wrong to conclude that thermodynamics requires the more stringent condition of subadditivity in addition to concavity. Thermodynamics offers more of a variety of concave functions to represent the entropy,

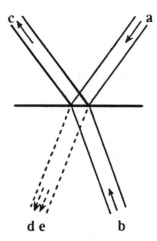

Figure 1.4: Inversion and recombination of pencils.

and in particular, it can have the form $S = \ln X$. It will be appreciated immediately that $X^{-1} \ln X$ is not a strictly decreasing function for all $X > 0$ so that we do not have a strictly subadditive entropy. With this aside on the property of subadditivity of the entropy, let us return to our discussion of the entropy of radiation.

Since Maxwell's equations are invariant under time reversal, $t \to -t$, which means that they are invariant under a reversal of propagation, we would be led to believe that the inversion of the direction of the propagation of the light rays should be an entirely reversible process. However, let us consider what happens when two pencils of light combine. For the process to be reversible, the two pencils a and b, in Fig. 1.4, would combine to give a single pencil c. But only part of pencil a will be reflected and we must also consider that a part of it, d, will be transmitted. Likewise, a part, e, of the pencil b will be reflected. In order that only the combined pencil c results, the transmitted part d and the reflected part e must cancel each other exactly. This can only happen if the amplitudes are equal and they have *exactly* opposite phases.

There is also an uncertainty in the frequency of the pencils because they are not made up of strictly monochromatic waves. And even if we had mirrors sensible enough to undo the splitting of one ray of the pencil, it would not be able to undo the splitting of the other rays which comprise the pencil. Here we have an example of a not impossible, but highly improbable, state of affairs which we referred to in §1.2 in connection with the second law. If we could be so precise as to make up a pencil with exactly monochromatic waves, we could in principle combine the split pencils in a completely reversible fashion. But the entropy would vanish! The presence of a finite entropy attests to the fact that such accuracy is not attainable so there will be deviations of the rays from a common frequency and this is responsible for the apparent irreversibility of the process. The finest spectrum line still has some breadth!

Moreover, since the slope of the solid curve in Fig. 1.3 is the inverse temperature, it is apparent that the temperatures of the split pencils are less than the combined one. In the process of splitting the beam, energy has passed from a higher temperature to a lower one, and this is inevitably accompanied by an increase in entropy. In the hypothetical reverse process, the temperature of the combined pencil would be equal to the temperatures of the split beams and the entropy would not increase. In no way can we envision the situation in which the temperatures of the split beams would be greater than the temperature of the combined one. This is guaranteed by the concavity of the entropy as a function of the energy,

$$S(E_i) - S(\bar{E}) - \left(\frac{\partial S}{\partial \bar{E}}\right)(E_i - \bar{E}) \leq 0, \tag{1.60}$$

obtained from (1.47) by setting $\bar{X} = \bar{E}$.

1.10   The Principle of Maximum Work Done by a System to Restore Equilibrium

In this section we will illustrate the concavity property of the entropy by deriving the principle of maximum work done by a system in order to restore it to a state of complete equilibrium. The thermodynamic principle states that the maximum work done by the system in contact with a thermal reservoir is equal to the decrease in the free energy of the system.

We construct a supersystem comprised of our system in contact with a thermal reservoir, which we shall refer to as the medium, and an external agent upon which the system performs work, or has work done on it. All three are isolated from the external world, and the external agent is thermally isolated from both the medium and our system. Without the presence of the medium, there would be no uncertainty in the amount of work that is done on the external agent since it is a problem of mechanics. The possibility of heat transfer between the medium and our system makes the resulting work less than certain. Moreover, we consider the medium to be much larger than our system so changes in our system will not appreciably alter the state of the medium. Thus, we can consider the medium's temperature, T_0, and pressure, P_0, to be essentially constant. The index 0 will refer to the properties of the medium while the unindexed quantities will pertain to those of our system.

Thus far, we have interpreted deviations from the equilibrium state, characterized by the average values of the energy, $\bar{E}$, and volume, $\bar{V}$, to be due to spontaneous fluctuations which alter the state of the system. However, there is nothing to prevent us from considering the deviations to be the result of the action of an external agent which performs work on the system.[11] The removal of the work source will lead to spontaneous irreversible processes tending to restore the system to a state of complete equilibrium. Even without the removal of the external agent, the system is able to regain equilibrium by performing work on it. In a reversible process, thermodynamics tells us that the system will perform maximum work on the external agent.

In the case under consideration, the concavity criterion is

$$2S(\bar{E}, \bar{V}) \geq S(E_0, V_0) + S(E, V). \tag{1.61}$$

Due to the fact of the enormous difference in size between our system and the medium, the total energy $2\bar{E}$ will not be equally shared between the two. Let us introduce a parameter λ, which is a proper fraction, to indicate how the extensive quantities are partitioned between our system and the medium. The energy of the medium is $(1+\lambda)\bar{E}$ while that of our system is $(1-\lambda)\bar{E}$. Likewise, $(1+\lambda)\bar{V}$ is the volume of the medium and $(1-\lambda)\bar{V}$ denotes the volume of our system. The fact that our system is much smaller than the medium requires $\lambda \approx 1$.

Our system has been displaced from the state in which it is in complete equilibrium with the medium. If the agent is not subsequently and suddenly removed, our system can perform work on the external agent in order to return to a state of complete equilibrium with the medium. The energy change of our system and medium will be, respectively, $\Delta E = (1 - \lambda)\bar{E} - E$ and $\Delta E_0 = (1+\lambda)\bar{E} - E_0$. Since the average energy $\bar{E} = \frac{1}{2}(E + E_0)$, we have the "closure," or conservation, condition $\Delta E = -\Delta E_0$. Likewise, the change in the volume of the system, $\Delta V = (1 - \lambda)\bar{V} - V$, is equal and opposite to the change in the volume of the medium, $\Delta V_0 = (1 + \lambda)\bar{V} - V_0$. This means that the total volume of the medium and our system remains constant.

Since the temperature and pressure of the medium remain constant, we apply the concavity criterion to the entropy of the medium. According to our convention regarding deviations, we write the concavity criterion

$$S(E_0, V_0) - S[(1 + \lambda)\bar{E}, (1 + \lambda)\bar{V}] + \frac{\partial S}{\partial E_0}\Delta E_0 + \frac{\partial S}{\partial V_0}\Delta V_0 \geq 0$$

as

$$-T_0\,\Delta S_0 + \Delta E_0 + P_0\,\Delta V_0 \geq 0. \tag{1.62}$$

The entropy change of the medium is

$$\Delta S_0 = (1 + \lambda)S(\bar{E}, \bar{V}) - S(E_0, V_0)$$

---

[11] By confounding a state which has been produced by an external constraint, or work source, and a state produced by a spontaneous fluctuation from equilibrium, we are really considering the process to be Markov. The Markovian assumption essentially states that the system does not "remember" how it got into the given (nonequilibrium) state. This assumption will be used in our formulation of the regression and growth of fluctuations to and from equilibrium in Chapter 8.

on account of the first-order homogeneity property of the entropy, (1.57). Using the closure property of the energy, (1.62) can be written as

$$\Delta E \leq -T_0 \Delta S_0 + P_0 \Delta V_0. \tag{1.63}$$

The term $-T_0 \Delta S_0$ represents the heat given up by the medium to our system and $P_0 \Delta V_0$ is the work done on it by the medium. By taking into account the work W that our system does on the external agent, we may convert (1.63) into the equality

$$\Delta E = -T_0 \Delta S_0 + P_0 \Delta V_0 - W. \tag{1.64}$$

We now express the concavity criterion (1.61) as

$$- \Delta S_0 \leq \Delta S, \tag{1.65}$$

where

$$\Delta S = (1 - \lambda)S(\bar{E}, \bar{V}) - S(E, V)$$

is the entropy change of our system. Introducing (1.65) into (1.64) and using the closure condition for the volume lead to

$$W \leq -\Delta(E - T_0 S + P_0 V), \tag{1.66}$$

where we have extracted the Δ from the sum of terms since T_0 and P_0 are constants.

Consequently, the maximum work that can be performed by our system on the external agent is

$$W_{\mathrm{max}} = -\Delta(E - T_0 S + P_0 V), \tag{1.67}$$

which obviously will occur for a *reversible* process. In the event that the external agent is removed, spontaneous irreversible processes will develop in the system which restore the system to equilibrium. Setting $W = 0$ in (1.66), these irreversible processes satisfy the inequality

$$\Delta(E - T_0 S + P_0 V) \leq 0. \tag{1.68}$$

This is to say that spontaneous and irreversible processes, which occur in our system, cause the quantity $E - T_0 S + P_0 V$ to decrease until it reaches complete equilibrium with the medium.

Two special cases are of primary interest. If the temperatures of the medium and system are equal, $T_0 = T$, and the volume of the body remains constant, $\Delta V = 0$, then the maximum work done by the system is equal to the decrease in the Helmholtz potential,

$$W_{\mathrm{max}} = -\Delta A \geq 0,$$

while if the temperature and-pressure of our system are equal to those of the medium, we have

$$W_{\mathrm{max}} = -\Delta G \geq 0,$$

where G is the Gibbs potential. The inequality asserts that the Gibbs potential will decrease until complete equilibrium has been achieved. The negative of the Gibbs potential is equal to the maximum available work by our system at constant temperature and pressure.

Now, it is often claimed that by interchanging initial and final states in a reversible process, the negative of (1.67) is the minimum work that can be done on our system by our external agent to bring it out of equilibrium and into the given nonequilibrium state. However, we cannot use the concavity property to establish this.

We shall appreciate in Chapter 8 that thermodynamic criteria of evolution refer only to the evolution *toward* equilibrium and not away from equilibrium. Thermodynamic criteria remain mute on the evolution away from equilibrium. Nonthermodynamic or stochastic criteria are required to pronounce on the most likely form of evolution that the system may choose to evolve from equilibrium. It is clear that the concavity criterion can be of no help in digging into these extremely rare events. In fact, it goes against the grain of the second law as well as the general spirit of probability theory, which attach a small probability to such events. Yet, certain forms of evolution are more improbable than others, and it is a problem of the stochastics to settle the type of behavior that will be displayed when the system has been subjected to small thermal perturbations for an unspecified length of time.

1.11   Einstein's Formula Vindicated

The equivalence of the two definitions of concavity, (1.47) and (1.51), can be established by expanding the entropy in a truncated Taylor series

$$S(X_i) = S(\bar{X}) + (X_i - \bar{X})S'(\bar{X}) + \tfrac{1}{2}(X_i - \bar{X})^2 S''(\hat{X}) \qquad (1.69)$$

about the average value $\bar{X}$, where the prime means differentiation and $\hat{X}$ lies between X_i and $\bar{X}$. It is now easy to see that either inequality, (1.47) or (1.51), follows from the other.

If we restrict ourselves to small fluctuations, then we have a bonus in store. In this case, we can set $\bar{X} = \hat{X}$ in (1.69) without committing any appreciable error. Then (1.46) reduces to the Gaussian probability density function

$$f(X_i) = \sqrt{\frac{-S''}{2\pi k}} \exp\left\{\frac{1}{2k}S''(X_i - \bar{X})^2\right\}. \qquad (1.70)$$

In order for (1.70) to be a proper probability density, the concavity condition (1.47) must be satisfied. We have reintroduced Boltzmann's constant to make the following point.

Einstein made brilliant use of (1.70) in his studies of spontaneous fluctuations from equilibrium, especially in his quantitative analysis of the scattering of light in the presence of strong density fluctuations in critical opalescence.

The Polish physicist Marian Smoluchowski had already considered opalescence generated by strong density fluctuations slightly above the critical point. He did not give, however, the decrease in the intensity of light scattered by opalescence and the intensity of the light source. This Einstein accomplished by considering the entropy as a function of some internal set of parameters with zero average values. This avoided the problem concerning the first-order derivatives in the series expansion of the entropy about equilibrium. The entropy difference was then strictly a negative quadratic form which Einstein considered to be equal to the work necessary to bring the system into the nonequilibrium state. In order to obtain an expression of the average work, he applied equipartition of energy. He then identified the set of internal parameters with the coefficients in the Fourier decomposition of the density. The average of their square together with Maxwell's equations are used to calculate the ratio of the intensity of a ray of the light source to the intensity of the scattered ray. Expressing this ratio in terms of experimentally measurable quantities allowed Einstein's theoretical results to be corroborated experimentally.

Einstein also gave the expression for the mean square fluctuation in any extensive variable, X, as

$$\overline{(X_i - \bar{X})^2} = -k/S''. \tag{1.71}$$

In the case where $\bar{X}$ represents the average energy and $S'' = -1/T^2 C_V$, where C_V is the heat capacity at constant volume, Einstein concluded:

> The absolute constant k therefore determines the thermal stability of the system. The relationship just found is particularly interesting because it no longer contains any quantity that calls to mind the assumptions underlying the theory.

Since the universal constant k is the ratio of the gas constant to Avogadro's number, (1.71) brought out a way of determining molecular dimensions through measurements in the fluctuations. Although no such measurement had been made at the time, Einstein did use (1.71) to determine the characteristic wavelength of black radiation. We shall give a critique of his derivation in §2.3.2.

1.12   Gauss' Principle for Reversible Processes

In equilibrium thermodynamics, the entropy and energy representations stand on equal footing. Instead of determining equilibrium by maximizing the entropy subject to the given constraints, the energy is to be minimized subject to the appropriate constaints. The equivalence of the two representations is summarized rather nicely in Herbert B. Callen's geometric analogy in which a circle can be characterized by either a closed curve with minimum perimeter for a fixed area or by a closed curve of maximum area for a given perimeter.

Once fluctuations are taken into account, the two representations are no longer equivalent.

An energy representation is available only for *reversible* processes. The entropy now figures among the set of independent variables whose mean value is given by

$$\bar{S} = \frac{1}{r} \sum_{i=1}^{r} S_i. \tag{1.72}$$

With the aid of the substitutions $S(E_i) \to S_i$, $S(\bar{E}) \to \bar{S}$, $E_i \to E(S_i)$, and $\bar{E} \to E(\bar{S})$ that effectuate a rotation of Gibbs space, the concavity criterion for the entropy (1.47) is transformed into the convexity criterion

$$E(S_i) - E(\bar{S}) - \left(\frac{\partial E}{\partial \bar{S}}\right)(S_i - \bar{S}) \geq 0 \tag{1.73}$$

of the energy. By virtue of the second law, $\partial E / \partial \bar{S} = \hat{T}$, Gauss' error law in the energy representation is

$$f(S_i; \bar{S}) = B \exp\left\{ -\frac{1}{\hat{T}} \left[E(S_i) - E(\bar{S}) - \hat{T}(S_i - \bar{S})\right] \right\}. \tag{1.74}$$

It is important to bear in mind that inequality (1.73) cannot be interpreted in terms of a principle of maximum work, as we did with the corresponding entropy inequality (1.61) in §1.10, precisely because the entropy is additive according to (1.72). Hence there is no criterion of evolution, meaning that the process is completely reversible. This implies that when (1.73) is summed over all i, we get the strict equality, $(1/r)\sum_{i=1}^{r} E_i = \bar{E}$, for otherwise, the total energy would not be conserved.

To illustrate the use of the error law (1.74), consider an ideal gas, where $E(\bar{S}) = \exp(\bar{S}/m)$ is the energy per molecule having $2m$ degrees-of-freedom. Introducing this fundamental relation together with $E(\bar{S}) = m\hat{T}$ into Gauss' error law (1.74) results in

$$f(S_i; \bar{S}) = B \exp\left\{ m\left(1 - e^{(S_i - \bar{S})/m}\right) + S_i - \bar{S} \right\}. \tag{1.75}$$

For small fluctuations, we may expand the exponent in a series and to second-order we have

$$f(S_i; \bar{S}) = B \exp\left\{ -\frac{1}{2m}(S_i - \bar{S})^2 \right\}.$$

But this must be equal to the truncated form of Gauss' principle,

$$f(S_i; \bar{S}) = B \exp\left\{ -\frac{1}{2}\frac{\partial^2 E}{\partial \bar{S}^2}(S_i - \bar{S})^2 \right\}.$$

Their equivalence requires $C_V = m$, which is the heat capacity at constant volume per molecule of an ideal gas with $2m$ degrees-of-freedom.

1.13   Gibbs' Paradox

The history of physics contains a number of paradoxes relating to the concept of entropy. Undoubtedly, the best known of these is the so-called Gibbs paradox. However, others of no less importance exist. The cause for such paradoxes is due to the involuntary interchange of one form of concave representation of the entropy to another. To the list of such paradoxes, we can add Einstein's combination of the entropy of the Wien law for thermal radiation with the volume dependence of the entropy of an ideal gas. Although this allowed Einstein to arrive at his light-quantum hypothesis, the association cannot be made because the two expressions for the entropy belong to different classes of concave functions. Since we need some prior knowledge of thermal radiation, we defer the discussion of this entropy paradox till §2.3.1.

Thermostatistics selects out two classes of concave functions for the entropy: $\ln X$ and $-X \ln X$. The classes $\ln X$ and $-X \ln X$ belong, respectively, to the canonical and grand-canonical ensembles. Anticipating the results of §5.2.1 and §5.2.2, respectively, we note that the two canonical ensembles are the temperature and pressure ensembles. In the temperature ensemble, the entropy has the form $m \ln E$, where m is one-half the number of degrees-of-freedom, while in the pressure ensemble, the entropy $N \ln V$ is a function of the volume V and the extensive parameter N is the number of particles. In these ensembles, the entropy is a concave function of the energy or the volume. Finally, in the grand-canonical ensemble, the entropy has the form $-N \ln N$.

It is commonly believed that the Gibbs paradox led to the prediction of the *indistinguishability* of particles, well in advance of quantum statistics. This conclusion is drawn from the fact that the empirical remedy, offered by Gibbs, makes the particles indistinguishable since permutations do not increase the number of states of the system.

Applying statistical mechanics, Gibbs obtained the expression

$$S = \bar{N} \left(s_0 + \ln V \right)$$

for the entropy of an ideal gas, where s_0 is a constant. Let us consider, as Gibbs did, the situation in which two ideal gases are mixed. In volume V_1 there are, on the average, $\bar{N}_1$ particles of gas while in volume V_2 there are, on the average, $\bar{N}_2$ particles. Both systems are initially at the same temperature so that when the two containers are brought together and the particles in one system are allowed to diffuse into the other, there will be no increase in entropy due to thermal interaction. However, Gibbs' expression for the entropy does, in fact, predict that there will be an entropy increase by the amount

$$\Delta S = \bar{N}_1 \ln \left(\frac{V_1 + V_2}{V_1} \right) + \bar{N}_2 \ln \left(\frac{V_1 + V_2}{V_2} \right). \tag{1.76}$$

If the gases are identical and occupy equal volumes, then (1.76) reduces to

$$\Delta S = 2 \bar{N} \ln 2,$$

which cannot be since the gases are identical and there should be no increase in entropy due to the mixing of identical gases.

If $\bar{N}_1$ are "red" and $\bar{N}_2$ are "black" balls, then one could argue that the state in which they are randomly mixed is a state of greater disorder than before mixing so that there should be an increase in the entropy. However, it is not difficult to see that, even when the particles are of the same color, there will be a definite increase in entropy upon mixing. This was considered a disaster on two counts: first, the entropy is not an extensive quantity because when two volumes containing the same gas are brought into contact and a partition is removed allowing the particles to intermix, the entropy is greater than the sum of the entropies of the separate containers. Second, we can continue this process and continuously remove the partitions which initially divided the container with the effect that the entropy could be made to increase without limit. This certainly would have catastrophic consequences! However, such a catastrophy is spurious since it amounts to using an incorrect form of concave representation for the entropy.

The volume of the composite system is $V = V_1 + V_2$ in which there are $N = N_1 + N_2$ particles. Since the gas is uniform, the probability that any given particle is in V_1 is obviously given by the fraction V_1/V. If we do not take into consideration the physical dimensions of the particles, the probability that there are N_1 particles present simultaneously in that subvolume is simply $(V_1/V)^{N_1}$. We have to add the clause about neglecting the volume that the particles occupy or, equivalently, that V should be much greater than the excluded volume; otherwise, the probabilities of one particle being found in the volume element will not be independent of the number of particles that are actually present in the volume element. In other words, the a priori probabilities will not all be equal to V_1/V [cf. §3.3.2 and §7.7].

With this proviso, we can likewise argue that V_2/V will be the probability that a particle is not in the volume V_1. Then the probability that N_2 particles will not be in V_1 is $(V_2/V)^{N_2}$. The probability of finding N_1 particles in the subvolume V_1 is therefore the product of the a priori probabilities multiplied by the number of ways N_1 particles can be chosen from the entire lot N, or

$$f(N_1) = \binom{N}{N_1} \left(\frac{V_1}{V}\right)^{N_1} \left(\frac{V_2}{V}\right)^{N-N_1}, \qquad (1.77)$$

subject to the closure conditions that $V = V_1+V_2$ and $N = N_1+N_2$. Expression (1.77) is none other than the binomial distribution.

Although N_1 is the random variable and V_1 is a mere parameter, we can still ask for the likelihood of a value of V_1; the true value will coincide with the maximum likelihood value in the sense of "degree-of-belief" rather than in the "frequency" sense. Anticipating a formal discussion of the method of maximum likelihood which will be undertaken in §3.3.2 and §4.3, we proceed to make r observations, N_{1i}, and construct the likelihood function as the product of binomial distributions,

$$e^{\mathcal{L}(V_1;N_1)} = \prod_{i=1}^{r} f(N_{1i}) \propto \left[\left(\frac{V_1}{V}\right)^{N_1} \left(\frac{V_2}{V}\right)^{N-\bar{N}_1} \right]^{r}$$

since each observation is independent of the other and each N_{1i} is a binomial variate.

Consider the log-likelihood function $\mathcal{L}$ as a function of V_1 for a fixed value of $\bar{N}_1$. Although this inversion is very common in mathematical statistics and leads to a subjective formulation in the sense of degree-of-belief rather than in the frequency sense, we must be careful *not* to interpret $\exp(\mathcal{L})$ as a probability density for V_1 since V_1 is not a random variable. We cannot therefore talk about the probabilities of different values of V_1 but rather content ourselves with comparing their likelihoods.

The most likely value of the subvolume V_1 is obtained from the likelihood equation

$$\frac{1}{r}\frac{\partial \mathcal{L}}{\partial V_1} = \frac{\bar{N}_1}{V_1} - \frac{N - \bar{N}_1}{V - V_1} = 0.$$

This expresses the most likely value of V_1, in the sense of degree-of-belief, in terms of the arithmetic mean $\bar{N}_1$ which is the most likely value of N_1, in the frequency sense. The likelihood equation is easily seen to require the condition of spatial homogeneity,

$$\frac{\bar{N}_1}{V_1} = \frac{N}{V}, \tag{1.78}$$

which comes as no surprise.

Gauss' law of error determines the entropy of the binomial distribution (1.77) as

$$S = N \ln\left(\frac{V}{V - V_1}\right) - \bar{N}_1 \ln\left(\frac{V_1}{V - V_1}\right), \tag{1.79}$$

although Gibbs would have referred to it as an entropy *difference* on the strength of (1.76). However, its functional dependency only becomes clairvoyant after we insert the maximum likelihood estimate (1.78). This is because V_1 in (1.79) has the status of a parameter rather than the average value of a random, extensive quantity so that the entropy is not a function of the volume, independent of the average number of particles. It is only when the maximum likelihood estimate (1.78) is introduced into (1.79) that we come out with

$$S(\bar{N}_1) = -\bar{N}_1 \ln\left(\frac{\bar{N}_1}{N}\right) - (N - \bar{N}_1) \ln\left(\frac{N - \bar{N}_1}{N}\right). \tag{1.80}$$

Therefore, the concave representation of the entropy belongs to the class $-X \ln X$ of the grand-canonical ensemble. If we equate the entropy given in (1.79) with the entropy difference in (1.76), then the individual entropies will have the form $N \ln V$. But we have no right to switch horses in midstream since it is the binomial distribution (1.77) which describes the process and whose entropy is given by (1.80). Hence, there is no paradox at all!

The usual way of eliminating the paradox is to divide the N_1-particle partition function by $N_1!$. This asserts that in writing the N_1-particle partition function as a product of single particle partition functions, the number of states has been overcounted exactly $N_1!$ times. The division by $N_1!$ is often referred to as "corrected" Boltzmann counting, although Boltzmann never fell into such a trap. There is nothing that concerns chance about the N_1-particle partition function to warrant such a statement, and it is not here that one should quibble over whether the particles are identical or not.

The introduction of such a correction means simply subtracting the entropy (1.80) from (1.79) to obtain

$$\Delta S = -\bar{N}_1 \ln \left(\frac{N}{V} \frac{V_1}{\bar{N}_1} \right) + (N - \bar{N}_1) \ln \left(\frac{N}{V} \frac{V - V_1}{N - \bar{N}_1} \right).$$

It is then argued that this entropy of mixing vanishes because the specific volume V/N is the same before and after mixing. However, the two expressions for the entropy, (1.79) and (1.80), are *numerically* equal under the maximum likelihood condition (1.78).

Bibliographic Notes

§1.2

A great deal of insight can be gained from Boltzmann's popular lectures,

- L. Boltzmann, *Theoretical Physics and Philosophical Problems*, B. McGuiness, ed. (D. Reidel, Dordrecht, 1974), esp. *Populäre Schriften*, Essay 3, pp. 13-32,

concerning his probabilistic reasoning.

Einstein began his scientific career by attempting to free statistical mechanics from its mechanical origins and obtain results of a completely general nature that would be independent of a particular model. This period covers the years 1902–1904, and although he considered it to be inferior to Gibbs' formulation, which he was unaware of at the time, it did set the way for his famous paper on light-quanta. A nice account of Einstein's work on fluctuation theory can be found in

- M.J. Klein, "Fluctuations and statistical physics in Einstein's early work," in *Proceedings of the Einstein Centennial Symposium* (Addison-Wesley, Reading, MA, 1979), pp. 39-58.

Einstein's reinterpretation of Boltzmann's principle first appeared in

- A. Einstein, "Zur allgemeinen molekularen theorie der Wärme," *Ann. d. Phys.* **14**, 354-362 (1904), in connection with the determination of the energy dispersion in black radiation. Einstein used a composite system in which two parts can exchange energy and determined the entropy difference.

Formula (1.2) is a stronger statement in that it says that all deviations from equilibrium, whether they are caused by a spontaneous fluctuation or an external perturbation, will inevitably lead to a diminution in the entropy. This form appears in Eq. (373) in

- M. Planck, *Theory of Heat*, transl. by H.L. Brose (Macmillan, London, 1932), Part 4, Ch. 1.

Both Planck's treatise and

- H.A. Lorentz, *Lectures on Theoretical Physics*, Vol. II, transl. by L. Silberstein and A.P.H. Trivelli (Macmillan, London, 1927), §16 of "Entropy and Probability"

discuss Boltzmann's principle in the context of composite systems.

A good critical analysis of Boltzmann's principle can be found in

- R.H. Fowler, *Statistical Mechanics*, 2nd ed. (Cambridge Univ. Press, Cambridge, 1936), §6.8 and §6.9.

Fowler brings out the important point that the definition of entropy in terms of the number of microscopic complexions is found wanting. He raises the question of how do we know that our analysis of complexions is really complete? "At one stage we may omit all references to nuclear spin and its orientation. Then we generalize the count to include it. But why should this be the last step?" Fortunately, things are not so bad as all this since the thermodynamic probability will always be given by a binomial or multinomial coefficient so that it is necessary to rely solely on probability theory.

§1.3 and §1.4

An exposition of Boltzmann's use of combinatorial analysis in his probabilistic interpretation of the H-function can be found in:

- L. Boltzmann, *Lectures on Gas Theory*, transl. by S.G. Brush (Univ. California Press, Berkley, 1964), Part I, §6.

A more detailed account is given in

- P. Ehrenfest and T. Ehrenfest, *The Conceptual Foundations of the Statistical Approach in Mechanics*, transl. by M.J. Moravcsik (Cornell Univ. Press, Ithaca, NY, 1959), §12.

The Ehrenfest article which appeared in the German *Encyclopedia of Mathematical Sciences* in 1912 gave a certain plausibility to Boltzmann's not always clear ideas and was widely disseminated during its day. It also contains a discussion of Gibbs' approach which is summarized in his treatise

- J.W. Gibbs, *Elementary Principles in Statistical Mechanics* (Yale Univ. Press, New Haven, CT, 1902).

We have argued that the energy constraint destroys the *a priori* argument for populating the cells in phase space in

- B.H. Lavenda and C. Scherer, "Statistical inference in equilibrium and nonequilibrium thermodynamics," *Rivista del Nuovo Cimento* # 6 (1988);

- B.H. Lavenda, "Particle distinguishability," *Nature* **331**, 308-309 (1988).

§1.5

An accurate historical account of Maxwell's derivation of his error law can be found in

- S.G. Brush, *The Kind of Motion We Call Heat*, Vol. 1 (North-Holland, Amsterdam, 1976), §5.1.

§1.6

An excellent account of Gauss' derivation of his error law and the timely connection with the method of least squares can be found in

- S.M. Stigler, *The History of Statistics: The Measurement of Uncertainty Before 1900* (Harvard Univ. Press, Cambridge, MA, 1986), pp. 140-143.

A clear and detailed exposition of the more modern derivation of Gauss' error law leading to the average value as the most probable value of the quantity measured, as well as other error laws, is given in

- J.M. Keynes, *Treatise on Probability* (St. Martin's Press, New York, 1921), Ch. 17.

McBride, in

- W.J. McBride, "A natural law," *Proc. IEEE* **56**, 1713-1715 (1968),

showed that Gauss' principle leads to density functions which belong to the exponential family of distributions.

§1.7

The discrepancy between Gauss' and Boltzmann's principles is discussed in

- B.H. Lavenda and J. Dunning-Davies, "Kinetic derivation of Gauss' law and its thermodynamic significance," *Z. Naturforsch.* **45A**, 873-878 (1990).

§1.8

The classic treatise on convex functions is

- G.H. Hardy, J.E. Littlewood, and G. Pólya, *Inequalities*, 2nd ed. (Cambridge Univ. Press, Cambridge, 1952), Ch. 3.

We have shown in

- B.H. Lavenda, "Black-body versus black hole thermodynamics," *Z. Naturforsch.* **45A**, 879-882 (1990),

- B.H. Lavenda and J. Dunning-Davies, "The essence of the second law is concavity," *Found. Phys. Lett.* **3**, 435-441 (1990)

that concavity is the definitive property of the entropy and it cannot be waived without leading to a contradiction with the second law. In contrast,

- L. Galgani and A. Scotti, "On subadditivity and convexity of thermodynamic functions," *Pure Appl. Chem.* **22**, 229-235 (1970);

- P.T. Landsberg, in *Proceedings of the Intern. Conf. on Thermodynamics, Cardiff, 1970*, P.T. Landsberg, ed. (Butterworth, London, 1970), p. 215

have advocated that it is superadditivity, and not concavity, which is at the root of the second law. They reason that since the entropy is a concave function in *all* its variables,

$$2S\left(\frac{E_1 + E_2}{2}, \frac{V_1 + V_2}{2}, \frac{N_1 + N_2}{2}\right) \geq S(E_1, V_1, N_1) + S(E_2, V_2, N_2),$$

and it is a first-order homogeneous function [i.e., condition (1.57)], it is also superadditive:

$$S(E_1 + E_2, V_1 + V_2, N_1 + N_2) \geq S(E_1, V_1, N_1) + S(E_2, V_2, N_2).$$

However, as we have shown, if the entropy is a first-order homogeneous function, it cannot be concave in *all* three variables simultaneously so that the concavity and superadditivity criteria coincide where only the equality sign applies.

A confrontation between Gauss' principle and classical ensembles can be found in

- B.H. Lavenda and J. Dunning-Davies, "Statistical thermodynamics without ensemble theory," *Ann. d. Physik* (in press).

§1.9

The illustration showing the relation between the concavity property of the entropy and irreversibility has been taken from Lorentz's Leiden *Lectures* [cf. §1.2 of Bibliographic Notes], §40 and §41 in "The Theory of Radiation." In view of the proposed superadditivity property of the entropy, Lorentz's result is a devastating blow: The concavity property of the entropy in one variable implies that it is "subadditive" rather than "superadditive." Since we have a counterexample, superadditivity can never be the characterizing property of the entropy since it is not universal.

The subadditive property of the information definition of entropy can be found in

- J. Aczél and Z. Daróczy, *On Measures of Information and their Characterizations* (Academic Press, New York, 1975).

§1.10

Undoubtedly, the best statement of the principle of maximum work, that done by a system on an external agent when it is in thermal contact with a medium which it can exchange heat and work, can be found in

- L.D. Landau and E.M. Lifshitz, *Statistical Physics*, 2nd ed. (Pergamon Press, Oxford, 1969), §20.

Landau and Lifshitz derive the principle of maximum work by first determining the minimum work that an external agent need do on the system to bring it into the given nonequilibrium state. They express the law of increase in entropy as $\Delta S + \Delta S_0 \geq 0$, where ΔS is the entropy difference between the system (which they refer to as the "body") and its value in a state of complete equilibrium. In regard to our analysis, they are interchanging initial and final states. Thus, the concavity of the entropy requires $\Delta S + \Delta S_0 \leq 0$ and their principle minimum work does not follow. In Chapter 8, it will be made apparent that thermodynamics cannot provide any criteria for evolution *away from* equilibrium since this is shunned at by the second law. For more details see

- B.H. Lavenda, *Nonequilibrium Statistical Thermodynamics* (Wiley, Chichester, 1985), Ch. 7.

We also failed to appreciate this in

- B.H. Lavenda, *Thermodynamics of Irreversible Processes* (Macmillan, London, 1978), §1.5.

§1.11

In

- A. Einstein, "Theorie der Opaleszenz von homogenen Flüssigkeiten und Flüssigkeitsgemischen in der Naehe des kritischen Zustandes" (Theory of the opalescence of homogeneous fluids and fluid mixtures near the critical state), *Ann. d. Physik* **33**, 1275-1298 (1910),

Einstein applies his theory of fluctuations to the phenomenon of critical opalescence, in which fluctuations are enhanced, to obtain an estimate of Avogadro's number. This method complements Einstein's other inventions for determining Avogadro's number—notably his theory of Brownian motion. This and other fluctuating phenomena are exponded upon in §26 of

- H.A. Lorentz, *Les Théories Statistique en Thermodynamique* (B.G. Teubner, Leipzig, 1916).

Actually, Einstein was very critical of Boltzmann's principle. In the 1910 article he claims that Boltzmann's principle is without any meaning unless one has, at hand, a complete molecular–mechanical theory of the elementary processes. Although Boltzmann's principle "seems without content, from a phenomenological point of view, without giving in addition such an *Elementartheorie*," Einstein's "hybrid" theory of mixing the microscopic with the macroscopic was, and still is, a very successful theory—perhaps too successful for it to have been a good guess!

This paper complements the work of Smoluchowski [*Ann. d. Physik* **25**, 205-226 (1908)], who studied strong density fluctuations of two fluids near the critical state of a mixture which produce opalescence or a single fluid near the critical state of condensation on the basis of the kinetic theory of heat. Einstein deals with the scattering of light that is produced by these phenomena. Instead of using composite systems, Einstein assumes that the entropy depends upon additional macroscopic variables whose average values are zero. In other words, these parameters measure the deviations from equilibrium where they vanish. For small deviations, the entropy difference can be represented as a negative quadratic form in these variables so that the thermodynamic probability can be represented by a normal law.

§1.12

The clearest and undoubtedly the best formulation of thermodynamics is the postulational one given in

- H.B. Callen, *Thermodynamics and an Introduction to Thermostatistics*, 2nd ed. (Wiley, New York, 1985).

§1.13

The paradox raised in this section of the mixing of two gases upon the removal of a partition is contained in Gibbs' memoir, "On the Equilibrium of Heterogeneous Substances," which can be found in

- J. Willard Gibbs, *Collected Works*, Vol. 1 (Yale Univ. Press, New Haven, 1948), p. 167.

Gibbs phrased it in these terms:

> Now we may without violence to the general laws of gases which are embodied in our equations suppose other gases exist than such as actually do exist, and there does not appear to be any limit to the resemblance which there might be between two such kinds of gases. But the increase of entropy due to the mixing of given volumes of the gases at a given temperature and pressure would be independent of the degree of similarity or dissimilarity between them.

He concludes his discussion with the famous phrase, "In other words, the impossibility of an uncompensated decrease in entropy seems to be reduced to improbability."

An interesting variation of the Gibbs paradox is given by Klein in

- M.J. Klein, "Note on a problem concerning the Gibbs paradox," *Am. J. Phys.* **26**, 80-81 (1958)

- M.J. Klein, "Remarks on the Gibbs paradox," *N.T. Natuurk.* **25**, 73-76 (1959)

where he analyses the following model. Two monatomic gases are considered with the only difference being that one is prepared in an isomeric, or excited, metastable state. If one waits a long enough time compared to the lifetime of the isomer, one will find that the isomeric state has decayed to the ground state and the system is comprised solely of one gas. Klein shows that the increase in entropy given up to the heat bath by the decay of the isomer more than compensates the entropy of mixing.

Chapter 2

Black Radiation:
A Case History

2.1 Historical Perspective

It has often been said that Max Planck hypothesized the existence of discrete quanta so as to "discourage' radiant energy from flowing into the high frequency part of the spectrum. Yet, Planck saw no compelling reason for a real *quantum* theory of radiation, and in the years that followed his radical step he steadily retreated from such an abrupt departure from classical physics. Planck's original goal was to justify his empirically proposed formula for black radiation and paid little attention to the radical idea that the energy of his oscillators, which were supposedly buried in the walls of a black body, were always a discrete multiple of an energy unit. In this sense, Planck was a reluctant revolutionary, and little of his original derivation and motivation can be found in texts on statistical mechanics. All that remains is the final result with some artifice to justify it. Yet, his thermodynamic thinking is at the basis of the different physical forms of statistics so that it would indeed be a shame to neglect it. Nothing is better than a lucid historical account of what drove Planck to attack the problem of black radiation and how the role of entropy played a fundamental role in justifying his empirical formula. To understand Planck's line of reasoning, we have to begin at the beginning.

The thermodynamic treatment of electromagnetic radiation goes back to Gustav Robert Kirchhoff (1860), Josef Stefan (1879), Ludwig Boltzmann (1884), and Wilhelm Wien (1896) which finally culminated in the birth of the quantum theory through Planck's formula for black radiation. Assuming an equilibrium between a black and a nonblack body at the same temperature, Kirchhoff related the rate of energy emitted by the nonblack body to its absorptivity. Radiation emitted by black bodies is called "thermal" radiation because both its intensity and spectrum are functions of the emitter's temperature only. In contrast, the spectrum of "nonthermal" radiation, noted for its high intensity, is characterized by an energy gap in the distribution of states as a function of energy. A perfect black body consists of a container with perfectly reflecting walls that are maintained at a constant temperature into which small bodies are introduced that can absorb and re-emit radiation at all possible wavelengths. This allows for energy exchange among the different modes of vibration of the electromagnetic field. Ordinary black bodies, such as charcoal, possess this property at least in the visible part of the spectrum; "idealistic" black bodies behave in the same way at all possible frequencies. Nonthermal radiation is far too large for the temperature of the body which emits it to be in thermal equilibrium. Some examples of this luminescent form of radiation are glow discharges, luminescent diodes, and lasers.

Kirchhoff formulated his law for thermal radiation by showing that any violation of it would result in a perpetual motion machine of the second kind.[1] Kirchhoff's law,

$$E_\nu = A_\nu \mathcal{J}(\nu, T),$$

for the radiation within "a space enclosed by bodies of equal temperature T, through which no radiation can penetrate" (Hohlraumstrahlung), represents the origin of the quantum theory of radiation. The emissive power, E_ν, of a body is the amount of energy emitted by it per unit area per unit frequency interval. Radiation impinging on a body may be refracted or absorbed. Assuming that all the absorbed energy is transformed into heat, this fraction of the total incident radiation is called the absorptive power, A_ν, of the body in the frequency interval from ν to $\nu + d\nu$. Kirchhoff's theorem then states that the ratio E_ν/A_ν for any body is independent of the nature of the body and depends only on the frequency ν and temperature T. If one body that absorbs red rays begins to glow at a given temperature, then any other body which absorbs the same radiation, no matter how little, will begin to glow at the same temperature.

For a perfect black body, $A_\nu = 1$: it absorbs and transforms into heat all radiation incident on it. What Kirchhoff showed was that the emissive power of one black body is equal to that of another and hence cannot depend upon the properties of the body. He made use of a general thermodynamic law which asserts that radiation cannot upset a state of thermal equilibrium. Consider a black body B in a box with perfectly black walls. We introduce two screens with apertures a and b as shown in Fig. 2.1. The aperture a has a center and the aperture b can either be fitted by a black plate or a "perfect" mirror which reflects all incident waves and does not radiate itself. For if such a mirror could radiate, thermal equilibrium would not be possible because it would cool down. Between the hole in the wall and the black body we put an aperture.

Now if everything has come to a common temperature, thermal equilibrium will persist whether we shut the hole with a black plate or a perfect mirror. In the former case, the radiation sent through the intermediate aperture from the black body will be absorbed on the black plate, while in the latter case,

---

[1] A continuous extraction of useful work from the heat of our environment is said to be a "perpetual motion of the second kind." Producing work from nothing is called a "perpetual motion of the first kind."

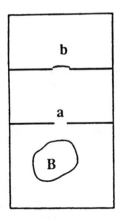

Figure 2.1: A black body B and two screens, with apertures a and b, enclosed in a box with perfectly black walls.

the rays will be reflected from the mirror back to the black body. Since thermal equilibrium prevails in both cases, the black body radiates through the intermediate aperture as much as the black plate would radiate through it. Replacing the black body by another black body does not change things so the two must have the same *total* emissive power. By means of a thin plane-parallel plate which reflects rays of different frequencies with different intensities, due to interference, Kirchhoff extended his result that the emissivities of two black bodies are equal *at each frequency*.

The challenge that Kirchhoff launched to the scientific world was to find the form of his function $\mathcal{J}(\nu, T)$ which represents the emissive power of the body. All further developments lay in discovering the form of Kirchhoff's function. In the case of *Hohlraumstrahlung*, or cavity radiation, where the radiation is homogeneous, isotropic, and unpolarized, Kirchhoff's function takes the form

$$\mathcal{J}(\nu, T) = (c/8\pi)\rho(\nu, T),$$

where $\rho(\nu, T)$ is the spectral density or the energy density per unit volume and c is the speed of light in a vacuum.

The next advance came in 1879 in a paper published by Stefan in which he conjectured that the total energy density,

$$u(T) = \int_0^\infty \rho(\nu, T)\, d\nu, \tag{2.1}$$

is proportional to T^4. His proposal came from the experimental evidence that

from weak red heat (about 525^0) to complete white heat (about 1200^0), the intensity of radiation increases from 10.4 to 122, thus nearly twelvefold (more precisely 11.7). The ratio of the absolute temperatures $273 + 1200$ and $273 + 525$ gives when raised to the fourth power 11.6.

Theoretical support for Stefan's empirical law came from Boltzmann in 1884. Boltzmann's reasoning involved the application of the laws governing a Carnot cycle to an engine in which radiation played the role of the working substance. Boltzmann wrote the second law in the form

$$\frac{\delta Q}{T} = dS(uV, V) = \frac{1}{T}[d(uV) + P\,dV], \tag{2.2}$$

where δQ is the heat that must be supplied when the volume V undergoes an alteration by an amount dV at constant temperature. This is where thermodynamics ends and the specification of a model begins.

Boltzmann had to introduce an equation of state between the pressure, P, and the energy density, u, that characterizes the working substance. He modified Maxwell's result for electromagnetic waves, where the radiation pressure acting on a surface is numerically equal to the energy density of the incident waves, to apply to thermal radiation. Unlike electromagnetic radiation, there is no preferential direction so that waves will be moving in all three directions with equal probability. Boltzmann therefore assumed that the radiation pressure in any one direction is one-third of the total energy density,

$$P = \tfrac{1}{3}u. \tag{2.3}$$

Boltzmann did not note the fact that the thermal radiation pressure is only one-half that of a gas of material particles. For an ideal gas $P = NT$ and $u = 3NT/2$ so that $P = 2u/3$ instead of (2.3). Boltzmann therefore did not draw the analogy between a gas of material particles and a "radiation" gas. It was not until 1905 that Einstein considered such a possibility, which we shall discuss in greater detail in §2.3.1.

Now, since the energy density, u, is a function only of the temperature, T, the exactness condition for a perfect differential is

$$\frac{\partial}{\partial V}\left(\frac{V}{T}\frac{du}{dT}\right) = \frac{\partial}{\partial T}\left(\frac{4}{3}\frac{u}{T}\right),$$

which is equivalent to

$$\frac{du}{dT} = 4\frac{u}{T}. \tag{2.4}$$

Integration gives

$$u = \sigma T^4, \tag{2.5}$$

which is precisely Stefan's law, and σ is $4/c$ times Stefan's constant, whose numerical value is 0.567×10^{-7} W/m²-K⁴.

An even neater proof of Stefan's law that does not make any assumption about the energy density but only involves the constitutive relation (2.3) is based on the Gibbs–Duhem relation

$$d\left(\frac{P}{T}\right) + u\,d\left(\frac{1}{T}\right) = 0.$$

Introducing (2.3) leads immediately to (2.4) and, by integration to Stefan's law, (2.5). Greater use of this relation will be made in our discussion of a "Stefan" gas in §6.2 which has the characteristic feature of not conserving the particle number.

The derivation of the Wien formula for the dependency of the energy density upon frequency and temperature makes brilliant use of the invariancy condition of the entropy and the energy balance relation which results when a pencil of rays impinges on a mirror. The rays, which comprise the pencil, cannot be strictly monochromatic. For otherwise, we would not be able to define an entropy or a temperature since the entropy and temperature of a single state cannot be defined. We thus allow for an interval in the frequency, $d\nu$, of the pencil.

Suppose we have a wave front of length ℓ and cross section A which impinges on a mirror; it will therefore have a volume $V = A\ell$. We can consider the process to be carried out adiabatically and reversibly. After the pencil is reflected by the mirror, the frequency interval will be changed into $d\nu'$ and the volume will undergo a change to V'. The invariancy of the entropy can be expressed as[2]

$$s'V'\,d\nu' = sV\,d\nu.$$

The energy, however, will not be invariant since there is work done on the mirror. As a result of the impinging radiation pressure, the mirror will be set in motion with a velocity v which we will consider small compared with the velocity of light. This will allow us to neglect higher powers in the ratio v/c. Now, the radiation pressure is equal to twice the energy density of the impinging pencil, $2\rho\,d\nu$. Since the time the pressure acts is $\ell/(c-v) \simeq \ell/c$, the work done on the mirror is

$$2\rho\,d\nu\frac{\ell}{c}Av.$$

This work needs to be subtracted from the incident energy so that the energy balance equation is

$$\rho'V'\,d\nu' = \rho V\,d\nu - 2\rho V\,d\nu\frac{v}{c}.$$

---

[2]In Lorentz's presentation, allowance is also made for the fact that the rays are not strictly parallel so that one must consider intervals in both the solid angle and the frequency. Nonetheless, we should expect that entropy invariancy and energy balance will only manifest themselves when the solid angle is integrated over the hemisphere. In other words, we cannot control the directions of the individual processes but, rather, consider some average value over all directions. The distinction appears when we consider that the volume changes proportionally with the cube of the wavelength and not the wavelength itself as Lorentz did.

The change in the frequency is obtained from Doppler's effect. The incident ray in the x-direction, say, will be described by

$$A \cos \nu \left(t - \frac{x}{c} + \varphi \right)$$

while

$$A' \cos \nu' \left(t + \frac{x}{c} + \varphi' \right)$$

describes the reflected ray, where A and φ are the amplitude and phase of the incident wave, respectively. At the surface of the mirror, $x = vt$, a relation exists between the two waves which necessitates that the coefficients of t in both terms be equal for $x = vt$, namely,

$$\nu' \left(1 + \frac{v}{c} \right) = \nu \left(1 - \frac{v}{c} \right).$$

But since we are interested in terms up to first-order in v/c, we have

$$\nu' = \nu \left(1 - 2\frac{v}{c} \right), \tag{2.6}$$

which is the change in frequency due to the Doppler effect.

Since the motion is in a single dimension, so that the transverse dimensions do not change, the volume should vary with the linear dimension. We must then investigate how the element of the solid angle changes. As noted by Lorentz, there is a complication owing to the fact that the ordinary laws of reflection are not applicable to a moving mirror. We can, however, avoid such complications by reasoning in a completely general way. We know that the volume of phase space occupied by the system must always be invariant. This is due to the fact that the phase space volume element is related to the probability of finding the system in a given state which should evidently be independent of the velocity of the observer relative to it. By introducing spherical coordinates, the volume element of momentum space $dp_x \, dp_y \, dp_z$ becomes $p^2 \, dp \, d\Omega$, where $d\Omega$ is the element of solid angle around the direction of the vector $\mathbf{p}$. The probability of finding the system in the volume element should not depend upon the direction of this vector, and integrating over angles gives the invariancy condition,

$$V'p'^2 \, dp' = Vp^2 \, dp,$$

of the phase space volume element. If the solid angle was retained in the invariancy condition, the volume would change proportionally with respect to the wavelength instead of the cube of the wavelength. The latter is the more natural situation since the momenta are proportional to the frequencies and the frequencies undergo a red-shift, (2.6).

Therefore, in order that the phase space volume element remains invariant, the volume must change in proportion to the cube of the wavelength so that

$$V' = \left(1 + 2\frac{v}{c} \right)^3 V.$$

Introducing this into the condition that the entropy remain invariant leads to

$$s' = \left(1 - 2\frac{v}{c}\right)^2 s = \left(\frac{\nu'}{\nu}\right)^2 s,$$

while the energy balance condition becomes

$$\rho' = \left(1 - 2\frac{v}{c}\right)^3 \rho,$$

because of the extra factor of $1 - 2v/c$ in that relation. We thus find the invariants are

$$\frac{s'}{\nu'^2} = \frac{s}{\nu^2} \quad \text{and} \quad \frac{\rho'}{\nu'^3} = \frac{\rho}{\nu^3}$$

under reflection of a pencil.

The entropy density s is some function of the energy density ρ and the frequency ν such that when ρ/ν^3 remains constant, so too does s/ν^2. We can express this functional dependency as

$$s = \nu^2 F\left(\frac{\rho}{\nu^3}\right).$$

Then inverting the second law,

$$\frac{1}{T} = \frac{\partial s}{\partial \rho} = \frac{1}{\nu} F'\left(\frac{\rho}{\nu^3}\right),$$

we obtain

$$\rho(\nu, T) = \nu^3 f\left(\frac{\nu}{T}\right), \tag{2.7}$$

which is Wien's formula.

If ν is the emitted frequency and ν' is the received frequency, then we will have a "red-shift" since $\nu > \nu'$. The fact that ν/T is an invariant implies that the observed temperature T' will undergo a red-shift by an amount

$$T' = T(1 - \Delta),$$

where $\Delta = (\nu - \nu')/\nu$. This will have important consequences on ways in which the radiation laws can be modified (cf. §2.4).

For graphical purposes, it is advantageous to switch from frequency to wavelength in Wien's formula. Wien's formula can then be written as

$$\Phi(\lambda, T) = \frac{1}{\lambda^5} \psi(\lambda T),$$

where

$$\psi(\lambda T) = c^4 f\left(\frac{c}{\lambda T}\right).$$

The form of the function Φ coincided with the experimental findings of Lummer and Pringsheim, who published their results in 1897. Plotting the curve Φ as a function of λ for a given temperature T will allow it to be determined at any other temperature T'. Select that wavelength λ at temperature T and another wavelength λ' at the other temperature T' such that the condition

$$\lambda'T' = \lambda T$$

is satisfied. Since ψ depends only on the product λT, it will be the same at this value as at $\lambda'T'$. To obtain Φ, we would have to divide ψ by λ^5, which is equivalent to magnifying the old Φ by the factor $(\lambda'/\lambda)^5$. Thus, at the temperature $T' > T$, the abscissa has been contracted by T/T' times the old abscissa while the ordinate undergoes a magnification proportional to $(T'/T)^5$ times the ordinate of the original curve.

The function ψ, being proportional to the energy density, will surely increase with the temperature. But the shape of the energy curve, as a function of the wavelength for a given temperature, is more complicated. Although the function ψ still increases as a function of wavelength, the energy density is cut down by the factor $1/\lambda^5$ so that one can expect it to exhibit a maximum. Denote by λ_m the wavelength at which the energy density has a maximum. It then follows from Wien's law that $\lambda_m T = \lambda'_m T'$, showing that the maximum wavelengths are inversely proportional to the temperatures. As the temperature is increased, the maximum of the energy density is shifted to shorter wavelengths. This is known as *Wien's displacement law*.

The functional form of Wien's formula gives the dependency of the energy density on the frequency and temperature. But it was as far as one could go on the basis of thermodynamic and electromagnetic considerations without the explicit introduction of a model. This step was taken by Wien three years later in 1896.

To find the form of $f(\nu, T)$, Wien drew an analogy between black radiation and a gas in which the velocities of the molecules obey Maxwell's distribution

$$v^2 e^{-v^2/2\Theta} \, dv.$$

By equipartition, the constant Θ is proportional to the absolute temperature. Now, Wien reasoned that the frequency ν of radiation emitted by any molecule is a function of its velocity, v, so that the relation can be inverted to give v as a function of λ. The spectral density of radiation whose frequency lies between ν and $\nu + d\nu$ is proportional to the number of molecules which emit radiation at this frequency and also a function of the velocity or frequency. Hence, Wien claimed that the spectral density had to have the form

$$\rho(\nu, T) = g(\nu)e^{-h(\nu)/T},$$

where $g(\nu)$ and $h(\nu)$ are two unknown functions. But Wien knew that the spectral density had to have the form (2.7), and this led to

$$\rho(\nu, T) = \alpha \nu^3 e^{-\gamma \nu/T} \quad \text{(Wien)}, \tag{2.8}$$

where α and γ are constants that could be determined from the experimental data.

Wien's law fitted the data remarkably well–at least at high frequencies–and Planck set out to give Wien's law a sound derivation based on the fundamental theory of the thermodynamics of radiation which he was about to create. In fact, had Wien's formula proved to be the final answer, classical physics would have predominated, but it was precisely the region in which classical theory holds that led to its ultimate downfall. The new technological advances made in experimental techniques in the far-infrared made it evident that $\rho(\nu, T)$ was proportional to T for small ν.

By November of 1899 technological advances were such that permitted the temperature range to be extended up to 1646 C. Also there was better screening of the bolometer, which measures radiant energy, from extraneous radiation sources as well as stable heating of the radiation source in order to maintain a constant temperature. These advances permitted Otto Lummer and Ernst Pringsheim to extend black radiation measurements into the region of large λT. In Fig. 2.2 they compared the experimental values against the Wien law (broken curve) and found that in order to apply Wien's law, the "constant" γ had to increase with increasing wavelength and hence could no longer be a constant. The time was ripe for Planck.

2.2   Planck's Law

Planck was teethed on the writings of Clausius, being fascinated by the second law and its implications about the irreversible tendency of the world to progress in a given direction. As a disciple of Clausius, Planck took the second law as one of the two commandments: processes in which the total entropy decreases simply do not exist. As such, he was at odds with Boltzmann, who, as we have seen, considered the second law a statistical law: the increase in entropy of all natural processes left to themselves was highly probable rather than being absolutely certain.

Writing in the preface to his *Treatise on Thermodynamics* in 1897, Planck expressed the hope that the mechanical interpretation of heat, with the obvious need for some dissipation to occur, would be replaced by an electromagnetic explanation in which there is neither heat transfer nor dissipation. What Planck was alluding to was an explanation of irreversible processes based on conservative forces, such as radiation damping, which involve no transformation of energy into heat. His basic mechanism for irreversibility was to be the scattering of electromagnetic radiation by charged oscillators in which an incident plane wave is converted into a spherical outgoing wave.

Not only was Planck sceptical of Boltzmann's ideas but Boltzmann was equally as sceptical about Planck's program–since Maxwell's equations are invariant under time reversal, there would have to be a reverse of the scattering process that would also be a legitimate solution to the equations of electro-

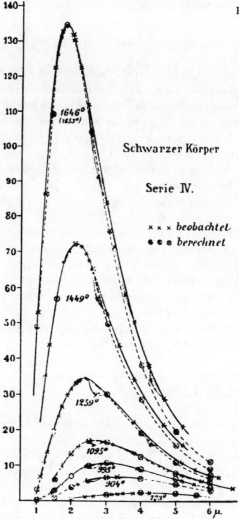

Figure 2.2: Spectral energy distribution as a function of the wavelength in micrometers (μm) at various temperatures taken from the published results of Lummer and Pringsheim in 1899. The higher the temperature and the greater the wavelength, the more apparent are the deviations of the theoretical curves (broken lines) from the experimental curves (solid lines). The hatched areas are due to the absorption of water vapor and carbon dioxide in the air.

dynamics. Just as mechanics cannot describe the irreversible approach of gas molecules to their equilibrium distribution, so too, electrodynamics could not offer a proof of why charged oscillators come to equilibrium with thermal radiation in a black body cavity. What was missing was some sort of statistical assumption concerning the randomness of the elementary building blocks of the theory and Planck proposed the assumption of "natural" radiation, mimicking Boltzmann's hypothesis of molecular chaos in the case of gas molecules, which essentially meant that there are to be no correlations between the different modes of the radiation field.

2.2.1   Step 1: Electrodynamics

With the aid of this hypothesis, Planck was able to derive an important relation between the spectral density, $\rho(\nu, T)$, and the average energy, $\bar{E}(\nu, T)$, of an oscillator at frequency ν and temperature T. He did so by considering a harmonic oscillator–or as he called it a *resonator*–capable of electrical oscillations. The resonator would undergo radiation damping through the interaction with an electric field of frequency ν_0 in the direction of its motion.

Let us recall that Planck wanted to replace the role of dissipation of energy into heat by radiation damping, which is a conservative mechanism. According to classical electrodynamics, a moving point charge emits radiation. This energy has to be contributed by the forces that keep the charge in motion. The kinetic energy of the charged particle therefore decreases in time, and we must consider not only the external forces acting on the particle but also the reaction of the field produced by the charge on its own motion. This reaction force is proportional to the time derivative of the acceleration and is given by $F = (2e^2/3c^3)\dddot{v}$, where e is the charge. In general, this reaction force of the radiation is small when compared to other forces that act on the particle, and to a first approximation, for a particle moving with a harmonic motion, $\ddot{v}$ can be replaced by $-(2\pi\nu)^2 v$.

The equation of motion for a Planck oscillator, which can be looked upon as a bound electron, that can oscillate in the x-direction, say, will be

$$\ddot{x} + \gamma\dot{x} + (2\pi\nu)^2 x = \frac{e}{M}E_0 e^{-i2\pi\nu_0 t},$$

where M and e are the mass and charge of the electron and E_0 is the x-component of the amplitude of the electric field. The radiative damping coefficient is $\gamma = 8\pi^2 e^2 \nu^2/3Mc^3$. Looking for a solution in the form $x = x_0 \exp(-i2\pi\nu_0 t)$ and inserting this into the equation of motion leads to the determination of the unknown amplitude x_0 as

$$x_0 = \frac{E_0 e/M}{4\pi^2 \left(\nu^2 - \nu_0^2\right) - i2\pi\nu_0\gamma}.$$

Since we are interested in the energy, $E(\nu) = M(2\pi\nu)^2 |x_0|^2/2$, of the oscillator at the frequency ν, we must take the square of the modulus of x_0. We then obtain

$$E(\nu) = \frac{e^2 |E_0|^2}{2M} \frac{\nu^2}{4\pi^2(\nu^2 - \nu_0^2)^2 + \nu_0^2\gamma^2}.$$

In the neighborhood of resonance, we may put $\nu \sim \nu_0$, giving the approximate expression

$$E(\nu) = \frac{e^2 |E_0|^2}{2M} \frac{1}{4\pi^2(\nu - \nu_0)^2 + \gamma^2}.$$

At this point, Planck introduced his random hypothesis that the electric field is made up of an incoherent superposition of modes in thermal equilibrium at a temperature T. This enabled him to obtain the average energy, $\bar{E}(\nu, T)$, of the oscillator, in thermal equilibrium with the modes of the electric field, by replacing the electric field density, $E_0^2/2$, by $4\pi\rho(\nu_0, T)\,d\nu_0/3$. Then integrating over all frequencies ν_0, he obtained

$$\bar{E}(\nu, T) = \frac{4\pi e^2}{3M} \int_0^\infty \frac{\rho(\nu_0, T)\,d\nu_0}{4\pi^2(\nu - \nu_0)^2 + \gamma^2}.$$

The possibility of performing this integration is conditional on the fact that the coefficient of damping of the free vibrations of the resonator, γ, is exceedingly small. The smaller the value of this coefficient, the narrower the interval into which the frequency of the external field must fall in order to excite vibrations of the resonator that are comparable in intensity to that which would be obtained at resonance. The weak damping condition means that our resonator will only be sensitive to external fields whose frequencies differ slightly from ν.

The function to be integrated has such a sharp maximum that we can safely put $\nu_0 = \nu$ in all terms except $\nu - \nu_0$. We can thus replace $\rho(\nu_0, T)$ by $\rho(\nu, T)$. By extending the limits of the integral from $-\infty$ to ∞, we are able to use the fact that

$$\int_{-\infty}^\infty \frac{d\omega_0}{(\omega - \omega_0)^2 + \gamma^2} = \frac{\pi}{\gamma}.$$

Then dividing by 2 since only one-half of the integral is wanted, we obtain Planck's celebrated relation

$$\rho(\nu, T) = \frac{8\pi\nu^2}{c^3}\,\bar{E}(\nu, T), \tag{2.9}$$

relating the spectral density to the average energy of an oscillator at thermal equilibrium.

2.2.2   Step 2: Thermodynamics

To complete his program, Planck had to show that radiation processes take place irreversibly and this entailed finding the form of the entropy. Since the entropy density was not given, Planck *defined* it to be

$$s(\bar{E}) = -\frac{\bar{E}}{\gamma\nu}\left[\ln\left(\frac{\bar{E}}{\alpha'\nu}\right) - 1\right]. \tag{2.10}$$

But where on earth did Planck get such an expression for the entropy density? Now Planck was *sure* that Wien's formula (2.8) was correct. And by eliminating the spectral density between Eqs. (2.8) and (2.9), he obtained the expression

$$\bar{E}(\nu, T) = \alpha'\nu e^{-\gamma\nu/T} \tag{2.11}$$

between the average energy of an oscillator at frequency ν and the temperature T, where $\alpha' = \alpha c^3/8\pi$. Solving for $1/T$, he invoked the second law in the form

$$\frac{\partial s}{\partial \bar{E}} = \frac{1}{T} = -\frac{1}{\gamma\nu}\ln\left(\frac{\bar{E}}{\alpha'\nu}\right). \tag{2.12}$$

Finally, integrating (2.12) Planck obtained (2.10) to within a constant of integration. From the fact that

$$\frac{\partial^2 s}{\partial \bar{E}^2} = -\frac{1}{\gamma\nu\bar{E}} < 0, \tag{2.13}$$

Planck concluded that the "total entropy of the system increases monotonically with time."

By March of 1900, deviations from Wien's law had already been observed and Planck had to weaken this conclusion by claiming that even if the second derivative of the entropy did not have the special form (2.13), it was only necessary for it to be negative in order to ensure its monotonic increase with time.

By October of that year, Planck seriously doubted the validity of Wien's formula. It became apparent that deviations were noted in the long wavelength region where it appeared that $\rho \propto T$. Too much knowledge would have been disastrous for Planck for such a dependency was predicted by the classical equipartition theorem, where each degree-of-freedom has associated with it an energy $T/2$. Since Planck was unaware of this result and had little esteem of kinetic theory in general, he was not aware of any great conflict.

The deviations were observed by the Berlin group of Rubens and Kurlbaum. As the story goes, Rubens and his wife paid a visit to the Plancks on the afternoon of Sunday, October 7, in which he mentioned that $\rho \propto T$ for long wavelengths, and it is most probable that Planck went to work shortly after they left and discovered his formula in the early evening of that same day.

Rather than abandoning Wien's formula, which worked so well in the high frequency part of the spectrum, Planck set out to find an interpolation between the new result and Wien's formula. But how to proceed? The second law could obviously not be violated, and this meant that $-(\partial^2 s/\partial\bar{E}^2)^{-1}$ would certainly have to be positive. According to Wien's law, it was proportional to the average energy. The next best guess was to set it proportional to a sum of linear and quadratic terms in $\bar{E}$. In place of (2.13), Planck proposed

$$\frac{\partial^2 s}{\partial \bar{E}^2} = -\left[\gamma\left(\nu\bar{E} + \frac{\bar{E}^2}{\alpha'}\right)\right]^{-1}. \tag{2.14}$$

Upon integrating he found

$$\frac{\partial s}{\partial \bar{E}} = \frac{1}{T} = \frac{1}{\gamma\nu}\left[\ln\left(\frac{\alpha'\nu + \bar{E}}{\bar{E}}\right)\right], \tag{2.15}$$

and solving for $\bar{E}$ as a function of T, Planck obtained

$$\bar{E}(\nu, T) = \frac{\alpha' \nu}{e^{\gamma \nu / T} - 1}. \tag{2.16}$$

Then using the relation between the average oscillator energy and the spectral density, Planck discovered

$$\rho(\nu, T) = \frac{\alpha \nu^3}{e^{\gamma \nu / T} - 1}. \tag{2.17}$$

For high frequencies, the exponential in the denominator of (2.17) dominates over unity and Wien's law (2.8) is recovered. Furthermore, for small ν, the exponential in the denominator is approximately $1 + \gamma \nu / T$ and the Rubens–Kurlbaum observation is predicted. Planck's law (2.17) fitted the experimental data far too well to be a lucky guess, but there did not seem any way to place it on a firm theoretical foundation that would replace the experimental evidence in its favor.

2.2.3   Step 3: Statistics

Planck tried everything–or almost everything–and all failed. Planck knew the key to resolving his problem lay not with the expression for the spectral density (2.17), but rather with the resulting entropy density

$$s(\bar{E}) = \frac{\alpha'}{\gamma} \left[\left(1 + \frac{\bar{E}}{\alpha' \nu} \right) \ln \left(1 + \frac{\bar{E}}{\alpha' \nu} \right) - \frac{\bar{E}}{\alpha' \nu} \ln \left(\frac{\bar{E}}{\alpha' \nu} \right) \right]. \tag{2.18}$$

The only thing Planck had not tried was Boltzmann's logarithmic relation between entropy and what Planck would later refer to as the "thermodynamic" probability. Planck's expression for the entropy in terms of the average energy of an oscillator, (2.18), too had the kind of logarithmic structure that Boltzmann's principle suggested. Since the thermodynamic entropy is the number of microscopic complexions compatible with a given macroscopic state, the average energy of an oscillator had somehow to be related to the average number of "particles." As Planck later recalled, "after a few weeks of the most strenuous work of my life, the darkness lifted and an unexpected vista began to appear."

The problem was to determine Ω in Boltzmann's principle,

$$S = \ln \Omega, \tag{2.19}$$

without any additive constant. Planck reasoned that Ω represented the number of ways in which the total energy, equal to the m times the *average* energy per oscillator $\bar{E}$, could be shared among m oscillators. In order to bridge the gap with Boltzmann's particle interpretation, Planck was forced to keep $\bar{E}$ discrete, being made up of a finite number of parts n, each having an "element" of energy ε:

$$m\bar{E} = n\varepsilon, \tag{2.20}$$

where n is an integer–a very large integer indeed.

Then Planck, in a rather nonchalant way, set the number of complexions, Ω, equal to the number of ways n energy units could be distributed among m oscillators,

$$\Omega = \binom{m+n-1}{n}, \tag{2.21}$$

which he undoubtedly got from reading Boltzmann. Although Planck was working in the dark, he knew that his final result must be his entropy expression (2.18), and this was more than sufficient motivation for choosing (2.21). Also, it should have worried Planck that the n in the binomial coefficient (2.21) is a random number which he had considered previously as some fixed number in (2.20).

Employing Boltzmann's principle, Planck obtained

$$s(n) = n \ln\left(\frac{n+m}{n}\right) + m \ln\left(\frac{n+m}{m}\right). \tag{2.22}$$

In order to obtain (2.22), Planck had to assume that both m and n are large numbers so that Stirling's approximation could be employed.

If $s(n)$ is truly the entropy density in (2.22), then n must be the *most probable* value that it can take on and which just happens also to be the *average* value. This is the crucial step that salvaged Planck's analysis for he should have written

$$m\bar{E} = \bar{n}\varepsilon \tag{2.23}$$

instead of (2.20) and

$$s(\bar{n}) = \bar{n} \ln\left(\frac{\bar{n}+m}{\bar{n}}\right) + m \ln\left(\frac{\bar{n}+m}{m}\right) \tag{2.24}$$

in place of (2.22). What Planck was interested in was not (2.24), but rather the entropy per oscillator. This is just $s(\bar{n})$ divided by m:

$$\frac{s(\bar{n})}{m} = \left(1 + \frac{\bar{n}}{m}\right) \ln\left(1 + \frac{\bar{n}}{m}\right) - \left(\frac{\bar{n}}{m}\right) \ln\left(\frac{\bar{n}}{m}\right), \tag{2.25}$$

or using (2.23),

$$\frac{s(\bar{E})}{m} = \left(1 + \frac{\bar{E}}{\varepsilon}\right) \ln\left(1 + \frac{\bar{E}}{\varepsilon}\right) - \left(\frac{\bar{E}}{\varepsilon}\right) \ln\left(\frac{\bar{E}}{\varepsilon}\right), \tag{2.26}$$

Planck was finally in a position to compare his statistical expression for the entropy with (2.18)–the expression he obtained from his radiation law.

Now the expression Planck obtained for the entropy (2.18) obviously satisfied Wien's displacement law (2.7), since on account of (2.9)

$$\bar{E}/\nu = (c^3/8\pi)f(\nu/T).$$

The same had to be true of (2.26). Hence, ε had to be a linear function of ν,

$$\varepsilon = h\nu, \tag{2.27}$$

where h is some constant of proportionality. In fact, comparing Planck's expression for the entropy (2.18) with expression (2.26) gave

$$\frac{\bar{E}}{\alpha'\nu} = \frac{\bar{n}}{m}.$$

And since the product $m\bar{E}$ is precisely the spectral density,

$$\rho = \bar{n}\alpha'\nu. \tag{2.28}$$

Let us observe the parallelism between (2.9), which gives the spectral density as the number of oscillators per unit volume per unit frequency interval times the average energy of the oscillator (2.9), and (2.28), which states that the spectral density is equal to the product of the average number of quanta and the energy of one quantum $\alpha'\nu$, or $h\nu$, substituting h for α'. The two ways of looking at the spectral density—either in terms of "quanta" or in terms of "oscillators"—are, in fact, equivalent.

Planck now deviated from Boltzmann, who always took the limit as $h \rightarrow 0$ at the end of the calculation. To Boltzmann, the combinatorial method was a mathematical artifice, in the same way that one counts cards or rolls dice. The physical world is a continuous one and so the limit must be taken at the end of the calculation. To Planck, however, the use of discrete energy units was "the most essential point of the whole calculation." But Planck went little beyond this to justify, or even discuss, his revolutionary idea and, in later years, he was even to shy away from it!

Planck's mark on science would have been made had he stopped here, but his analytical mind drove him further to evaluate the two coefficients α' and γ that appear in both Wien's formula, (2.8), and his own, (2.17), for the spectral density. From the numerical data available to him at the time from black radiation, Planck found $\alpha' = 6.885 \times 10^{-27}$ erg-sec and $\gamma = 4.818 \times 10^{-11}$ sec-K. Planck was quick to point out that their ratio $k = \alpha'/\gamma = 1.429 \times 10^{-16}$erg-K^{-1} had to equal the ratio of the universal gas constant, R, to Avogadro's number. And determining k from the radiation data, he was able to calculate the size of an atom and how many atoms there are in a mole (i.e., a molecular weight in grams). According to Avogadro's hypothesis, this volume is the same for all ideal gases, namely 22,414 cm^3 at standard temperature and pressure. The number of atoms in this volume is Avogadro's number, and knowing it is the same as knowing the mass of an atom, since the latter is simply a mole of the gas divided by Avogadro's number.

Then, from Faraday's law for univalent electrolytes, Planck determined the charge of an electron. The Faraday, $F = Ne$, is the amount of charge needed to plate out one gram mole of univalent ions, which is equal to the product of Avogadro's number N and the charge of an electron, e. Like R, the molar quantity F was also well known, even in those times, but was not so well known was Avogadro's number, N. Planck obtained the value $e = 4.69 \times 10^{-10}$ esu. J. J. Thomson had already measured the charge of the electron and obtained $e = 6.5 \times 10^{-10}$ esu.

However, the significance of his second constant, h, was not immediate. In a private correspondence in July of 1905, Planck drew the analogy between e, the elementary unit of electric charge, and h, an elementary "quantum" of energy. Planck, and later Einstein, noted the dimensional equivalence of h and e^2/c; the latter prophetically remarked that the "same modification of the theory which contains the elementary charge as one of its consequences will also contain the quantum structure of radiation."

2.3 Einstein on Quanta

The clash of Planck's theory with classical theory occurred unknowingly with Lord Rayleigh's suggestion to apply "the Maxwell–Boltzmann doctrine of the partition of energy" to thermal radiation in a cavity. He calculated the spectral density as the number of modes in a given frequency interval multiplied by the average energy of an oscillator, $\bar{E} = T$, that is predicted from the law of equipartition of energy. If this value is introduced into (2.9), there results

$$\rho(\nu, T) = (8\pi\nu^2/c^3)T. \tag{2.29}$$

Although it is known as the Rayleigh–Jeans law, Einstein had a definite hand in it, especially in pointing out the disastrous consequences in the high frequency range where it diverges. Paul Ehrenfest later gave it the picturesque name "ultraviolet catastrophy" and the name has stuck.

As it appears from his June 1900 note in the *Philosophical Magazine*, Rayleigh undermined his conclusion that the Boltzmann–Maxwell doctrine "fails in general" although "it seems possible that it may apply to the graver modes" by subsequently introducing an exponential factor so that "the complete expression will be $\rho(\nu, T) \propto \nu^2 T \exp(-\beta\nu/T)$." Rayleigh then urged that this expression for the spectral density be compared with experimental findings. This is precisely what Lummer and Pringsheim did, and their comparison of the Rayleigh and Wien laws with the experimental results are shown in Fig. 2.3.

What has filtered down to us today is not so much Rayleigh's conclusions but, rather, his normal mode calculation of the density of states since, oddly enough, it still forms part of the discussion on black radiation found in almost all texts on the structure of matter. Hindsight allows a one-line derivation. The ratio of the volume of phase space is expressed as

$$\frac{\mathcal{V}}{\mathcal{V}_0} = \frac{4\pi p^3}{3h^3} V,$$

where $\mathcal{V}_0 = h^3$ is the minimum possible phase space volume in which a system can be located. Using the de Broglie relation, $p = h/\lambda = h\nu/c$—which was

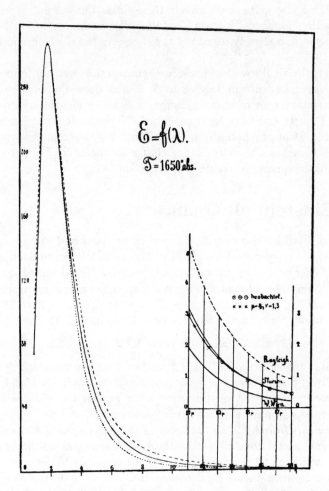

Figure 2.3: Spectral energy curves as a function of wavelength. The solid line represents the experimental curve; above it lies Rayleigh's law (broken curve) while below it lies Wien's law. The authors conclude that the "ascending branch of Wien's isotherm nearly coincides with that drawn from the observational data." The inset is a 20-fold magnification of the comparison in the long wavelength region. The experimental data is now represented by a thin continuous curve and an additional formula due to Thiesen is shown, which is yet a further variant of Wien's law.

still 18 years away—and multiplying by 2 for the two directions of polarization of an electromagnetic wave give $8\pi\nu^2/c^3\,d\nu$ as the number of modes per unit volume in the interval $d\nu$.

Einstein in 1905 was faced with the dilemma that although (2.17) was infinitely more successful in describing black radiation over all frequencies ranges than the Rayleigh–Jeans law (2.29), it was at odds with classical theory. He chose to avoid any analysis based on the generation and propagation of radiation, thereby avoiding the use of Eq. (2.9), and concentrated on a thermodynamic analysis of the high frequency region of black radiation, where the Wien law could be applied unreservedly.

2.3.1 Light-Quantum Hypothesis

What Einstein did was to compare the entropy of the Wien formula with the entropy difference of an ideal gas. Out of this came his famous light-quantum hypothesis.

Consider a system of volume V, pressure P, and temperature T. According to the second law (2.2), the change in the entropy S can be expressed in terms of the corresponding changes in the energy $\bar{U}$ and volume V as

$$dS = \frac{1}{T} d\bar{U} + \frac{P}{T} dV.$$

Furthermore, we are told that $\bar{U} = \bar{U}(V, T)$. The Gibbs relation can then be written more explicitly, and less generally, as

$$dS = \frac{C_V}{T} dT + \frac{1}{T} \left(\frac{\partial \bar{U}}{\partial V} + P \right) dV$$

where C_V is the heat capacity at constant volume. The exactness condition for the entropy is

$$\frac{\partial}{\partial V} \left(\frac{\partial S}{\partial T} \right) = \frac{\partial}{\partial T} \left(\frac{\partial S}{\partial V} \right),$$

implying that

$$T \frac{\partial P}{\partial T} = \frac{\partial \bar{U}}{\partial V} + P.$$

However, for an ideal gas, $PV = NT$, and this implies $\partial \bar{U} / \partial V = 0$ so that the heat capacity can be at most a function of temperature—which it turns out not to be. Hence,

$$dS = C_V \, d\ln T + N \, d\ln V.$$

Now for a system that undergoes a compression from volume V_0 to a smaller volume V, at constant temperature, the entropy change is

$$\Delta S = N \ln \left(\frac{V}{V_0} \right). \tag{2.30}$$

Rather than the thermodynamic derivation of (2.30), Einstein chose another road to it that was based on Boltzmann's principle (2.19). For according to this principle, the entropy difference between states having volumes V and V_0, while at the same temperature, must be

$$S - S_0 = \ln \left(\frac{\Omega}{\Omega_0} \right).$$

Einstein then asked: What is the probability that N particles of the gas will be found in the subvolume V assuming that they are ideal? His answer was:

$$\frac{\Omega}{\Omega_0} = \left(\frac{V}{V_0}\right)^N,$$

which again gives (2.30).

Einstein then switched from a system of homogeneously distributed particles of an ideal gas to radiation in an enclosure. Einstein wrote down the entropy change when the volume containing the radiation is changed from V_0 to V as

$$S(\nu, V, T) - S(\nu, V_0, T) = \frac{\bar{E}}{h\nu} \ln \left(\frac{V}{V_0}\right), \tag{2.31}$$

while *keeping the energy of the (monochromatic) radiation fixed at the value* $\bar{E}$. Capitalizing on the fact that (2.31) displays the same dependence on the volume as (2.30), Einstein concluded that in the domain of validity of Wien's formula, "monochromatic radiation behaves as if it consists of mutually independent energy quanta of magnitude $\gamma\nu$."

Three years earlier, Lord Rayleigh published a paper on the pressure of vibrations in which he commented on the behavior of a normal mode under a sufficiently slow variation of the dimensions of the enclosure. Ehrenfest later extracted the essence of this and converted it into the following theorem: If the reflecting walls of the enclosure are made to approach each other infinitely slowly, the energies of all proper vibrations change in exactly the same proportion.

This means that the ratio $\bar{E}(\nu, T)/\nu$ is *invariant* in such an "adiabatic" process, during which ν *varies inversely as the dimensions of the enclosure.* The change in the Wien entropy is not (2.31) but rather

$$S(\nu, V, T) - S(\nu_0, V_0, T) = -\frac{\bar{E}}{\gamma\nu} \left[\ln\left(\frac{\bar{E}}{\alpha'\nu}\right) - 1\right] + \frac{\bar{E}_0}{\gamma\nu_0} \left[\ln\left(\frac{\bar{E}_0}{\alpha'\nu_0}\right) - 1\right],$$

which vanishes on account of the Rayleigh–Ehrenfest theorem

$$\frac{\bar{E}}{\nu} = \frac{\bar{E}_0}{\nu_0}. \tag{2.32}$$

Moreover, since the frequency varies as the inverse of the volume,

$$\nu > \nu_0 \quad \text{for} \quad V_0 > V.$$

In the light of the invariancy of (2.32), Einstein's demonstration of the existence of quanta appears to be contrived.

But surely Einstein's conclusion regarding the magnitude of energy quanta is sound, so there must be something to his argument. The crux of the problem lies in Einstein's comparison of one concave function (2.30) with another one (2.31). In other words, the volume dependency of the concave function

(2.30) cannot be compared with the Wien entropy (2.10). We must, therefore, look for a process whose entropy has the same form as the Wien entropy. We have already considered such a process in §1.13 in connection with the Gibbs paradox.

There, we considered a volume V_0 that contains N_0 gas particles. A sub-volume V is singled out, and we ask for the probability of finding N particles in it. This probability is simply the binomial distribution

$$f(N) = \binom{N_0}{N} \left(\frac{V}{V_0}\right)^N \left(\frac{V_0 - V}{V_0}\right)^{N_0 - N}.$$

Suppose that the total volume V_0 and N_0 are allowed to increase indefinitely or, what is equivalent, to consider $V \ll V_0$ and $N \ll N_0$. Then the approximation

$$N! \simeq (N_0 - N)! N_0^N$$

together with the omission of N in the exponent of $N_0 - N$, allow us to write the binomial distribution as

$$f(N) \simeq \frac{1}{N!} \left(\frac{N_0 V}{V_0}\right)^N \left(1 - \frac{V}{V_0}\right)^{N_0},$$

where we have dropped N in the exponent of the last term since it is small compared to N_0. Using the fact that the system is homogeneous, $\bar{N}/V = N_0/V_0$, we have

$$f(N) \sim \frac{\bar{N}^N}{N!} \left(1 - \frac{\bar{N}}{N_0}\right)^{N_0},$$

which is beginning to look more like a Poisson distribution. In fact, if we allow $N_0 \to \infty$ and avail ourselves of the formula

$$\lim_{N_0 \to \infty} \left(1 - \frac{\bar{N}}{N_0}\right)^{N_0} = e^{-\bar{N}},$$

we come out precisely with the Poisson distribution,

$$f(N) = \frac{\bar{N}^N}{N!} e^{-\bar{N}}, \tag{2.33}$$

in the limit.

The logarithm of the Poisson distribution (2.33) can be cast in the form

$$\ln f(N) = (N - \bar{N}) \ln \left(\frac{\bar{N}}{N_0}\right) + \bar{N} \ln \left(\frac{\bar{N}}{N_0}\right) - \bar{N} - N \ln \left(\frac{N}{N_0}\right) + N,$$

where we have added and subtracted terms in $\ln N_0$ so as to make it extensive. Comparing this expression with Gauss' error law,

$$f(N) = A \exp \left\{ S(N) - S(\bar{N}) - \left(\frac{\partial S}{\partial \bar{N}}\right)_V (N - \bar{N}) \right\}, \tag{2.34}$$

we can read off the entropy as

$$S(\bar{N}) = -\bar{N}\left[\ln\left(\frac{\bar{N}}{N_0}\right) - 1\right].$$ (2.35)

On the strength of the homogeneity condition, this is equivalent to

$$S(\bar{N}) = -\bar{N}\left[\ln\left(\frac{V}{V_0}\right) - 1\right].$$ (2.36)

We may now compare this expression with the Wien entropy of the *field*,

$$S(\rho) = -V_0\frac{\rho}{h\nu}\left[\ln\left(\frac{\rho c^3}{8\pi h\nu^3}\right) - 1\right],$$ (2.37)

rather than the entropy of the *oscillator* given by (2.10). Equating the prefactors, we obtain

$$\rho V_0 = m\bar{E} = \bar{N}h\nu,$$ (2.38)

while equating the arguments of the logarithms gives

$$\frac{\bar{N}}{N_0} = \frac{V}{V_0} = \frac{\rho c^3}{8\pi h\nu^3}.$$

Then eliminating $\rho/h\nu$ with the aid of (2.38) results in

$$N_0/V_0 = m = 8\pi\nu^2/c^3,$$

which is precisely the number of oscillators per unit volume per frequency interval $d\nu$. In this way not only do we obtain Einstein's sought after relation (2.38) but, in addition, we get the correct expression for the number of Planck oscillators per unit volume per unit frequency interval.

Now the odd thing about the two identifications is that in the first instance N is identified as the number of photons while in the second instance N_0 is the number of oscillators. However, from our derivation N should be the same species as the total number N_0. This is our first indication of a breakdown in the distinction between "particles" and "oscillators" which comprise the field. We will be dealing with complementarity more fully in §2.5 and Chapter 6. Let it suffice here to say that they are different manifestations of the same phenomenon.

2.3.2   Fluctuations: Synthesis of Waves and Particles

There is probably no one that wielded statistical arguments better than Einstein to obtain insights into quantum theory. Einstein realized that energy fluctuations in a volume the size of a cavity filled with thermal radiation would certainly be too small to be observed. However, he reasoned, if one were to consider a very small volume whose linear dimension is the size of a wavelength, then the fluctuation should be of the order of the energy itself.

Thus, over such a dimension, the dispersion in energy should be given by

$$\overline{(\Delta U)^2} = \bar{U}^2. \tag{2.39}$$

Using the fact that the dispersion is related to the heat capacity by

$$\overline{(U - \bar{U})^2} = T^2 \frac{d\bar{U}}{dT} \tag{2.40}$$

and identifying the mean energy $\bar{U}$ with the product of the total energy density given by Stefan's law (2.5) and the volume, namely,

$$\bar{U} = \sigma V T^4, \tag{2.41}$$

Einstein obtained the characteristic dimension as

$$V^{1/3} = \left(\frac{4}{\sigma}\right)^{1/3} T^{-1}.$$

Setting $V^{1/3}$ equal to a characteristic wavelength, λ, led him to the relation

$$\lambda = \frac{0.418}{T} \qquad \text{[cm]}. \tag{2.42}$$

Experimentally, it was known that the maximum in the black radiation curve occurred at

$$\lambda_m = \frac{0.29}{T} \qquad \text{[cm]} \tag{2.43}$$

—a coincidence that was too close to be mere chance. In Einstein's words,

> one sees that the temperature dependence of λ_m as well as its order of magnitude can be determined correctly by means of the general molecular theory of heat, and I believe that because of the great generality of our assumptions, this agreement ought not be ascribed to chance.

As enticing as the argument may appear, it is inaccurate. It is clear from (2.40) that the dispersion in energy is an extensive quantity. The same cannot be said of the square of the average energy, (2.39). Hence, the two cannot be equated. A feasible expression would be to replace the right side of (2.39) by $\bar{E}^2/V$, analogous to the Rayleigh–Jeans limit, where $\bar{E}$ is the energy in the frequency interval $d\nu$. As a result, the volumes would cancel out in the final result so that no expression for the maximum wavelength could be obtained by this procedure.

Einstein later returned to the analysis of fluctuations to argue for a theory of radiation which would see the synthesis of the particle and wave aspect. This he did by again turning to the expression for the dispersion in energy (2.40).

When Planck's radiation law, (2.16), is used for $\bar{E}$, the resulting expression for the dispersion is

$$\overline{(E - \bar{E})^2} = \frac{(h\nu)^2}{e^{h\nu/T} - 1}\left(1 + \frac{1}{e^{h\nu/T} - 1}\right)$$
$$= \left(h\nu\bar{E} + \bar{E}^2\right). \qquad (2.44)$$

From this expression, Einstein concluded that there are two independent sources of fluctuations which are additive. The linear term is what would be obtained from, say, "shot" noise or a random sequence of points that are distributed according to a Poisson distribution. The dispersion in the number of particles is obtained from (2.44) by dividing both side by $(h\nu)^2$. However, we must first show that the Wien entropy is, in fact, the entropy of the Poisson distribution.

The entropy of the field in the Wien approximation is

$$S(\rho V) = -V\frac{\rho}{h\nu}\left[\ln\left(\frac{\rho}{h\nu m}\right) - 1\right]$$
$$= -V\bar{n}\left[\ln\left(\frac{\bar{n}}{m}\right) - 1\right]. \qquad (2.45)$$

Introducing this expression for the entropy into Gauss' principle (2.34), with $\bar{X} = \bar{N}$, and simplifying terms, we obtain precisely the Poisson distribution (2.33). Now, in order to calculate the dispersion, we differentiate the series

$$e^{\bar{N}} = \sum_{N=0}^{\infty} \frac{\bar{N}^N}{N!}$$

to obtain

$$e^{\bar{N}} = \frac{1}{\bar{N}}\sum_{N=1}^{\infty} N\frac{\bar{N}^N}{N!}.$$

A second differentiation yields

$$e^{\bar{N}} = \sum_{N=1}^{\infty} N(N-1)\frac{\bar{N}^{N-2}}{N!} = \frac{1}{\bar{N}^2}\left\{\sum_{N=1}^{\infty} N^2\frac{\bar{N}^N}{N!} - \sum_{N=1}^{\infty} N\frac{\bar{N}^N}{N!}\right\}.$$

Rearranging we get

$$\overline{(N - \bar{N})^2} = e^{-\bar{N}}\sum_{N=1}^{\infty} N\frac{\bar{N}^N}{N!} = \bar{N}, \qquad (2.46)$$

showing that the mean and dispersion of the Poisson distribution are equal. Multiplying (2.46) by $(h\nu)^2$, we get the first term in (2.44).

Einstein argued on dimensional grounds that the second term in the dispersion relation (2.44) is characteristic of the wave properties of light. The term $\bar{E}^2$ is proportional to the intensity of the radiation in exactly the same way that the square of the electric field is proportional to the energy.

In the high frequency, low temperature region of the spectrum only the particle nature of light would subsist and the dispersion in energy would be that of Wien's law. Alternatively, in the low frequency, high temperature region, the second term is dominant. This would correspond to the Rayleigh–Jeans law. If only the second term were present, we would have

$$\frac{\partial^2 s}{\partial \bar{E}^2} = -\frac{1}{\bar{E}^2}.$$

Integrating and using the second law in the form $\partial s/\partial \bar{E} = 1/T$ gives the equipartition law $\bar{E} = T$ for a harmonic oscillator. A second integration gives

$$s = \ln \bar{E} + \text{const.}$$

This is the entropy density of one oscillator. The entropy density of m oscillators per unit volume is

$$s(\bar{u}) = m \ln \left(\frac{\bar{u}}{m} \right) + \text{const.},$$

where we have set $\bar{u} = m\bar{E}$.

Introducing this expression for the entropy into Gauss' principle (2.34), with $\bar{X}$ representing the average energy, we have

$$f(u; \bar{u}) = \left(\frac{u}{\bar{u}} \right)^m e^{-mu/\bar{u}+m+\psi(u)},$$

where $\psi(u)$ is an undetermined function of integration that will be used to normalize the density. Introducing equipartition of energy in the form $\bar{u} = m/\beta$ and setting $\psi(u) = m/u$ give

$$f(u; \beta) = \beta \frac{(\beta u)^{m-1}}{\Gamma(m)} e^{-\beta u} \qquad (2.47)$$

provided m is sufficiently large to warrant Stirling's approximation. Hence, u is a continuous random variable which has a gamma distribution (2.47). We will later identify m with one-half the number of degrees-of-freedom. This is what we would expect *classically*, and we shall return to the discussion in §2.5.

2.4   Einstein's Radiation Theory: The Second Coming of Planck's Law

We jump ahead in time to the year 1916. Einstein attempted a derivation of Planck's law based on the physical processes of absorption and emission of radiation which where still "obscure" processes. The processes of absorption and spontaneous emission were known from classical electrodynamics. However, in the process of deriving Planck's law, Einstein found it necessary to postulate a second type of emission process, referred to as "stimulated" or "induced" emission, in order to obtain thermal equilibrium in a gas emitting and absorbing radiation. In Einstein's own words, by postulating some hypotheses on the emission and absorption of radiation by molecules, I was able to show that molecules with a quantum-theoretical distribution of states in thermal equilibrium, were in dynamical equilibrium with the Planck radiation; in this way, Planck's formula [(2.17)] could be derived in an astonishingly simple and general way.

To begin with, Einstein assumed that the molecules absorbing and emitting radiation obeyed classical Boltzmann statistics. The probability of finding a molecule in its nth (discrete) state is

$$p_n = g_n e^{-\varepsilon_n/T},$$

where g_n is the statistical weight of this quantum state. Without loss of generality, he focused his attention on two states m and n such that $\varepsilon_m > \varepsilon_n$. Transitions between these two states could be effectuated by the emission or absorption of radiation.

Concerning the process of emission, Einstein assumed that the emission process could occur spontaneously, in agreement with classical theory, without any external influence of the radiation field. In an exactly analogous way to radioactive decay, the probability dW_{sp} that a radiative transition occurs in time dt is

$$dW_{sp} = A_{m \to n}\, dt \qquad \text{(spontaneous emission)},$$

where $A_{m \to n}$ is some constant which characterizes this process and which is supposed to be *temperature independent*. The idea of spontaneous emission was to subsequently trouble Einstein, for it was essentially a probabilistic mechanism. Einstein's thoughts, in later years, were to turn to a completely causal universe in which "God does not play dice with the world." Such a probabilistic mechanism would have no role in such a world.

A classical, charged oscillator will absorb radiation, depending on the phase of its motion, when it interacts with radiation in the neighborhood of its own frequency. This depends upon the spectral density ρ and the frequency ν of the radiation. The probability of transition from the state n to m in time dt caused by the absorption of radiation is proportional to the spectral density,

$$dW_{ab} = \rho B_{n \to m}\, dt \qquad \text{(absorption)},$$

where again the coefficient $B_{n \to m}$ is assumed to be *temperature independent*. The three types of transitions between the two levels is shown in Fig. 2.4.

This was all that was known at the time, and surely Einstein must have considered this in his initial attempt at deriving Planck's law. If the system is to achieve thermal equilibrium, the rate of absorption must balance the rate of spontaneous emission. The condition for such a dynamical equilibrium is

$$\rho B_{n \to m} g_n e^{-\varepsilon_n/T} = A_{m \to n} g_m e^{-\varepsilon_m/T}, \qquad (2.48)$$

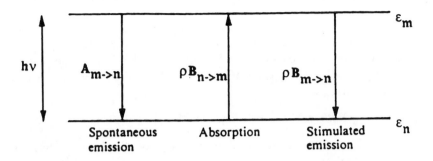

Figure 2.4: The three basic forms of radiative processes.

which on solving for ρ gives

$$\rho = \frac{g_m}{g_n} \frac{A_{m \to n}}{B_{n \to m}} e^{-(\varepsilon_m - \varepsilon_n)/T}.$$

This resembles the Wien distribution (2.8), and in fact, if one were to impose the Bohr frequency condition,

$$\varepsilon_m - \varepsilon_n = h\nu, \qquad (2.49)$$

and set

$$\frac{g_m}{g_n} \frac{A_{m \to n}}{B_{n \to m}} = \alpha \nu^3,$$

it would certainly coincide with it. Therefore, it is quite possible that based on these considerations, Einstein realized there was something missing from the possible radiation processes. Certainly, what was missing is not a classical effect, and it therefore must be related to the quantum nature of radiation.

By the very fact that the dynamic equilibrium condition must yield Planck's law, Einstein was probably led to suppose the existence of a reverse transition, having the same structural form as the absorption process. This meant that the field could induce emission at the frequency ν present before the emission process. The probability for this to occur is

$$dW_{st} = \rho B_{m \to n}\, dt \qquad \text{(stimulated emission)},$$

where, again, the coefficient characterizing this process, $B_{m \to n}$, is *temperature independent*.

We have repeatedly emphasized the temperature independence of the Einstein coefficients in order to bring out the fact that the overall rates were split up into a product of terms: a term independent of the temperature, which could be subsequently calculated from quantum mechanical perturbation theory, and a factor which depended on the temperature, based on the assumption that the molecules, which are emitting and absorbing radiation, obey *classical* statistics.

Now Einstein reasoned that in a state of dynamical equilibrium the rates of processes going in one direction will be exactly balanced by those going in the other direction. This condition of "detailed balancing" of all elementary processes requires that the rates of the two processes balance one another at equilibrium. Hence, Einstein equated

$$\rho B_{n \to m} g_n e^{-\varepsilon_n/T} = (\rho B_{m \to n} + A_{m \to n}) g_m e^{-\varepsilon_m/T}, \qquad (2.50)$$

where the rates at which the processes occur are the products of the probability per unit time and the probability of finding a molecule in the given state. Faced with the dilemma that there are three unknowns but only one equation, Einstein proceeded to the asymptotic limit where "ρ tends to infinity with T." This led to the condition that

$$g_n B_{n \to m} = g_m B_{m \to n},$$

relating the coefficients of absorption and induced emission. With this condition, one of the two Einstein coefficients in (2.50) could be eliminated, leaving only a ratio of the other two that could be identified on comparison with Planck's radiation law. For on solving for ρ he found

$$\rho = \frac{A_{m \to n}/B_{m \to n}}{e^{(\varepsilon_m - \varepsilon_n)/T} - 1}. \qquad (2.51)$$

Einstein was quick to conclude that (2.51) has the same form as Planck's formula (2.17), and in fact, if one equated

$$\frac{A_{m \to n}}{B_{m \to n}} = \alpha \nu^3$$

and employed the Bohr frequency condition, it would be a perfect match.

Although quantum mechanics appears to have fully vindicated Einstein's radiation mechanism, it nevertheless leads one to ask whether his assumptions are both necessary and sufficient to obtain Planck's formula. Could less restrictive assumptions be made and still obtain the radiation law? In particular, why must the molecules obey *classical* statistics? Let us consider the assumption concerning temperature independence of the Einstein coefficients.

Three years earlier, Einstein and his collaborator Otto Stern wanted to discriminate between Planck's radiation law with and without a "zero-point energy." The zero-point energy arose for the following reason. For large values of the temperature, the radiation law (2.51) becomes

$$\rho = \frac{A_{m \to n}}{B_{m \to n}} \left(\frac{T}{h\nu} - \frac{1}{2} \right) \qquad \text{(Rayleigh–Jeans limit).} \qquad (2.52)$$

Thus, it appears that Planck's formula in the classical limit does *not* give the equipartition of energy result. However, the addition of the factor $A_{m \to n}/2B_{m \to n}$

precisely cancels the second term to make it agree with the Rayleigh–Jeans formula (2.29).

The formula

$$\rho = \alpha\nu^3 \left\{ \frac{1}{e^{h\nu/T} - 1} + \frac{1}{2} \right\} \tag{2.53}$$

made its first appearance in Planck's "second theory" in which he attempted to modify his earlier theory so that it would appear to be a less radical departure from classical theory. In this theory, a molecule continuously absorbs quanta and only the emission is discontinuous. Planck's second theory has long since passed into oblivion, except for the vestige of the zero-point energy which he obtained by averaging the energies of successive rings in the phase plane of a harmonic oscillator, i.e., the average of $(n-1)h\nu$ and $nh\nu$ give $(n-1/2)h\nu$.

In contrast to the first radiation formula, (2.17), the second one, (2.53), does give the equipartition result

$$\rho = \frac{A_{m\to n}}{B_{m\to n}} \frac{T}{h\nu}$$

in the high temperature limit. The scope of the Einstein–Stern article was to discriminate between the two expressions by analyzing the data obtained for rotating gas molecules. According to Hermann Walther Nernst, such a system would behave very similarly to a monochromatic system whose frequency could be changed independently of the temperature. The "effective" frequency of rotation was determined by setting the kinetic energy of rotation equal to the oscillator energy and it was temperature dependent. On the basis of the specific heat data supplied by Arnold Eucken, Einstein and Stern concluded that only expression (2.53) could account for the measurements.

However, as far as black radiation is concerned, the second form of the spectral density, (2.53), would again lead to the ultraviolet catastrophe in the high frequency, or low temperature, region of the spectrum. So we conclude that Planck's radiation law must be given by the original expression (2.17). In the Rayleigh–Jeans limit, it becomes (2.52), which must be equated with the Rayleigh–Jeans formula, which, as we recall, was obtained from purely classical considerations, i.e., equipartition of energy. Therefore, upon equating (2.29) with the asymptotic expression (2.52), we obtain

$$\frac{A_{m\to n}}{B_{m\to n}} = \alpha\nu^3 \frac{T}{T - h\nu/2},$$

in contrast to Einstein's result.

Hence, for the same reason that Einstein and Stern, in their 1913 paper, found it necessary to distinguish between the two proposed forms of Planck's law, (2.17) and (2.53), we see here that the Einstein coefficients are *not* temperature independent, albeit, the temperature dependency is a weak one. We emphasize that of the two expressions for the radiation law, it is only the original expression (2.17) that is acceptable for black radiation, although it does not give the classical equipartition limit result. We will leave till §3.3.1 the demonstration that there is no need to decompose the rates into temperature dependent and temperature independent parts. It is precisely the second law that will provide automatically the correct temperature dependency as well as the stability of the radiation process.

We can avoid the Einstein coefficients entirely by introducing some information about the electrons which are absorbing and emitting radiation. Surely, Einstein had no knowledge of the exclusion principle, and much less knew of Fermi–Dirac statistics, so his derivation cannot be undermined by facts which were only to be learned afterward. Nevertheless, it is an interesting exercise to show that one form of statistics implies the other.

Let φ_m be the probability (a proper fraction!) that an electron is in state m and $1-\varphi_m$ that it is not. The probability that an electron will make a transition $n \to m$ is the product $\varphi_n (1 - \varphi_m)$. Einstein's dynamical equilibrium condition can now be expressed as

$$\bar{n}\varphi_n (1 - \varphi_m) = (\bar{n} + m)\varphi_m (1 - \varphi_n),$$

where $m = 8\pi\nu^2/c^3$ is the density states for photons in the cavity and $\bar{n}$ is the average density of photons in the same frequency interval. This condition of dynamical equilibrium between the rates of absorption and emission of radiation can be rearranged to read

$$\frac{\varphi_n}{1 - \varphi_n}\bar{n} = \frac{\varphi_m}{1 - \varphi_m}(\bar{n} + m).$$

Now, if we know that the electrons obey "Fermi–Dirac" statistics [cf. §3.3.3],

$$\frac{\varphi_n}{1 - \varphi_n} = e^{-\varepsilon_n/T}, \tag{2.54}$$

we come out with

$$\bar{n} = \frac{m}{e^{h\nu/T} - 1}, \tag{2.55}$$

provided the Bohr frequency condition (2.49) holds.

Alternatively, if we know that photons obey Bose–Einstein statistics, we obtain (2.54), again provided the Bohr frequency condition is satisfied. Thus we see that one statistics implies the other. Even if you do not know the exclusion principle and assume that the φ's are given by Boltzmann statistics with the same statistical weights for the two states in question, you come out with the average photon density, (2.55).

Although everything seems to be consistent, there is a small subtlety: since electrons are conserved and photons are not, the former do not have to have a vanishing chemical potential. For example, if we consider an equilibrium between electrons and photons in a cavity with perfectly reflecting walls which are composed of a homogeneous semiconducting material, we could have electronic transitions between occupied states in the conduction band and unoccupied states in the valence band. These transitions would occur much more infrequently than transitions between states of different energies in the conduction band or those in the valence band. The latter processes establish a uniform chemical potential throughout the conduction band, μ_m, and the valence band, μ_n. But because of the rarity of the transitions between electrons in the conduction band and those in the valence band, chemical equilibrium may not exist between the two bands with the consequence that $\mu_c \neq \mu_v$. This necessitates the introduction of a factor $+\mu_n/T$ in the exponent of (2.54) and leads to the modification

$$\bar{n} = \frac{m}{e^{(h\nu - \mu_\gamma)/T} - 1}$$

of Planck's law (2.55), where the photon chemical potential is

$$\mu_\gamma = \mu_m - \mu_n. \tag{2.56}$$

Equation (2.56) establishes an equilibrium between excitations in the semiconductor, characterized by a chemical potential difference $\mu_m - \mu_n$, and a photon chemical potential μ_γ. Only in the case of thermal black radiation, in which the two sets of states have the same chemical potential, will their difference, μ_γ, vanish. For "nonthermal" radiation, which we shall discuss in §7.3, we would expect a nonvanishing photon chemical potential, μ_γ.

We can derive an interesting property about this type of chemical potential from the invariancy of ρ/ν^3. We recall from §2.1 that the temperature must undergo a red-shift if the frequency undergoes one by an amount $T' = (1-\Delta)T$, where T' is the observed temperature and T is the temperature at emission. In order that this invariancy not be destroyed upon introducing the chemical potential into Planck's law, μ_γ/T must also be an invariant. This implies that the chemical potential also undergoes a red-shift by an amount

$$\mu'_\gamma = \mu_\gamma(1 - \Delta).$$

The invariancy of the ratio μ_γ/T leads immediately to the relation

$$\frac{d\mu_\gamma}{\mu_\gamma} = \frac{dT}{T},$$

which, as we will see in §6.3, reduces the Gibbs–Duhem equation to the Clapeyron equation.

The Clapeyron equation describes a phase equilibrium in which the pressure depends only on the temperature, independent of the volume. Interestingly enough, black radiation has the same functional dependency. When the volume is varied at constant temperature, the density of energy, and consequently the pressure, remain constant. This is analogous to the behavior of the pressure of a saturated vapor, as we shall see in §7.3. When the volume is varied at constant temperature, the liquid makes sure that the pressure is not altered.

2.5   Underlying Probability Distributions

It is always intriguing to wonder if a great scientific discovery could have been overlooked had the discoverer used all the knowledge that was available at the time. Knowing too much can prevent the step that is necessary to break with the past. Planck's discovery of the black radiation law (2.17) is such a case. When he wrote down his formula, he was not at all concerned about the fact that it contradicted the law of equipartition of energy in all but the long wavelength limit. He probably did not know the law existed. As late as April 1897, Planck argued in his preface to *Vorlesungen über Thermodynamik* that irreversible processes in radiation cannot be explained satisfactorily by kinetic theory. Surely, Rayleigh was well aware of the law of equipartition but could not pass judgment on Planck's formula because he could not follow its derivation. As we have seen, it was several years afterward that Einstein fully grasped the undermining of the laws of classical physics that went into Planck's derivation of his radiation law.

Planck's search for his radiation law could have begun by the observation that the fundamental relation, which expresses the entropy as a function of the extensive variables, is really not that "fundamental" at all since it does not give rise to an error law for the distribution in the energy. The fundamental relation for black radiation is obtained by introducing the pressure (2.3) and the temperature $T = (\bar{U}/\sigma V)^{1/4}$ obtained from Stefan's law (2.41) into the Euler relation,

$$S = \frac{\bar{U} + PV}{T}. \tag{2.57}$$

We are immediately led to the fundamental relation

$$S = \tfrac{4}{3}\sigma^{1/4}\bar{U}^{3/4}V^{1/4} \tag{2.58}$$

for the entropy as a function of the energy $\bar{U}$ and volume V. Although (2.58) is thermodynamically admissible, since the entropy is extensive and a concave function of either the energy or volume, it does not give rise to an error law.

Consider the entropy as a function of $\bar{U}$ only, holding the volume constant. The law of error leading to the average energy as the most probable value of the energy measured is

$$f(U;\bar{U}) = A\exp\left\{S(U) - S(\bar{U}) - S'(\bar{U})(U - \bar{U})\right\}, \tag{2.59}$$

where the prime means differentiation. It is quite clear that (2.58) does not give rise to any error law at all. So, Planck could have used this as a motive to search for a more fundamental relation that would replace (2.58). This is actually what he did, for Planck replaced (2.58) by the entropy of a spectral component within the frequency range from ν to $\nu + d\nu$.

We can use Gauss' principle (2.59) as a guide in our search for this more fundamental relation. In §4.1 we will show that Gauss' error law (2.59) can be cast in the so-called canonical form

$$f(U; \hat{\beta}) = \frac{e^{-\hat{\beta}U}}{\mathcal{Z}(\hat{\beta})} \Omega(U),$$ (2.60)

where

$$S'(\bar{U}) = \hat{\beta}$$ (2.61)

is the inverse temperature. For continuous values of U, (2.60) is the probability density of the canonical ensemble. Leaving till Chapters 4 and 5 the full development of the canonical ensemble, we will use here only those properties that are vital to the discussion. The norming factor, $\mathcal{Z}(\beta)$, is known as the partition function and the prior density, or "structure" function, $\Omega(U)$, completely determines the mechanical structure of the isolated system before it is converted into a closed system by putting it in thermal contact with a heat reservoir. At most, $\Omega(U)$ increases as a finite, fixed power of U which requires a fundamental relation of the logarithmic form like that of an ideal gas. That this is not fulfilled by the "fundamental" relation of black radiation (2.58) should have been a cue to search for another, more fundamental relation of the logarithmic form for the entropy.

2.5.1   Statistics of Black Radiation

Planck's derivation of his radiation law can be portrayed as an attempt to convert the fundamental relation (2.58) into a logarithmic form so that it would correspond to a law of error for the energy, though it will not be the total radiant energy, but rather the energy in a given frequency interval. Let us recall that Planck had two leads to go on: Wien's displacement law, (2.7), and Wien's radiation law, (2.8), for the spectral distribution. From Wien's displacement law it follows that the spectral distribution is determined for all temperatures once it is known for a single temperature. This function must be such that when it is integrated over all frequencies, it reproduces Stefan's law (2.5).

In what follows, we will designate by $\bar{u}_\nu$ what we have previously called $\rho(\nu, T)$ so as to distinguish between the fluctuation, u, and its average value, $\bar{u}_\nu$. Since modes of the electromagnetic field do not interact with each other, it is safe to assume that the Euler relation, (2.57), is valid in each frequency interval of the electromagnetic field. Therefore, the entropy density in the frequency interval from ν to $\nu + d\nu$ is

$$s(\bar{u}_\nu) = \frac{\bar{u}_\nu + P_\nu}{T}.$$ (2.62)

The radiation pressure, P_ν, in the same frequency interval can be determined by comparing the local Euler relation (2.62) with the Wien entropy of the field,

$$s(\bar{u}_\nu) = -\frac{\bar{u}_\nu}{\gamma \nu} \left[\ln\left(\frac{\bar{u}_\nu}{\alpha \nu^3}\right) - 1 \right].$$ (2.63)

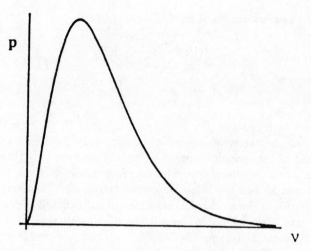

Figure 2.5: The Wien pressure as a function of the frequency.

From the expression for the Wien temperature (2.12) or its field counterpart,

$$s'(\bar{u}_\nu) = \frac{1}{T} = -\frac{1}{\gamma\nu}\ln\left(\frac{\bar{u}_\nu}{\alpha\nu^3}\right), \qquad (2.64)$$

we know that Wien's law only holds for weak intensities, $\bar{u}_\nu < \alpha\nu^3$. Comparing (2.62) and (2.63), we obtain

$$P_\nu = T\frac{\bar{u}_\nu}{\gamma\nu}. \qquad (2.65)$$

Integrating the Wien pressure, (2.65), over all frequencies gives $P = 2\alpha T^4/\gamma^4$. while integrating the Wien energy density (2.8) over the same range gives $u = 3!\alpha T^4/\gamma^4$ so the pressure–energy relation (2.3) is, in fact, obeyed with the value of the Stefan–Boltzmann constant given by $\sigma = 3!\alpha/\gamma^4$. The Wien pressure per mode as a function of frequency is shown in Fig. 2.5.

Expression (2.10) is a fundamental relation and should correspond to a law of error identifying the average value of the energy as the most probable value of the energy which is measured. On substituting the entropy (2.63) into the error law (2.59), we obtain the expressions

$$\Omega(u) = \left(\frac{u}{\alpha\nu^3}\right)^{-u/\gamma\nu} e^{u/\gamma\nu} \qquad (2.66)$$

and

$$\ln \mathcal{Z}(\beta) = \frac{\alpha\nu^2}{\gamma}e^{-\beta\gamma\nu} \qquad (2.67)$$

for the structure function and logarithm of the partition function, respectively. It thus appears that we are no closer to a fundamental relation for black radiation than our original relation, (2.58).

But let us not be too hasty in our judgment! The structure function, (2.66), would look like Stirling's approximation for a factorial if only u could somehow be treated as a *discrete* variable. The logarithm of the partition function, (2.67), would then be some average number density. In fact, by defining,

$$\bar{n}_\nu \equiv \bar{u}_\nu/\gamma\nu, \quad n \equiv u/\gamma\nu, \quad \text{and} \quad \bar{m}_\nu \equiv \alpha\nu^2/\gamma, \tag{2.68}$$

the entropy density per mode, (2.63), can be written as

$$s(\bar{n}_\nu) = \ln\left(\frac{\bar{m}_\nu^{\bar{n}_\nu}}{\bar{n}_\nu!}\right), \tag{2.69}$$

provided $\bar{n}_\nu$ is sufficiently large to warrent Stirling's approximation. Now, (2.69) is precisely the entropy of the Poisson distribution [cf. (2.33)],

$$f(n) = \frac{\bar{n}_\nu^n}{n!}e^{-\bar{n}_\nu}. \tag{2.70}$$

The revolutionary step is contained in (2.68)—the assumption of discrete "particles"—and its justification resides in the derivation of the Poisson distribution as the law of error identifying the average number of "particles" as the most probable value of the quantity measured.

Further support would have been forthcoming had (2.68) been introduced into Wien's formula, (2.7), to obtain

$$\bar{n}_\nu = \bar{m}_\nu e^{-\varepsilon_\nu/T}, \tag{2.71}$$

where $\varepsilon_\nu = \gamma\nu$. Expression (2.71) is the most probable "distribution" of Boltzmann statistics, where $\bar{n}_\nu$ is the most probable, or equivalently, average number of particles of the group ν that is found in $\bar{m}$ cells [cf. §3.6.1]. Surely, the law of equipartition of energy cannot be applied locally to each mode. Instead, it applies to the average energy density

$$\bar{u} = 3\bar{n}T \tag{2.72}$$

for a system with $6\bar{n}$ degrees-of-freedom per unit volume.

Having gotten this far, Planck could have introduced the first relation in (2.68) into the Wien expression for the pressure per mode, (2.65). He would have then realized that the pressure

$$P_\nu = \bar{n}_\nu T$$

is that of an ideal gas in each frequency interval $d\nu$. Integrating over all frequencies gives $P = \bar{n}T$, which, in view of (2.72), gives the radiation pressure (2.3).

Hence, had Planck attempted to place Wien's formula on firm theoretical ground, he would have certainly been led to the discreteness hypothesis contained in (2.68) and found that its justification was simpler than the path he was to follow since it involved only *classical* Boltzmann statistics. Planck, however, was not aware of the limiting nature of the Poisson distribution, and it was not apparent that (2.71) is only valid under the condition that $\bar{n}_\nu \ll \bar{m}_\nu$. This was to emerge from the experimental investigations of Rubens and Kurlbaum, who showed that Wien's formula, (2.7), did in fact breakdown in the far-infrared region, where $\bar{u}_\nu \propto T$, in accordance with the law of equipartition of energy. It is rather ironic that quantum theory was discovered only after "classical" deviations from the quantum regime had been observed in the far-infrared!

The short wavelength regime of black radiation points to the discreteness of "quanta" of energy,

$$\bar{u}_\nu = \bar{n}_\nu \varepsilon_\nu, \tag{2.73}$$

that obey classical, or Boltzmann, statistics. As we have seen in §2.3.2, this was emphasized by Einstein in his 1905 paper on light-quanta. It is precisely in the long wavelength regime—where the law of equipartition of energy is obeyed—that the statistics becomes *non*-classical. Recalling our discussion in §2.2.2, Planck worked from the expression for $-(\partial^2 s/\partial \bar{E}^2)^{-1}$ in which he added a quadratic term in the energy. Then integrating once and using the second law, he obtained his celebrated formula. In order to place his formula on firm theoretical ground, he had to take recourse to Boltzmann's probabilistic formulation of the entropy. We will now proceed in a way that employs the actual probability distributions rather than their binomial coefficients.

Boltzmann statistics, like the Poisson distribution, is a limiting form that is valid in a well defined limit. That limit was unknown to Planck at the time, and it took another quarter of a century before it became clear what was happening in the passage to that limit. Historically, the Poisson distribution was known as "Poisson's limit law" to which the binomial distribution converges in the limit as the number of Bernoulli trials increases without bound while the probability of success tends to zero in such a way that their product is of moderate magnitude. This, together with the particle nature that emerges in the Wien limit, could have led Planck to explore the possibility of obtaining the actual probability distribution. However, Boltzmann's principle, relating the entropy to the thermodynamic probability in contrast to a proper probability distribution, does not deal with probability distributions at all, so Planck, following Boltzmann, could not have imagined the connection [cf. §1.2].

Classically, the probability that a "particle" is present is given by the Boltzmann factor, $q = e^{-\varepsilon_\nu/T}$. The probability of there not being one is obviously $p = \left(1 - e^{-\varepsilon_\nu/T}\right)$. Thus, the probability that there are n particles is given by the geometric distribution

$$f(n) = e^{-(n-1)\varepsilon_\nu/T}\left(1 - e^{-\varepsilon_\nu/T}\right), \tag{2.74}$$

which holds in the frequency interval from ν and $\nu + d\nu$. Now suppose we have $\bar{m}_\nu$ oscillators lying in this frequency interval. Since each of these oscillators is independent (Planck's assumption of "natural radiation"), the moment generating function will be $G^{\bar{m}_\nu}(z)$, where

$$G(z) = \frac{pz}{1 - qz}$$

is the moment generating function of the geometric distribution, (2.74). By virtue of the binomial series expansion,

$$(1 - qz)^{-\bar{m}_\nu} = \sum_{n=0}^{\infty} \binom{\bar{m}_\nu + n - 1}{n} (qz)^n,$$

we have

$$
\begin{aligned}
G^{\bar{m}_\nu}(z) &= \sum_{n=0}^{\infty} \binom{\bar{m}_\nu + n - 1}{n} p^{\bar{m}_\nu} q^n z^{\bar{m}_\nu + n} \\
&= \sum_{k=\bar{m}_\nu}^{\infty} \binom{k - 1}{\bar{m}_\nu - 1} p^{\bar{m}_\nu} q^{k - \bar{m}_\nu} z^k.
\end{aligned}
\tag{2.75}
$$

Therefore, the probability of finding n *indistinguishable* particles among the $\bar{m}_\nu$ oscillators, with empty oscillators being admissible, is given by the negative binomial distribution:

$$f(n) = \binom{\bar{m}_\nu + n - 1}{n} e^{-n\varepsilon_\nu/T} \left(1 - e^{-\varepsilon_\nu/T}\right)^{\bar{m}_\nu}.
\tag{2.76}$$

If we impose the constraint that particle number can never be inferior to the number of oscillators, the negative binomial distribution (2.76) is converted into the Pascal distribution:

$$f(n) = \binom{n - 1}{\bar{m}_\nu - 1} e^{-(n - \bar{m}_\nu)\varepsilon_\nu/T} \left(1 - e^{-\varepsilon_\nu/T}\right)^{\bar{m}_\nu}, \qquad n \geq \bar{m}_\nu,
\tag{2.77}$$

as shown by formula (2.75). The radiation laws that are derived from (2.76) and (2.77) differ by an integral zero-point energy term, and this excludes the high frequency end of the spectrum in (2.77). We will have more to say about this in §7.2 where we propose an alternative radiation mechanism.

Casting the negative binomial distribution, (2.76), as an error law,

$$f(n) = \exp\left\{s(n) - s(\bar{n}_\nu) - s'(\bar{n})(n - \bar{n})\right\},$$

gives the expression

$$s(\bar{u}_\nu) = \frac{\bar{u}_\nu}{T} - \bar{m}_\nu \ln\left(1 - e^{-\varepsilon_\nu/T}\right)
\tag{2.78}$$

for the entropy density in the frequency interval $d\nu$. Comparing (2.78) with the Euler relation, (2.62), gives the pressure as

$$P_\nu = -T\bar{m}_\nu \ln\left(1 - e^{-\varepsilon_\nu/T}\right).
\tag{2.79}$$

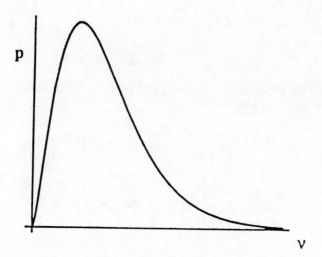

Figure 2.6: The Planck pressure as a function of the frequency.

This transforms into the Wien expression for the pressure, (2.65), in the limit $\varepsilon_\nu \gg T$. The pressure as a function of frequency is shown in Fig. 2.6.

Since the chemical potential of thermal radiation is identically equal to zero, $-P_\nu$ concides with the Helmholtz potential density per mode. Differentiating $-P_\nu/T$ with respect to $1/T$ yields Planck's formula (2.17). Then consulting the first definition in (2.68) leads to the average number of photons given by (2.55). In the limit $\varepsilon_\nu \gg T$, it transforms into the Boltzmann distribution, (2.71).

Solving (2.17) for $1/T$ and either integrating the second law, (2.61), or substituting it directly into (2.78) gives Planck's expression for the entropy,

$$s(\bar{u}_\nu) = \frac{\bar{u}_\nu}{\varepsilon_\nu} \ln \left(\frac{\bar{u}_\nu + \bar{m}_\nu \varepsilon_\nu}{\bar{u}_\nu} \right) + \bar{m}_\nu \ln \left(\frac{\bar{u}_\nu + \bar{m}_\nu \varepsilon_\nu}{\bar{m}_\nu \varepsilon_\nu} \right), \qquad (2.80)$$

which can, according to (2.68), be expressed in terms the average particle density as given in (2.24).

2.5.2   Particles versus Oscillators

In October 1914, Lorentz received a note from Ehrenfest together with a short communication by Heike Kamerlingh Onnes and himself. Ehrenfest considered his analysis of the negative binomial coefficient to be highly amusing. This "joke," as he called it, can be found in almost every text of statistical mechanics, with no recognition of the authors who created it.

What Ehrenfest and Kamerlingh Onnes did was to consider a sequence of balls separated by bars, as shown in Fig. 2.7. Two consecutive bars indicate a cell. If there are n balls and m cells, there will be $m+1$ bars, but since the first and last bars must always be fixed, only $m-1$ are movable. These bars together

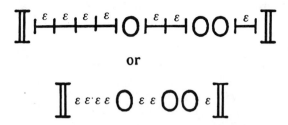

or

Figure 2.7: This is the original scheme given in the Ehrenfest–Kamerlingh Onnes paper of 1914. The first resonator on the left has 4 "energy grades" ε, the second resonator has 2ε, the third resonator has $0 \cdot \varepsilon$, and the fourth resonator has a single energy grade. The symbol $\bigcirc$ separates the different resonators and is movable. The $n + m - 1$ elements $\varepsilon \cdots \varepsilon \ \bigcirc \cdots \bigcirc$ may be arranged in $(n + m - 1)!$ distinguishable ways between the fixed ends. But the permutations of the ε's among themselves, or the $\bigcirc$'s among themselves, do not produce any new arrangements so that $(n + m - 1)!$ must be divided by $n!(m - 1)!$ in order to get the number of distinguishable arrangements.

with the n balls can appear in any order so that the number of distinguishable distributions equals the number of ways of selecting n places out of $m + n - 1$, namely, the negative binomial coefficient in (2.76). Let $\eta = m + n$ be the total number of balls and cells. For a fixed number of cells, η will vary because n does. But now suppose that we fix η and allow m to vary. In other words, the number of cells and balls may vary but in such a way that their sum remains constant.

First, suppose that $m \gg n$. Eliminating m in favor of η in (2.76) gives

$$f(n) = \binom{\eta}{n} e^{-n\varepsilon_\nu/T} \left(1 - e^{-\varepsilon_\nu/T}\right)^{\eta-n} \tag{2.81}$$

since $\eta \gg 1$. Expression (2.81) is the binomial distribution which, in the limit as $e^{-\varepsilon_\nu/T} \to 0$ and $\eta \to \infty$ such that their product,

$$\bar{n}_\nu = \eta e^{-\varepsilon_\nu/T}, \tag{2.82}$$

is of moderate size, transforms into the Poisson distribution (2.70). This is the Wien limit where (2.82) is essentially expression (2.71).

Second, consider the opposite limit, $n \gg m$. Eliminating n in favor of η in (2.76) gives

$$f(m) = \binom{\eta}{m} e^{-(\eta-m)\varepsilon_\nu/T} \left(1 - e^{-\varepsilon_\nu/T}\right)^m, \tag{2.83}$$

which is again a binomial distribution. In the limit as $1 - e^{-\varepsilon_\nu/T} \approx \varepsilon_\nu/T \to 0$ and $\eta \to \infty$ such that

$$\eta\varepsilon_\nu/T = \bar{m}_\nu \tag{2.84}$$

is of moderate size, the binomial distribution (2.83) transforms into the Poisson distribution

$$f(m) = \frac{\bar{m}_\nu^m}{m!} e^{-\bar{m}_\nu}, \tag{2.85}$$

whose entropy density is

$$s(\bar{m}_\nu) = \ln\left(\frac{\bar{n}_\nu^{\bar{m}_\nu}}{\bar{m}_\nu!}\right). \tag{2.86}$$

Expression (2.84) is essentially the law of equipartition of energy, and (2.86) has been derived from the distribution (2.83) in the limit $\varepsilon_\nu \ll T$.

The physical picture which emerges is that of an entity which can be in one of two forms: either a "particle" or an "oscillator." These are Bernoullian random variables which are governed by the binomial distributions (2.81) and (2.83), respectively. In both the Wien and Rayleigh–Jeans limits, one of the two forms predominates. In the former (latter) limit, the appearance of a particle (oscillator) is a rare event which is governed by the "law of small numbers," or the Poisson distribution (2.70) [(2.85)]. The granularity in the number of oscillators appears in the Rayleigh–Jeans limit in exactly the same way that the granularity in the number of photons occurs in the Wien limit.

We now return to our discussion in §2.3.2 of Einstein's argument in support of the particle nature of light in the Wien limit to show that the same holds true for the oscillators in the Rayleigh–Jeans limit. We need to determine the probability that in a small volume V, of a much larger volume V_0 in which there are N_0 gas particles which are uniformly distributed, there will be exactly N particles. This is given by the binomial distribution

$$f(N) = \binom{N_0}{N} \left(\frac{V}{V_0}\right)^N \left(1 - \frac{V}{V_0}\right)^{N_0-N}. \tag{2.87}$$

In the limit as $N_0 \to \infty$ and $V_0 \to \infty$ such that their ratio

$$\frac{N_0}{V_0} = \frac{\bar{N}}{V} \tag{2.88}$$

is finite, where $\bar{N}$ is the average number of particles in the volume V, the binomial distribution (2.87) transforms into the Poisson distribution

$$f(N) = \frac{\bar{N}^N}{N!} e^{-\bar{N}},$$

whose entropy is

$$S(\bar{N}) = -\bar{N} \left[\ln\left(\frac{\bar{N}}{N_0}\right) - 1\right]. \tag{2.89}$$

With the aid of the homogeneity condition (2.88), the entropy (2.89) can be written as

$$S(\bar{N}) = -\bar{N} \left[\ln\left(\frac{V}{V_0}\right) - 1\right]. \tag{2.90}$$

In the Wien limit, the entropy (2.90) must be the same as V_0 times the entropy density (2.69); this requires

$$\frac{\bar{N}}{N_0} = \frac{V}{V_0} = \frac{\bar{n}_\nu}{\bar{m}_\nu} \tag{2.91}$$

and

$$\bar{N} = \bar{n}_\nu V_0. \tag{2.92}$$

The average number of particles in the subvolume is thus identified as the average number of photons present in the cavity. Introducing (2.92) into (2.91) gives $N_0 = \bar{m}_\nu V_0$, which associates the total number of particles with the number of modes per unit frequency interval.

Alternatively, in the Rayleigh–Jeans limit, (2.90) must coincide with V_0 times the entropy density (2.86); this requires

$$\frac{\bar{N}}{N_0} = \frac{V}{V_0} = \frac{\bar{m}_\nu}{\bar{n}_\nu} \tag{2.93}$$

and

$$\bar{N} = \bar{m}_\nu V_0. \tag{2.94}$$

The average number of particles in the subvolume is now to be identified as the average number of oscillators per unit frequency interval in the cavity. Eliminating $\bar{N}$ between (2.93) and (2.94) leads to $N_0 = \bar{n}_\nu V_0$, which associates the total number of particles with the number of photons per unit frequency interval. It is quite evident from (2.92) and (2.94) that the roles of the photons and oscillators have been interchanged at the opposite ends of the black radiation spectrum.

The binomial distribution, (2.83), furnishes two expressions for the entropy density: the statistical entropy density, or the logarithm of the binomial coefficient evaluated at the average value,

$$s(\bar{m}_\nu) = -\eta \ln\left(\frac{\eta - \bar{m}_\nu}{\eta}\right) - \bar{m}_\nu \ln\left(\frac{\eta - \bar{m}_\nu}{\bar{m}_\nu}\right),$$

and the thermodynamic entropy density,

$$s(\bar{m}_\nu) = \frac{u}{T} - \bar{m}_\nu \ln\left(e^{\varepsilon_\nu/T} - 1\right), \tag{2.95}$$

where the total energy density, $\eta\varepsilon = u$, is constant. Equating their derivatives with respect to $\bar{m}_\nu$ gives the average density of oscillators per frequency interval as

$$\bar{m}_\nu = \eta\left(1 - e^{-\varepsilon_\nu/T}\right). \tag{2.96}$$

This is entirely equivalent to Planck's expression for the average density of photons given by (2.55). Comparing (2.95) to the Euler relation, (2.62), results in the expression

$$P_\nu = -T\bar{m}_\nu \ln\left(e^{\varepsilon_\nu/T} - 1\right) \tag{2.97}$$

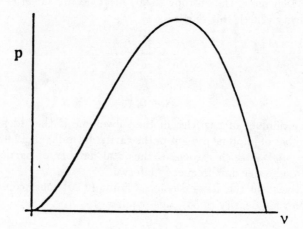

Figure 2.8: The Rayleigh–Jeans pressure as a function of frequency.

for the pressure shown in Fig. 2.8.

The area under the curve is one-third the total energy density. The pressure curve, shown in Fig. 2.7, stands on par with the pressure distribution, shown in Fig. 2.4. The underlying probability distributions are both discrete and describe, respectively, the distribution of the number of oscillators and particles in the low and high frequency ranges of the spectrum. The energy density per mode is an unbounded function of the frequency while the pressure per mode passes through a maximum and vanishes at

$$\nu_{\text{thres}} = \frac{T}{\gamma} \ln 2.$$

This is the threshold frequency that separates the wave from the particle nature of light.[3] Numerically, $\nu = 0.14 \nu_{\max}$, where $\nu_{\max}$ is the frequency at which $\bar{u}$ is maximum. For a temperature of 600 K, ν_{thres} lies on the border between the far- and near-infrared. The threshold frequency is related to the maximum uncertainty of finding a photon in a given mode,

$$e^{\gamma \nu_{\text{thres}}/T} = 1 - e^{-\gamma \nu_{\text{thres}}/T} = \tfrac{1}{2}.$$

Integrating the pressure (2.97) over frequencies up to ν_{thres} gives the radiation pressure (2.3).

---

[3]In communication theory it is known that a minimum energy of $0.693T$ must be used to send one bit of information. Communicating by means of a signal of frequency ν, the number of bits per photon is ν/ν_{thres}.

2.5.3 The "Classical" Limit of Rayleigh–Jeans

In the low frequency limit where $\varepsilon_\nu \ll T$, the exponential in (2.79) may be expanded to first-order. Together with the law of equipartition of energy

$$\bar{u}_\nu = \bar{m}_\nu T \tag{2.98}$$

we have

$$P_\nu = -T\bar{m}_\nu \ln\left(\frac{\bar{m}_\nu \varepsilon_\nu}{\bar{u}_\nu}\right). \tag{2.99}$$

The Rayleigh–Jeans limit is the high intensity limit, where $\bar{u} > \alpha\nu^3$. The inequality ensures that the pressure per frequency interval, (2.99), will be positive. Introducing both equations of state, (2.99) and (2.98), into the Euler relation (2.62) leads to the fundamental relation

$$
\begin{aligned}
s(\bar{u}_\nu) &= \bar{m}_\nu \left[\ln\left(\frac{\bar{u}_\nu}{\alpha\nu^3}\right) + 1\right] \\
&= \ln\left[\frac{(\bar{u}_\nu/\varepsilon_\nu)^{\bar{m}_\nu}}{\bar{m}_\nu!}\right],
\end{aligned}
\tag{2.100}
$$

provided $\bar{m}_\nu$ is large enough to permit the use of Stirling's approximation.

Expression (2.100) relates the entropy to the logarithm of the phase volume occupied by the system. With the aid of (2.73), it can be written as (2.86). A comparison with the Wien entropy, (2.69), shows that $\bar{m}_\nu$ and $\bar{n}_\nu$ have swapped roles! There is no classical analog to such an expression; nevertheless, it is a direct consequence of the classical limit where equipartition of energy, (2.98), applies. By contrast with the Wien limit, "particle" discreteness is replaced by "mode" discreteness. Moreover, Planck's calculation of the number of electromagnetic modes per unit volume, having frequencies in the interval between ν and $\nu + d\nu$, $\bar{m}_\nu d\nu = (8\pi\nu^2/c^3)d\nu$, should be interpreted as the *average* number of oscillators in this interval.

Introducing (2.100) into the error law (2.59) leads to

$$f(u) = \frac{(\beta u)^m}{m!}e^{-\beta u}. \tag{2.101}$$

But this cannot be the law of error for the energy because it is not normalized. Rather, it is the Poisson distribution, (2.85), where the scale parameter β is the expected number of oscillators in a given frequency interval per unit energy. What we have uncovered is a statistical equivalence principle that holds between nonconjugate quantities like oscillators and particles and energy and degrees-of-freedom. Where one is discrete the other is continuous.

The Poisson distribution, (2.101), for the discrete variable m can be contrasted with the gamma density, (2.47), for the continuous variable u. Both are characteristic of the Rayleigh–Jeans region and offer equivalent, though seemingly different, descriptions of the same phenomenon. In fact, the two distributions are related to one another depending upon which of the two variables we choose to let fluctuate at a fixed value of the other. We will have a great deal more to say about this type of *complementarity* in Chapters 5 and 6.

Bibliographic Notes

§2.1

The clearest exposition that we found of Kirchhoff's results together with separate derivations of Wien's, the Rayleigh–Jeans formula, and Planck's laws is given in

- H.A. Lorentz, *Lectures on Theoretical Physics*, Vol. II, transl. by L. Silberstein and A.P.H. Trivelli (Macmillan, London, 1927), pp. 209-274.

Lorentz's lectures take us up to 1911, the year in which he decided to give up his chair in Leyden and go to Haarlem, leaving the chair to Ehrenfest.

A more chronological account of the developments in black radiation, which includes a full account of the experimental results, can be found in

- H. Kangro, *Early History of Planck's Radiation Law* (Taylor and Francis, London, 1976).

§2.2

It is a real pity that modern-day texts on the structure of matter have extracted Planck's results from their thermodynamic setting so that it would be unrecognizable even to Planck himself. The development of Planck's ideas are usually left to the historians of science, and no better account could be given than that found in

- M.J. Klein, *Paul Ehrenfest: The Making of a Theoretical Physicist* (North-Holland, Amsterdam, 1970), Ch. 10.

Klein's presentation is really superb for its clarity and focalization of the important points without getting bogged down in mathematical details. In it, he has borrowed from his previous researches:

- M.J. Klein, "Max Planck and the beginnings of quantum theory," *Arch. Hist. Exact Sci.* **1**, 459-479 (1962);

- M.J. Klein, "Planck, entropy and quanta, 1901-1906," *Natural Philos.* **1**, 83-108 (1963);

- M.J. Klein, "Thermodynamics and quanta in Planck's work," *Phys. Today* **19**, 23-32 (1966);

- M.J. Klein, "The beginnings of the quantum theory," in *History of 20th Century Physics* (Proceedings of the International School of Physics "Enrico Fermi," Course 57) (Academic Press, New York, 1977), pp. 1-39.

Another account of Planck's formulation which includes his "second" theory as well as a reinterpretation of Planck's thoughts on the quantum discontinuity is given in

- T.S. Kuhn, *Black-body Theory and the Quantum Discontinuity 1894-1912* (Univ. Chicago Press, Chicago, 1987).

And, of course, there is the account given by Planck himself which can be found in

- M. Planck, *The Theory of Heat*, transl. by H.L. Brose (Macmillan, London, 1932).

Still other accounts of the development of the theory of black radiation are to be found in

- A. Pais, *Subtle is the Lord... The Science and Life of Albert Einstein* (Oxford Univ. Press, Oxford, 1982), Ch. 19.

It is to be recommended for its schematic character which, in part, we have followed.

§2.3

Lord Rayleigh's 1900 note is

- Lord Rayleigh, "Remarks upon the Law of Complete Radiation," *Phil. Mag.* **49**, 539 (1900).

Einstein became attracted to the powerful tool that the analysis of fluctuations provided. The first time he used fluctuation theory was in his 1904 in his paper

- A. Einstein, *Ann. d. Physik* (Leipzig) **14**, 354 (1904).

He was to return to it many times afterward to get physical insights into radiation phenomena. The incorrectness of Einstein's conclusion that $\overline{(\Delta U)^2} \sim U^2$ has also been noticed by Pais in his book cited above. The reason he notes this is that fluctuations are therefore not all that different for radiation as for an ideal gas. Pais also claims that this work brought to Einstein's attention the importance of the volume dependence of thermodynamic quantities, like the entropy, that was to appear in his very next paper on quanta.

That paper appeared the next year and dealt with his proposal of the photon concept:

- A. Einstein, "Concerning an heuristic point of view toward the emission and transformation of light," *Ann. Physik* **17**, 132 (1905); transl. by A.B. Arons and M.B. Peppard, *Am. J. Phys.* **33**, 367-374 (1965).

This paper also contains the analogy he made between the entropy of an ideal gas and the entropy of the Wien distribution. It is duly famous for his explanation of the photoelectric effect for which he won the Nobel prize in 1921. The background to this paper has been given by

- M.J. Klein, "Einstein's first paper on quanta," *Natural Philos.* **2**, 59-86 (1963);

- M.J. Klein, "Einstein on the wave-particle duality," *Natural Philos.* **3**, 3-49 (1964).

A good account of Einstein's analysis of fluctuations in black radiation leading to the particle-wave duality is given in

- H.A. Lorentz, *Les Théories Statistiques en Thermodynamique* (B.G. Teubner, Leipzig, 1916), §42.

Ehrenfest developed his "adiabatic" principle from Lord Rayleigh's observation concerning the invariancy of the ratio of energy to the frequency of a normal mode under sufficiently slow variations of the dimensions of the enclosure. Lord Rayleigh's published his work in

- Lord Rayleigh, "On the pressure of vibrations," *Phil. Mag.* **3**, 338-346 (1902).

According to Klein, Ehrenfest was unaware of the exchange between Lorentz and Einstein at the first Solvay conference where Einstein made the observation that the ratio of energy to the frequency for a pendulum would be unaffected if the length of the pendulum were slowly shortened. Apparently, Einstein did not see how this remark affected his analogy between the entropy of an ideal gas and the entropy of Wien's law. Ehrenfest's adiabatic principle,

- P. Ehrenfest, "On adiabatic changes of a system in connection with the quantum theory," *Proc. Acad. Amsterdam* **19**, 576-597 (1916); reprinted in *Collected Scientific Papers*, M.J. Klein, ed. (North-Holland, Amsterdam, 1959), pp. 378-399,

formed one of the cornerstones of the old quantum theory.

§2.4

Einstein's derivation of Planck's law, in which he proposed the process of stimulated emission, can be found in

- A. Einstein, "On the quantum theory of radiation," *Sources of Quantum Mechanics*, B.L. van der Waerden, ed. (North-Holland, Amsterdam, 1967), pp. 63-77.

Our critique of Einstein's radiation theory in regard to the decomposition of the Einstein coefficients into temperature dependent and independent parts is given in

- B.H. Lavenda, "Einstein's theory of quantum radiation," *Int. J. Theoret. Phys.* **28**, 391-406 (1989);

- B.H. Lavenda, "Comment on 'Using Einstein's method to derive both the Planck and Fermi–Dirac distributions,'" *Am. J. Phys.* **58**, 91-92 (1990).

The latter comment was in response to

- F.S. Crawford, "Using Einstein's method to derive both the Planck and Fermi–Dirac distributions," *Am. J. Phys.* **56**, 883-885 (1988).

Crawford used the high temperature, or long wavelength, limit to show that the ratio of the spontaneous to stimulated emission coefficients for a single atom is temperature independent. However, as we have shown in the text, this is not the case. The paper we refer to is

- A. Einstein and O. Stern, "Einige Argumente für die Annahme einer molekularen Agitation beim absolutem Nullpunkt," *Ann. d. Phys. (Leipzig)* **40**, 551-560 (1913).

A nonzero chemical potential for photons is discussed in

- P. Würfel, "The chemical potential of radiation," *J. Phys. C* **15**, 3967-3985 (1982).

§2.5

Our analysis of the underlying probability distributions to Planck's radiation law can be found in

- B.H. Lavenda, "Underlying probability distributions of Planck's radiation law," *Int. J. Theoret. Phys.* **29**, 1379-1392 (1990).

As Klein tells us, the joke was Ehrenfest's memorable combinatorial proof of the negative binomial coefficient which was published in

- P. Ehrenfest and H. Kamerlingh Onnes, "Simplified deduction of the formula from the theory of combinations which Planck uses as the basis of his radiation theory," *Proc. Acad. Amsterdam* **17**, 870 (1914); reprinted in P. Ehrenfest, *Collected Scientific Papers*, M.J. Klein, ed. (North-Holland, Amsterdam, 1959), pp. 870-873.

Chapter 3

From One to Infinity

3.1 Occupancy Problems

The traditional path to the different types of physical statistics involves models of placing randomly n balls into m cells. The "cells" correspond to small regions into which the phase space has been subdivided. The statistics is specified once we know how the particles are distributed among the cells.

The number of ways in which n *different* particles can be distributed into m *different* cells is m^n when blank cells are admissible. In other words, n particles can be successively placed in m cells and therefore there are m^n arrangements. Without any information to the contrary, it would appear that all m^n arrangements are equally likely. In this case we would speak about *Maxwell–Boltzmann statistics*. Yet, such a type of statistics does not apply to any known particle; all the m^n arrangements are not equally probable.

All known particles fit into one of two probability models in which the particles are treated as *indistinguishable*. This does not mean that a third probability model may not be introduced, and in fact, such an "intermediary" model has been proposed on various occasions stemming from the original idea of Giovanni Gentile junior in 1940. The lack of uniqueness of the two types of statistics is not unlike that of Boltzmann's principle.

In the statistics of deaths among different age groups or the distribution of anniversaries among the calendar days, we are concerned with the number of occurrences rather than the individuals involved. So, too, in the derivation of physical statistics, we should be concerned with the distribution of the particles among the different cells and not which particle is found in which cell. In other words, we should consider the particles to be *indistinguishable*.

Suppose we have m cells into which any number of the n particles at our disposition can be placed. The number of distinguishable distributions is

$$C_{m-1}^{m+n-1} = \binom{m+n-1}{m-1}. \tag{3.1}$$

We recall from §2.2.3 that it was the logarithm of this expression which Planck set equal to the entropy of a system composed of n "energy-grades" and m

"resonators." Planck did not give any justification for this but rather chose to refer to standard texts on combinatorial methods. A good one would have been W. A. Whitworth's *Choice and Chance*, which went through several editions in the last decade of the last century.

The explanation of (3.1), given by Whitworth, is that if blank cells are not admissible, then there are

$$C_{m-1}^{n-1} = \binom{n-1}{m-1} \tag{3.2}$$

number of ways that n particles may be placed in the m cells. With the number of particles placed in a row, this is the number of ways $m-1$ dividers can be distributed among the $n-1$ intervals. For the distribution of n particles, when blank cells are permissible, is the same as the distribution of $n+m$ things when they are not admissible, and hence (3.1) follows in the case where blank cells are permitted.

The proof physicists choose to remember is that given by Ehrenfest and Kamerlingh Onnes in 1914 [cf. §2.5.2], in which Planck's "formal" device of distributing n energy-elements over m resonators is contrasted with Einstein's light-quanta. They reached the conclusion that Planck could not interpret his energy-grades in the sense of Einstein's light-quanta, for instead of their being only (3.1) ways, Einstein would have found that there are m^n ways of distributing n *distinguishable* quanta over m resonators. But then again, Einstein's reasoning applies to Wien's, and not to Planck's, law. It was to take another decade to come to the realization that an assembly of quanta can be thought of as an ideal gas of *indistinguishable* particles, where each distinguishable arrangement has a probability $1/C_n^{m+n-1}$. Today we refer to this type of statistics as *Bose–Einstein statistics*, after the Indian physicist Satyendra Nath Bose, who acted as a catalyzer for Einstein's realization that radiation actually constitutes an ideal quantum gas.

In 1931 Ehrenfest together with J. Robert Oppenheimer showed that this statistics applies not only to photons but also to nuclei containing an *even* number of "elementary" particles (i.e., electrons and protons). Nuclei containing an odd number of particles were shown to obey a different kind of statistics in which the assignment of a particle to a cell can have only two values, 0 or 1.

This obviously requires that the number of particles be less than, or at most equal to, the number of cells, $n \leq m$. An arrangement consists of specifying which of the m cells are occupied. The number of ways the cells can be chosen is

$$C_n^m = \binom{m}{n}, \tag{3.3}$$

which is the number of ways that n things can be selected from a total of m things. Since each arrangement is equally probable, the probability of an arrangement is $1/C_n^m$. This type of statistics is referred to as *Fermi–Dirac statistics*.

At the time Enrico Fermi was meditating about the significance of the entropy constant and the not unambiguous counting procedure used in Boltzmann's principle, Wolfgang Pauli's exclusion principle appeared in print. Several weeks after reading Pauli's paper, in February 1926, Fermi presented to the Accademia dei Lincei a paper developing a new form of statistics in which a maximum of one particle can be admitted to any cell. This was Fermi's first fundamental contribution to physics, and it obtained almost immediate recognition. Although Fermi attempted to apply this type of statistics to the heat of vaporization in metals, he did not publish his results. A few months after the appearance of Fermi's paper Paul Adrien Maurice Dirac laid the foundations of the statistics based on quantum mechanics. Arnold Sommerfeld and his students applied successfully Fermi–Dirac statistics to "degenerate" phenomena caused by electrons in metals, such as the problem of the specific heat at low temperatures.

About 15 years later, another Italian, Gentile, was pondering over the modifications of the statistics that would be necessary when one no longer makes the particular hypotheses of there being a maximum of one particle per cell in the Fermi–Dirac case and an infinite number in the case of Bose–Einstein. A situation could be realized where an intermediate number of particles, d, were allowed to occupy each cell. Not only would such a type of statistics be able to bridge the gap between two known types of statistics, it would moreover provide a way of passing from one type of statistics to another without any discontinuity. A year later in 1941,[1] Gentile found an application of this type of *intermediate statistics* to the Bose–Einstein condensation.

In 1924 Einstein predicted that a sort of "condensation" would occur in momentum space where an *infinite* number of particles would accumulate in the zero momentum, or ground, state as the temperature is lowered below the critical value which is determined by the vanishing of the chemical potential. Admittedly, Einstein realized that only a certain number of particles of the gas would enter into this new "phase," and Gentile thought such a gas would obey a modified form of statistics where only d particles could be found in the ground state that would constitute the superfluid phase. Gentile found basically the same thermodynamic characteristics that Einstein did but claimed there would be no real division between the "liquid" and "gas" phases.

More recently, intermediate statistics has undergone a facial uplift together with a change of name. Now referred to as "parastatistics," it has been conjectured that quarks satisfy such a type of statistics rather than the usual Fermi–Dirac statistics. Instead of a state being occupied by one particle, it was assumed that a state could accommodate a maximum of three particles where each of the three types of quarks could carry three new degrees-of-freedom, referred to by the colors "red, white and blue." The parastatistics model was shown to be equivalent with the tricolored triplet model of quarks under the

---

[1] The note "Intermediate statistics and liquid helium" appeared posthumously in the April 1942 edition of *Il Nuovo Cimento*.

condition that the three triplets carry the same charges. So, at last, it seemed that intermediate statistics found itself a home in a physical phenomenon that could not be explained by the usual Fermi–Dirac statistics.

The same ambiguities that arise between Fermi–Dirac and Bose–Einstein statistics as opposed to intermediate statistics occurs in the specification of the number of microstates compatible with a given macroscopic state in Boltzmann's principle. It would seem that there is nothing that would, a priori, exclude intermediate statistics. The number of ways in which n particles can be distributed over m cells with each cell having a maximum capacity of d particles,

$$\vartheta(n; m, d) = \sum_{k=0}^{[n/(d+1)]} (-1)^k \binom{m+n-(d+1)k-1}{m-1} \binom{m}{k}, \qquad (3.4)$$

where $[n/(d+1)]$ is the largest integer less than or equal to $n/(d+1)$, appears to be just as good an expression as the negative binomial coefficient, (3.1), or the binomial coefficient, (3.3), to use in Boltzmann's principle.

Since the quantity (3.4) is the coefficient of x^n in the polynomial

$$f(x) = (1 + x + x^2 + \cdots + x^d)^m = \left(\frac{1 - x^{d+1}}{1 - x}\right)^m,$$

it appears more general than either the negative binomial or binomial coefficients since the latter can be obtained in the particular cases where $d = \infty$, in which case only the term $k = 0$ subsists in (3.4), and $d = 1$, respectively. But it was already apparent to several authors that it did not make any difference whether one took $d = n$ or $d = \infty$ for both were equivalent to Bose–Einstein statistics. Dirk ter Haar concluded that "any results peculiar to the case of the intermediate statistics must be spurious" on the basis that there are not enough particles present to make the restriction on the occupancy effective. The claim that if there were enough particles present, we would be able to distinguish between intermediate and Bose–Einstein statistics appears to go against the grain of certain limit theorems in probability theory which show that all distributions converge to the normal one in the limit as the number of indistinguishable particles increases without limit while the number of cells remains finite. Rather, we must look to something more fundamental to rule out the possibility of an intermediary form of statistics.

From the time of Boltzmann, one has always dealt with binomial or multinomial coefficients rather than bona fide probability distributions precisely because one did not know what to do with the unknown a priori probabilities. Admitting total ignorance, it was just as well to set them all equal to one another so that the probability to obtain a given set of occupancy numbers $n_1, n_2, \ldots, n_m$ equals

$$\frac{n!}{n_1! n_2! \cdots n_m!} m^{-n} \qquad (3.5)$$

subject to the condition

$$\sum_{i=1}^{m} n_i = n.$$

In other words, the fact that all the m^n placements are equally probable will have the effect of changing the entropy constant when the logarithm of the multinomial distribution (3.5) is identified with the entropy. The essential factor has always been the multinomial coefficient in (3.5) rather than the fact that it is a *bona fide* probability distribution. And it is precisely for this reason that the ambiguities turn up in Boltzmann's principle. For if *bona fide* probability distributions could be invoked, any and every form of *physical* statistics would have to be associated with probability distributions. The statistics— whether it be quantum or classical—are governed by the same probability distributions as the distribution of raisins in the dough, the number of misprints on the page, or the probability for a given number of successes to turn up.

Therefore, we will find it necessary, just as we did in §1.7, to insert Boltzmann's principle into another principle that deals directly with the probability distributions themselves rather than only a part of them. This will remove the ambiguity from Boltzmann's principle since it will now be derived from first principles rather than being postulated. In other words, Boltzmann's principle is a consequence of a statistical line of reasoning. Moreover, it will also rule out intermediary statistics because such a form of statistics is not governed by any probability distribution.

3.2   Markov Stochastic Processes

Following Einstein's lead, we will derive quantum statistics from stochastic processes that admit stationary probability distributions. Stochastic processes in which the future depends on the present, independent of the whole past history of the process, are known as Markov processes. We will be primarily interested in Markov processes which are continuous in time with discrete states. Such processes go under the heading of "birth and death" or "generation and recombination" processes. They are also frequently referred to as "one-step" processes. We have already treated an example of a one-step process, namely Einstein's radiation mechanism, discussed in §2.4.

Denote by $r(n)$ (recombination) the probability per unit time for the transition $n \rightarrow n-1$, where n is a positive integer representing the number of photons, particles, etc. The probability per unit time that the transition $n \rightarrow n+1$ will occur is $g(n)$ (generation). The probability for these events to happen is proportional to the interval in time, Δt. The probability to jump two or more units is $O(\Delta t^2)$, which can be neglected if Δt is small. The rate of change of the state n is governed by the so-called master equation

$$\dot{f}(n,t) = r(n+1)f(n+1,t) + g(n-1)f(n-1,t) - [r(n) + g(n)]f(n,t),$$

where $f(n, t)$ is the probability of finding n particles at time t or the probability of being in state n at that time.

Since we are concerned with positive values of n, one of the boundaries will always be $n = 0$. We can make the master equation hold there by defining

$$r(0) = g(-1) = 0.$$

At the other boundary $n = m$, where m may be infinity, we define

$$r(m + 1) = g(m) = 0.$$

Instead of having to write out all the different terms, there is a more compact way of writing the master equation in terms of what are called the "step" operators, $\mathbf{E}$ and $\mathbf{E}^{-1}$. When acting on an arbitrary function $\varphi(n)$, they have the following effects:

$$\mathbf{E}\varphi(n) = \varphi(n + 1) \qquad \text{and} \qquad \mathbf{E}^{-1}\varphi(n) = \varphi(n - 1).$$

Since $n \in [0, m]$, the $\mathbf{E}$'s are just shorthand notations and not real operators. They do possess the important property that

$$\sum_{n=0}^{m-1} \psi(n)\mathbf{E}\varphi(n) = \sum_{n=1}^{m} \varphi(n)\mathbf{E}^{-1}\psi(n),$$

where ψ is another arbitrary function. Using these step operators, the master equation can be written succinctly as

$$\dot{f}(n, t) = \left\{(\mathbf{E} - 1)\, r(n) + \left(\mathbf{E}^{-1} - 1\right) g(n)\right\} f(n, t). \qquad (3.6)$$

It is now an easy matter to calculate the rate equations for the moments of the distribution. For instance, the rate equation for the first moment is obtained by multiplying the master equation (3.6) by n and summing over all n. This gives

$$
\begin{aligned}
\dot{\bar{n}} \;\equiv\; \frac{d}{dt}\sum_{n} n f(n, t) &= \sum_{n} n \left\{(\mathbf{E} - 1)\, r(n) + \left(\mathbf{E}^{-1} - 1\right) g(n)\right\} f(n, t) \\
&= \sum_{n} f(n, t) \left\{r(n)\left(\mathbf{E}^{-1} - 1\right) + g(n)\left(\mathbf{E} - 1\right)\right\} n \\
&= \overline{-r(n)} + \overline{g(n)}. \qquad\qquad\qquad\qquad\qquad\qquad (3.7)
\end{aligned}
$$

We will be particularly interested in the stationary solution to the master equation (3.6). Setting $\dot{f} = 0$ results in

$$
\begin{aligned}
0 &= \left\{(\mathbf{E} - 1)\, r(n) + \left(\mathbf{E}^{-1} - 1\right) g(n)\right\} f(n) \\
&= (\mathbf{E} - 1)\left\{r(n)f(n) - \mathbf{E}^{-1}g(n)f(n)\right\}.
\end{aligned}
$$

This implies that the term within the curly brackets must be independent of n and, at most, equal to a constant,

$$r(n)f(n) - g(n - 1)f(n - 1) = -J,$$

where J is the stationary "flow" of probability from n to $n-1$.

In the case that detailed balancing holds, J vanishes and we get the iterative relations

$$r(n)f(n) = g(n-1)f(n-1),$$

from which the stationary distribution is obtained as

$$f(n) = \frac{g(n-1)g(n-2)\cdots g(0)}{r(n)r(n-1)\cdots r(1)} f(0).$$

The term $f(0)$ is determined by the condition that $f(n)$ be normalized. We now have the machinery to affront the different physical processes that give rise to the different types of statistics.

3.3 Path to Quantum Statistics

3.3.1 Kinetic Derivation of the Probability Distributions

In physics, the term "statistics" is used in a very peculiar way. Rather than referring to the probability distribution, it refers to the average number of particles as a function of the temperature, or what would be the first moment of the distribution. Two different probability models have been introduced to describe two types of particles each of which are indistinguishable. In order to distinguish between the two types of particles, consider n *indistinguishable* particles in m cells. Bose–Einstein statistics does not restrict the number of particles in any one cell while Fermi–Dirac statistics restricts the particle occupancy to a single particle.

Bose–Einstein statistics was around since Planck derived his radiation law, but it was not until 1924 that Bose showed that radiation could be conceived as an ideal quantum gas. In Einstein's words,

> if it is justified to conceive as radiation as a quantum gas, then the analogy between the quantum gas and a molecular gas must be a complete one.

Einstein then went on to derive the thermodynamic properties of such a gas including the so-called Bose–Einstein condensation, of which we will have more to say about in §7.5. As mentioned in the previous section, Fermi–Dirac statistics was motivated by Pauli's exclusion principle.

In order to derive the two forms of statistics, we set $r(n) = \alpha n$ and $g(n) = \beta(m - \eta n)$ for the rates of recombination and generation, respectively. The rate parameters, α and β, are independent of n and $\eta = 1$ for Fermi–Dirac statistics, while $\eta = -1$ for Bose–Einstein statistics.

As a physical picture for Fermi–Dirac statistics we can imagine an adsorption isotherm where the rate of evaporation of n molecules from a surface consisting of m sites is αn, while the rate of condensation is proportional to the surface not already covered, $m - n$, with β representing the rate at which it is taking place. Alternatively, for Bose–Einstein statistics, we may think in terms of Einstein's radiation mechanism, discussed in §2.4, where α, β, and $\gamma(= \beta m)$ are the coefficients of absorption, stimulated emission, and spontaneous emission, respectively.

Introducing these rate coefficients into the master equation (3.6) gives

$$\dot{f}(n, t) = \left\{ (\mathbf{E} - 1)\,\alpha n + \left(\mathbf{E}^{-1} - 1\right) \beta (m - \eta n) \right\} f(n, t). \qquad (3.8)$$

At equilibrium the rate of each process will be balanced exactly by the reverse of that process. The detailed balancing of the individual processes leads to the stationary solution,

$$f_\eta(n) = \frac{(m + \eta - \eta n) \cdots m}{n!} \left(\frac{\beta}{\alpha} \right)^n f_\eta(0), \qquad (3.9)$$

of this master equation, where

$$\frac{(m + \eta - \eta n) \cdots m}{n!} = \begin{cases} \dbinom{m}{n} & \text{if } \eta = 1, \\[2ex] \dbinom{m + n - 1}{n} & \text{if } \eta = -1. \end{cases}$$

The constants $f_\eta(0)$ in (3.9) can be found from the condition that $f(n)$ be normalized. For Fermi–Dirac and Bose–Einstein statistics, respectively,

$$\sum_{n=0}^{m} \binom{m}{n} \left(\frac{\beta}{\alpha} \right)^n = \left(1 + \frac{\beta}{\alpha} \right)^m$$

and

$$\sum_{n=0}^{\infty} \binom{m + n - 1}{n} \left(\frac{\beta}{\alpha} \right)^n = \left(1 - \frac{\beta}{\alpha} \right)^{-m}.$$

Fermi–Dirac statistics does not impose any condition on the rate coefficients while for Bose–Einstein statistics we must have

$$\alpha > \beta. \qquad (3.10)$$

Borrowing Einstein's terminology, this requires the coefficient of absorption to be greater than the stimulated emission coefficient. We shall soon see that this is guaranteed by the second law and how its violation can lead to the mechanism of lasing [cf. §7.4].

Consequently, the stationary distribution of Fermi–Dirac statistics,

$$f_1(n) = \binom{m}{n} \left(\frac{\beta}{\alpha + \beta} \right)^n \left(\frac{\alpha}{\alpha + \beta} \right)^{m-n}, \qquad (3.11)$$

is the binomial distribution while the negative binomial distribution,

$$f_{-1}(n) = \binom{m+n-1}{n} \left(\frac{\beta}{\alpha}\right)^n \left(\frac{\alpha-\beta}{\alpha}\right)^m, \qquad (3.12)$$

is the stationary probability distribution of Bose–Einstein statistics. Not only have we derived the probability distributions for the two types of statistics but we have, in addition, determined the *a priori* probabilities for "success" and "failure." We shall now pause to show that these values are their maximum likelihood estimates.

3.3.2 Statistical Interactions

The binomial and negative binomial distributions, (3.11) and (3.12), respectively, can be derived from parent distributions which manifest the particular nature of the particle statistics. The *a priori* probabilities in these distributions do not vary from one trial to the next so that, in this sense, they are idealizations of probabilistic processes which do not take into consideration that a streak of bad luck should diminish the *a priori* probability of success. In terms of physical particles, these distributions idealize them as points which occupy no volume and do not exert any forces on other point particles. However, their parent distributions do take such factors into account so that we may learn how these distributions arise in the appropriate limits and what is the physical nature of the interactions that give rise to the statistics.

Statistics of Repulsion

Let V_0 and n_0 be the volume of a gas and the number of particles in it, respectively. Consider a subvolume V and ask for the probability that any given particle is in this volume. Even though the gas is uniform, we cannot conclude that the probability for any particle to be found in the subvolume is V/V_0, regardless of how many particles may already be in the subvolume because particles occupy space. The space taken up by the particles, σ, is referred to as the "excluded volume." For a monatomic gas this is 4 times the "molecular volume." If this elementary volume is not negligible compared to the subvolume, the *a priori* probabilities, dependent upon observation will depend on n.

The product of *a priori* probabilities for finding n atoms in the volume V is

$$\prod_{i=0}^{n-1} p_i = \frac{V}{V_0} \frac{V-\sigma}{V_0-\sigma} \frac{V-2\sigma}{V_0-2\sigma} \cdots \frac{V-(n-1)\sigma}{V_0-(n-1)\sigma}$$

$$= \frac{(V/\sigma)!}{(V/\sigma-n)!} \frac{(V_0/\sigma-n)!}{(V_0/\sigma)!}.$$

Each particle added to the subvolume decreases the *a priori* probability of their being an additional particle. Likewise, the product of *a priori* probabilities of finding $n_0 - n$ particles in the complementary volume, $V_0 - V$, is

$$\prod_{i=0}^{n_0-n-1} q_i = \frac{V_0 - V}{V_0 - n\sigma} \frac{V_0 - V - \sigma}{V_0 - (n+1)\sigma} \cdots \frac{V_0 - V - (n_0 - n - 1)\sigma}{V_0 - (n_0 - 1)\sigma}$$

$$= \frac{[(V_0 - V)/\sigma]!}{[(V_0 - V)/\sigma - (n_0 - n)]!} \frac{(V_0/\sigma - n_0)!}{(V_0/\sigma - n)!}.$$

Consequently, the probability of finding n particles in the subvolume V will be the product of these two factors multiplied by the number of ways that n molecules can be chosen from n_0 or

$$f_h(n) = \binom{n_0}{n} \prod_{i=0}^{n-1} p_i \prod_{i=0}^{n_0-n-1} q_i$$

$$= \binom{n_0}{n} \binom{V_0/\sigma - n_0}{V/\sigma - n} / \binom{V_0/\sigma}{V/\sigma}. \qquad (3.13)$$

This is the hypergeometric distribution of sampling *without replacement*.

At low densities $V_0 \gg n_0\sigma$, the hypergeometric distribution transforms into the binomial distribution

$$f_1(n) = \binom{n_0}{n} \left(\frac{V}{V_0}\right)^n \left(\frac{V_0 - V}{V_0}\right)^{n_0-n} \qquad (3.14)$$

for sampling *with replacement*. All trace of the excluded volume has disappeared, and the *a priori* probabilities no longer depend upon the number of particles which are already in the subvolume. In §1.13 we showed that the most likely situation regarding the binomial distribution (3.14) is when the system is spatially homogeneous,

$$\frac{\bar{n}}{n_0} = \frac{V}{V_0}, \qquad (3.15)$$

where $\bar{n}$ is the average number of particles in the subvolume. We now show that this also holds true for the parent distribution, (3.13), in the case where the excluded volume is small compared to the volume available to each particle.

We again employ the method of maximum likelihood [cf. §1.13], where the functional dependency in (3.14) is now inverted: V is taken as the variable and the sample mean $\bar{n} = (1/r) \sum_i n_i$ is taken as the parameter. Data are obtained by making observations on the number of particles in the subvolume at fixed intervals in time. Each of the observations n_i are independent random variables having a common distribution, (3.13), so that the log-likelihood function is

$$\mathcal{L}(V|n_1, n_2, \ldots, n_r) = \sum_{i=1}^{r} \ln f_h(n_i).$$

Unlike the case of the binomial distribution, we cannot express the log-likelihood function in terms of the sample mean because the *a priori* probabilities also depend on the observations n_i. The likelihood equation, $(\partial \mathcal{L}/\partial V) = 0$, is explicitly given by

$$\sum_{i=1}^{r} \left\{ \ln \left[1 - \frac{(n_0 - n_i)\sigma}{(V_0 - V)} \right] - \ln \left(1 - \frac{n_i \sigma}{V} \right) \right\} = 0.$$

There would be no meaning to the requirement that each observation preserve spatial homogeneity, for then our observations would not be informative. The only other way to satisfy the likelihood equation is to require the excluded volume to be small compared to the volume per particle so that we can use the approximation $\ln(1 - x) \approx -x$. This gives the homogeneity condition, (3.15).

For the binomial distribution, (3.11), the condition of homogeneity, (3.15), implies

$$\frac{\bar{n}}{n_0} = \frac{\beta}{\alpha + \beta}, \tag{3.16}$$

which we shall soon appreciate as the stationary solution to the average equation of motion. This implies that we have used the maximum likelihood estimates for the a priori probabilities in the binomial distribution (3.11) without even knowing it! We shall return to this point once we have introduced the thermodynamics.

Statistics of Attraction

The negative binomial distribution, (3.12), can also be derived from a parent distribution in which the statistics can be interpreted as physical attractions among the particles.

Consider an urn with m_0 balls in it. There are m black and $m_0 - m$ red balls. Each time a ball of a given color is drawn, it is replaced and another one is added of the same color. This is sampling with replacement. The property that "like generates like" is a characteristic property of the particles that can be translated in saying that the particles attract one another.

The probability that a black ball will be drawn on the first drawing is m/m_0. The probability that a black ball will be drawn on the second trial is $(m+1)/(m_0+1)$ and so on. The probability that out of n_0 drawings the first n trials result in black balls is

$$\prod_{i=0}^{n-1} p_i = \frac{m}{m_0} \frac{m+1}{m_0+1} \cdots \frac{m+n-1}{m_0+n-1},$$

while the probability that the remaining $n_0 - n$ drawings result in red balls is

$$\prod_{i=0}^{n_0-n-1} q_i = \frac{m_0 - m}{m_0 + n} \frac{m_0 - m + 1}{m_0 + n + 1} \cdots \frac{m_0 - m + n_0 - n - 1}{m_0 + n_0 - 1}.$$

Thus the probability that out of n_0 drawings n will result in black and $n_0 - n$ red balls is simply the product of the a priori probabilities multiplied by the number of ways n things can be selected from n_0,

$$f_{-1}(n)$$
$$= \binom{n_0}{n} \frac{m}{m_0} \frac{m+1}{m_0+1} \cdots \frac{m+n-1}{m_0+n-1} \frac{m_0-m}{m_0+n} \cdots \frac{m_0-m+n_0-n-1}{m_0+n_0-1}.$$

In order to bring $f_{-1}(n)$ into canonical form, we first introduce factorials and then binomial coefficients:

$$
\begin{aligned}
f_{-1}(n) &= \frac{n_0!}{n!}(n_0-n)! \frac{(m+n-1)!}{(m-1)!} \frac{(m_0-1)!}{(m_0+n_0-1)!} \\
&\quad \times \frac{(m_0-m+n_0-n-1)!}{(m_0-m-1)!} \\
&= \binom{m+n-1}{n} \binom{m_0-m+n_0-n-1}{n_0-n} \Big/ \binom{m_0+n_0-1}{n_0}.
\end{aligned}
$$

This is known as the Pólya distribution, named after the person who invented it, George Pólya.

In the limit where the number of drawings and balls increase without bound in such a way that their ratio n_0/m_0 remains finite, the Pólya distribution transforms into

$$f_{-1}(n) = \binom{m+n-1}{n} \left(\frac{n_0/m_0}{1+n_0/m_0}\right)^n \left(\frac{1}{1+n_0/m_0}\right)^m. \tag{3.17}$$

But this is precisely the negative binomial distribution,

$$f_{-1}(n) = \binom{m+n-1}{n} p^m q^n. \tag{3.18}$$

The only trace left of the parent population, m_0, and total number of drawings, n_0, is to be found in the *a priori* probabilities, p and q. The most likely values of these *a priori* probabilities will again lead to a state of homogeneity.

Suppose we make r observations: on the first, we observe n_1 particles, on the second n_2, and so on. The log-likelihood function of the negative binomial distribution is

$$\mathcal{L}(p|n_1,\ldots,n_r) = \sum_{i=1}^{r} \ln f_{-1}(n_i) \propto r\,(m \ln p + \bar{n} \ln q),$$

where $\bar{n} = (1/r)\sum_i^r n_i$ is the arithmetic mean. The likelihood equation $\partial \mathcal{L}/\partial p = 0$ is given explicitly as

$$\frac{m}{p} - \frac{\bar{n}}{1-p} = 0.$$

Therefore, the most likely values of the *a priori* probabilities are

$$\hat{p} = \frac{m}{m+\bar{n}} \qquad \text{and} \qquad \hat{q} = \frac{\bar{n}}{m+\bar{n}}, \tag{3.19}$$

which also identify the most probable value of n as its average value, $\bar{n}$. Upon comparing (3.19) with their counterparts given in (3.17), we have

$$\frac{\bar{n}}{m} = \frac{n_0}{m_0}. \tag{3.20}$$

This says that the ratio of the average number of black balls drawn to the total number is equal to the ratio of the total number of drawings to the total number of balls. If the drawings are not biased, this is precisely the result that we should expect.

Introducing the homogeneity condition, (3.20), into the negative binomial distribution, (3.17), and comparing the *a priori* probabilities with those of (3.12) give

$$\frac{\bar{n}}{m} = \frac{\beta}{\alpha - \beta}. \tag{3.21}$$

The important lesson to be learnt is that we began with a distribution in which the *a priori* probabilities depend on the history of the process. If an event occurred, it would be more probable that it would be repeated. In terms of Bose–Einstein particles, this means that the presence of one particle would attract another and so on. In the limit where the total number of balls together with the number of drawings increase without limit, the *a priori* probabilities become equal to one another and the negative binomial distribution (3.17) results. The number of black balls, m, corresponds to the number of "cells" or "oscillators," and the number of drawings n in which black balls occur is analogous to the number of particles. When these quantities are not negligible with respect to the total numbers, we expect that the particles will show attractive behavior, like "photon bunching" in optics.

3.3.3   Thermodynamics

Multiplying the master equation (3.8) by n and summing over all possible values, we obtain the average equation of motion

$$\dot{\bar{n}}(t) = -\alpha \bar{n}(t) + \beta(m - \eta \bar{n}). \tag{3.22}$$

The stationary solution to this equation,

$$\bar{n} = \frac{\beta m}{\alpha + \eta \beta}, \tag{3.23}$$

corresponds to (3.16) when m is substituted for n_0 and $\eta = 1$, while for $\eta = -1$, it corresponds to (3.21). This shows that the maximum likelihood estimates coincide with the stationary solution to the average equation of motion, (3.22).

We have already observed that both (3.11) and (3.12) are the maximum values of the binomial and negative binomial probability distributions, respectively, since their *a priori* probabilities are given by the maximum likelihood values. As such, they are in the form amenable to Gauss' law of error leading to the arithmetic mean as the most probable value of the quantity measured. Moreover, since both the binomial and negative binomial distributions belong to the exponential family, the arithmetic mean will coincide with the mean of the distribution for any sample size.

The binomial and negative binomial distributions, evaluated at their maximum likelihood estimates, can be written as the Gaussian law of error,

$$\ln f_\eta(n) = S(n) - S(\bar{n}) - (n - \bar{n}) \left(\frac{\partial S(\bar{n})}{\partial \bar{n}} \right)_V + \text{const.}, \qquad (3.24)$$

for deviations in the particle number from its average or most probable value, $\bar{n}$. Comparing like terms in expressions (3.24) and the maximum likelihood values of the binomial, (3.11), and negative binomial, (3.12), distributions gives the expressions for the statistical entropies as

$$S(\bar{n}) = \bar{n} \ln \left(\frac{m - \eta \bar{n}}{\bar{n}} \right) - \frac{m}{\eta} \ln \left(\frac{m - \eta \bar{n}}{m} \right). \qquad (3.25)$$

For $\eta = 1$, (3.25) is the entropy of an ideal Fermi–Dirac gas, while for $\eta = -1$, it is the entropy of an ideal Bose–Einstein gas. It is also readily apparent that the stochastic entropies, $S(n)$, are the same function of n that the statistical entropies (3.25) are of its average value, $\bar{n}$. This will hold true any time that $\bar{n}$ and n are large enough to justify the use of Stirling's approximation. We will soon see that this is responsible for there being a single thermodynamics and not separate thermodynamics for each of the ensembles. This is so vital to the universality of thermodynamics that we will elevate it to a principle in §3.6.

Connection with the outside world is made by invoking the second law in the form

$$\left(\frac{\partial S(\bar{n})}{\partial \bar{n}} \right)_V = \left(\frac{\partial S(\bar{n})}{\partial \bar{n}} \right)_{V,\bar{U}} + \left(\frac{\partial S(\bar{n})}{\partial \bar{U}} \right)_{V,\bar{n}} \frac{d\bar{U}}{d\bar{n}}$$

$$= \frac{-\mu + \varepsilon}{T}, \qquad (3.26)$$

where $\varepsilon = \bar{U}/\bar{n}$ is the energy per particle. The average energy $\bar{U}$ is not an independent variable as it is in thermodynamics. With the statistical entropies given by (3.25), Eq. (3.26) can be solved for the average number of particles as a function of the absolute temperature T. We then obtain

$$\bar{n}(T) = \frac{m}{e^{(\varepsilon - \mu)/T} + \eta}, \qquad (3.27)$$

which are referred to as statistical distributions, a name peculiar to physics. Actually, (3.27) are the first moments of the distributions under the constraints (3.26).

For $\eta = -1$, it was first derived by Einstein in his 1924 paper on the ideal quantum gas. Einstein observed that the "degeneracy parameter" A must satisfy

$$A = e^{\mu/T} \le 1 \qquad (3.28)$$

in order that the "distribution" (3.27) does not blow up at arbitrarily small values of the energies, ε. This implies that the chemical potential be negative for Bose–Einstein statistics. Einstein also drew an analogy between the limiting case, where the degeneracy becomes unity, and a type of "phase" transition in which there appears to be a condensation in momentum, rather than configuration, space. We will have more to say about such a condensation in §7.5.

The equivalence between the statistical and thermodynamic definitions of the second law (3.26) can be written in the form

$$\frac{m - \eta \bar{n}}{\bar{n}} = \frac{\alpha}{\beta} = e^{(\varepsilon - \mu)/T} \tag{3.29}$$

with the aid of (3.23). We can now appreciate that the second law ensures that the Bose–Einstein condition (3.10) is complied with. For thermal radiation, the chemical potential $\mu = \mu_\gamma = 0$, while in the case of an ideal gas, it obeys condition (3.28).

3.4 Asymptotic Limit: Regaining Identity

Expression (3.27) must be compatible with classical statistics in the high temperature or low density limits. This is certainly apparent from the inequality

$$e^{(\varepsilon - \mu)/T} \gg 1. \tag{3.30}$$

But what is not so obvious is how *indistinguishable* particles apparently become *distinguishable* in the same limit. For a classical monatomic gas, inequality (3.30) is satisfied when

$$A = e^{\mu/T} = \lambda^3 \cdot \frac{N}{V} \ll 1, \tag{3.31}$$

where $\lambda \equiv h/\sqrt{2\pi MT}$ is the so-called thermal wavelength of a monatomic particle of mass M. Since the cube root of the volume per particle, V/N, is proportional to the mean distance between the particles, inequality (3.31) says that classical, or Maxwell–Boltzmann, statistics can be applied when the thermal wavelength is small compared to the mean distance between the particles.

Undoubtedly the "generating" function is the simplest and succinct method to obtain the asymptotic forms of probability distributions. Since it will appear over and over again, it is better that we pause here to introduce it.

3.4.1 Aside: Generating Functions

The generating function was invented by the mathematician Leonhard Euler as a mathematical gimmick to put all the information about the probability distribution into a capsule form which would be immediately accessible.

Suppose a random variable, n, can take on only nonnegative integer values with probability

$$f(n = k) = p_k, \qquad k = 0, 1, 2, \ldots .$$

If we multiply each of these probabilities by powers of a "dummy" variable, s, and add them, we get the power series

$$G(s) = p_0 + p_1 s + p_2 s^2 + \cdots = \sum_{k=0}^{\infty} p_k s^k,$$

which is called the *generating* function of the sequence of numbers p_k. Under the restriction $|s| < 1$ (which ensures the convergence of the power series), we can differentiate the generating function to obtain

$$G'(s) = \sum_{k=1}^{\infty} k p_k s^{k-1}, \qquad G''(s) = \sum_{k=2}^{\infty} k(k-1) p_k s^{k-2},$$

and so on. Now observe that if we set the dummy variable $s = 1$, we get precisely

$$G'(1) = \sum_{k=1}^{\infty} k p_k = \bar{n}, \qquad G''(1) = \sum_{k=2}^{\infty} (k^2 - k) p_k = \overline{n^2} - \bar{n},$$

and so on. These expressions are related to the moments of the distribution. A knowledge of all the moments,

$$\bar{n} = G'(1), \qquad \overline{n^2} = G''(1) + G'(1),$$

and so on, is tantamount to a knowledge of the probability distribution itself. Moreover, to obtain the variance of n, we have to add $\bar{n} - \bar{n}^2$ to $G''(1)$. These relations frequently provide the simplest means to calculate the central moments of the distribution.

An important property of generating functions is their so-called convolution property. If two random variables $\mathbf{N}_1$ and $\mathbf{N}_2$ are nonnegative, integral-valued, and mutually independent and have generating functions $G_1(s)$ and $G_2(s)$, then their sum $\mathbf{N}_1 + \mathbf{N}_2$ has the generating function

$$G_{12}(s) = G_1(s) G_2(s) = \sum_k q_k s^k,$$

where the sequence $\{q_k\}$ is the *convolution* of the sequences of the individual generating functions. This product rule allows us to buildup distributions from more elementary ones. For instance, the generating function of the negative binomial distribution is the product of generating functions of the geometric distribution. It will have even more important repercussions for us when the dummy variable, s, acquires a physical significance, as it will in §4.4.

Since it will be important for later developments that begin in Chapter 4, a word is in order concerning continuous random variables. For a positive, continuous random variable, $\mathbf{X}$, with density $f(x)$, it proves more convenient to set $s = e^{-\gamma x}$, where γ is another dummy variable. The generating function then takes the form of a Laplace transform

$$G(\gamma) = \int_0^\infty e^{-\gamma x} f(x) \, dx,$$

which is defined for $\gamma > 0$. The method is, in fact, due to Pierre Simon Laplace, who developed it in 1782. If we differentiate the logarithm of the generating function and evaluate it at $\gamma = 0$, we get $d \ln G(0)/d\gamma = -\overline{\mathbf{X}}$. The second derivative, evaluated at the same value, is $d^2 \ln G(0)/d\gamma^2 = -\overline{(\mathbf{X} - \overline{\mathbf{X}})^2}$. Hence, this appears as an extremely neat way to generate the central moments of a distribution associated with a continuous random variable. The technique will be essential to us in Chapter 4 when we deal with *improper* distribution functions whose integral over all values is not unity but infinity! The dummy variable, γ, will then acquire a physical significance, and we will not be able to set it equal to unity in order to derive the moments of a proper probability distribution which is "conjugate" to our original, improper distribution.

3.4.2   Classical Limit

The generating function of the binomial distribution, evaluated at the maximum likelihood value of the *a priori* probabilities (3.11), is

$$G_B(s) = \sum_{n=0}^{m} \binom{m}{n} (\hat{p}s)^n \hat{q}^{m-n} = (\hat{q} + \hat{p}s)^m, \tag{3.32}$$

while that of the negative binomial distribution is

$$G_{NB}(s) = \hat{p}^m \sum_{n=0}^{\infty} \binom{m+n-1}{n} (\hat{q}s)^n = \left(\frac{\hat{p}}{1 - \hat{q}s} \right)^m. \tag{3.33}$$

The maximum likelihood values of the *a priori* probabilities of the binomial distribution are

$$\hat{p} = \frac{\bar{n}}{m} \qquad \text{and} \qquad \hat{q} = \frac{m - \bar{n}}{m}, \tag{3.34}$$

while those for the negative binomial distribution are given in (3.19).

We are interested in the limit where the number of cells m increases without bound for fixed $\bar{n}$. In this limit, both the generating function of the binomial distribution,

$$G_B(s) = \left[1 - \frac{\bar{n}}{m}(1 - s) \right]^m,$$

and the generating function of the negative binomial distribution,

$$G_{NB}(s) = \left(\frac{1 - \bar{n}/m}{1 - s\bar{n}/m} \right)^m,$$

transform into

$$G_P(s) = e^{\bar{n}(s-1)}. \tag{3.35}$$

This is the generating function of the Poisson distribution. In deriving these results, we have made use of the well known formula

$$\lim_{m \to \infty} \left(1 - \frac{x}{m}\right)^m = e^{-x}.$$

One of the most elegant ways of deriving the Poisson distribution is via a limiting scheme whereby the binomial or negative binomial distribution is transformed into the Poisson distribution,

$$f(n) = \frac{\bar{n}^n}{n!}e^{-\bar{n}}. \tag{3.36}$$

In fact, historically, Poisson's distribution was known as Poisson's *limit law*. So we know that in the limit where $\bar{n} \ll m$, both the binomial and the negative binomial distributions, (3.11) and (3.12), go over into the Poisson distribution, (3.36). But this condition is precisely the condition (3.30) for the validity of classical statistics which we have applied to the probability distributions themselves rather than to their average values.

Casting the Poisson distribution (3.36) as a law of error of the form (3.24), we obtain the expression

$$S(n) = -n(\ln n - 1)$$

for the stochastic entropy. However, this expression is not admissible because it is not *extensive*. To render it extensive, we could add on a term $n \ln m$ which would give

$$S(n) = n - n \ln\left(\frac{n}{m}\right). \tag{3.37}$$

This has no effect on the Poisson distribution (3.36) since all the terms in m cancel out.

However, in view of Boltzmann's principle, this would correspond to a thermodynamic probability or statistical weight,

$$\Omega = \frac{m^n}{n!}, \tag{3.38}$$

assuming Stirling's approximation to be valid. Now, m^n is the number of ways in which n *distinguishable* particles can be put into m cells, and the division by $n!$ is the "corrected" Boltzmann counting to take into account that the particles are actually *identical* and that the $n!$ permutations of the particles do not lead to any new arrangements.

We recall from §1.13 that the division by $n!$ in (3.38) is known as the Gibbs paradox. It was necessary in order that the expression for the entropy turn out to be extensive. The antithesis has been the necessity of introducing the factor m^n into the statistical weight (3.38) for precisely the same reason. The basic probability distributions are the negative binomial (Bose–Einstein statistics) and binomial (Fermi–Dirac statistics) distributions. The Poisson distribution is only a limiting form of these distributions which, qualitatively speaking, is valid when there is an overwhelmingly large number of cells. Hence, if the particles are distinguishable, they have to be told so by writing the number of ways in which an arrangement of n distinguishable particles can be placed in m cells in the expression for the statistical weight (3.38). This leads to an extensive expression for the stochastic entropy, namely (3.37).

3.5   Intermediate Statistics

It has been relatively easy to derive Fermi–Dirac and Bose–Einstein statistics, but we have offered nothing in the way of demonstrating their uniqueness. In fact, Gentile, in his *Nuovo Cimento* article of 1940, wrote:

> Whoever considers the two quantum statistics of Bose–Einstein and Fermi–Dirac is naturally led to ask himself which properties remain and which are modified when one doesn't make the particular hypotheses (Fermi–Dirac) that in an elementary cell there cannot be more than one particle, or the other, none the less special (Bose–Einstein) case, that there can be any number, even an infinite number, of particles.
>
> Although it is clear that the most general hypothesis, that in a cell there can be a finite number, d, of particles, leaves essentially unaltered the specific character of the limitation imposed by Pauli's principle, it is evident that this hypothesis permits, with the increase of d from 1 to ∞, to pass without discontinuity from one to the other typical case by passing through all the intermediate cases. In any event, as we will show, proceeding in this manner one has a method to treat, in a unified and elementary way, the diverse quantum statistics which, as is well known, have in common that the particles must be considered indistinguishable.

On the contrary, our job will be to show that any value of d different from 1 or ∞ does not correspond to a stationary probability distribution. In other words, there is no stationary probability distribution which tends to the binomial and negative binomial distributions in the limits as $d \to 1$ and $d \to \infty$, respectively. Since we have derived the stationary distributions from physical stochastic processes, this will demonstrate conclusively that there are no physical processes which obey intermediate statistics. The supposition that quarks obey an intermediate form of statistics simply means that either all the degrees-of-freedom have not been properly accounted for or the number of states has not been properly enumerated or both.

Since there is no purely *thermodynamic* argument that can be advanced in favor or against intermediate statistics, we will have to employ a *statistical* argument. The statistical argument we choose is to show that the stationary probability distribution can only satisfy a recursion relation that is obtained from the time independent master equation when it coincides with the binomial, (3.11), negative binomial, (3.12), or Poisson, (3.36), distributions. As we have seen, these distributions govern Fermi–Dirac, Bose–Einstein, and classical statistics, respectively. In this way, we lay to rest the possibility of there being still yet other types of particles with spins different from semi-integral or integral values.

3.5.1   Derivation of the Distribution Function

The analysis is most easily performed using the method of the generating function that we described in §3.4.1. The generating function we will need is

$$G_I(s) = \left(p \sum_{k=0}^{d} (qs)^k \right)^m = \left(\frac{p[1 - (qs)^{d+1}]}{(1 - qs)} \right)^m \tag{3.39}$$

without necessarily requiring $p+q = 1$. Two limiting forms of (3.39) have direct physical meanings: for $d = 1$ and $q = 1/p - 1$, it reduces to the generating function of the binomial distribution, (3.32), which we now write as

$$G_B(s) = [p(1 + qs)]^m, \tag{3.40}$$

while for $d = \infty$ and $p + q = 1$, it becomes

$$G_{NB}(s) = \left(\frac{p}{1 - qs} \right)^m. \tag{3.41}$$

This is the generating function of the negative binomial distribution, (3.33).

On the strength of the binomial expansion, the numerator and denominator in the generating function (3.39) can be expanded, respectively, as

$$[1 - (qs)^a]^m = \sum_{k=0}^{m} (-1)^k \binom{m}{k} (qs)^{ak}$$

and

$$(1 - qs)^{-m} = \sum_{j=0}^{\infty} \binom{-m}{j} (-qs)^j,$$

where $\binom{-m}{j} = (-1)^j \binom{m+j-1}{j}$ and $a \equiv d + 1$. The product of these two series can be written as the convolution

$$G_I(s) = p^m \sum_{n=0}^{dm} \vartheta(n; m, d)(qs)^n \tag{3.42}$$

since only $dm + 1$ terms in the sum are different from zero. The coefficient $\vartheta(n; m, d)$ has been defined in (3.4). The probability that there are n particles is simply the coefficient of s^n in (3.42):

$$f(n; m, d) = \vartheta(n; m, d)p^m q^n. \tag{3.43}$$

Although the form of the coefficient (3.4) is generally complicated, it reduces to the binomial coefficient $\binom{m}{n} = \vartheta(n; m, 1)$ in the case $d = 1$ and to the negative binomial coefficient $\binom{m+n-1}{m-1} = \vartheta(n; m, \infty)$ in the case $d = \infty$. That is to say, the condition $d = \infty$ means that we must take only the $k = 0$ term in sum (3.4).

It is easy to show that the probability distribution (3.43) is properly normalized in the two cases: $d = 1$ and $d = \infty$. In the former case we have

$$\sum_{n=0}^{m} \binom{m}{n} q^n = (1+q)^m = p^{-m},$$

while in the latter case

$$\sum_{n=0}^{\infty} \binom{m+n-1}{n} q^n = (1-q)^{-m} = p^{-m}.$$

In either case, if we take the limit as $m \to \infty$ and $q \to 0$ such that $qm \to \lambda$, we find

$$\sum_{n=0}^{\infty} \frac{\lambda^n}{n!} = e^{\lambda} = p^{-m}$$

since $e^q \sim 1 + q$.

The fact that the generating function (3.39) is the mth power of the ratio $p[1 - (qs)^a]/(1 - qs)$ shows that the sequence $\{f(n; m, d)\}$ is the distribution of a sum, $\mathbf{S}_m = \mathbf{N}_1 + \cdots + \mathbf{N}_m$, of m independent random variables with the common generating function $p[1 - (qs)^a]/(1 - qs)$. In the case $d = 1$, each variable $\mathbf{N}_i$ assumes the value 0 with probability p and the value 1 with probability pq [i.e., $p(1+q) = 1$].

In the case $d = \infty$, the random variable $\mathbf{N}_i$ can be interpreted as the number of failures following the $(i-1)$st and preceding the ith success. The sum $\mathbf{S}_m$ is the total number of failures preceding the mth success. The probability that there are exactly n failures preceding the mth success is $\Pr\{\mathbf{S}_m = n\} = f(n; m, \infty)$, where $\mathbf{S}_m + m$ is the number of trials up to and including the mth success.

The analysis that we have been using can actually be traced back to Abraham De Moivre, who obtained (3.43) for the probability of getting a score of $m + n$ in a throw of m dice.

Suppose that each of the random variables, $\mathbf{N}_i$, can assume only a finite number of values, $0, 1, 2, \ldots, d$, and that each occurs with a probability $1/a$ where we recall that $a = d+1$. In terms of intermediate statistics, $\mathbf{N}_i$ represents the level i that can accommodate d particles. In the case $d = 1$, the probability of there being a particle is simply $\frac{1}{2}$. The probability that $\mathbf{S}_m = n$ is

$$\Pr\{\mathbf{S}_m = n\} = \frac{1}{a^m} \sum_{k=0}^{[n/a]} (-1)^k \binom{m}{k} \binom{m+n-ak-1}{m-1}, \qquad (3.44)$$

and

$$\Pr\{\mathbf{S}_m \le n\} = \frac{1}{a^m} \sum_{k=0}^{[n/a]} \binom{m}{k} \binom{m+n-ak}{m}$$

is the probability that the levels will contain no more than n particles. In the limit as n and a both tend to infinity in such a way that their ratio has a finite value, x,

$$\Pr\{\mathbf{S}_m \le n\} \to \frac{1}{m!} \sum_{k=0}^{[x]} (-1)^k \binom{m}{k} (x-k)^m.$$

This expression, originally derived by Joseph Louis Lagrange, states that "in the limit, the sum of m independent random variables is uniformly distributed in the interval $0,1$ with a constant density of 1." In the case $d = 1$, $\Pr\{\mathbf{S}_m \le n\} = (1/2^m)\binom{m-1}{n}$, while in the case $d = \infty$ with $n \to \infty$ such that their ratio tends to a constant, $x < 1$, $\Pr\{\mathbf{S}_m \le \infty\} = x^m/m!$

However, distributions having the form (3.44) are not compatible with stochastic processes which generate the known forms of physical statistics. In other words, we intend to show that there is no stationary probability distribution which can transform from one statistics to the other through a continuous variation of a characteristic parameter like d.

3.5.2   Criterion for Stationary Distributions

In §3.3.1, we have shown how the stationary probability distributions governing the two forms of quantum statistics can be derived from dynamical equilibria. In the case of Fermi–Dirac statistics, the dynamical equilibrium is between the rates of desorption, αn, and absorption, $\beta(m - \eta n)$, on a lattice of m sites, while for Bose–Einstein statistics, the dynamical equilibrium is between the absorption and emission (spontaneous + stimulated) of radiation by an atom.

In the intermediate case, where $\eta \in [-1, 1]$, an intermediate form of statistics should be applicable. Our aim is to show that any intermediate form of statistics is, itself, spurious in that the stationary probability distribution (3.43) provides for only three constant values of η: $-1, 0$, and 1. All other values of η depend on the number of particles n and consequently do not correspond to any physical process. To put it picturesquely, we may say that the values of η are "quantized": they correspond to stationary probability distributions that belong to the exponential family.

In order for (3.43) to be the stationary solution of the master equation (3.8), $q = \beta/\alpha$. Setting equal to zero the coefficients of the different powers in β/α give

$$\eta = \frac{m}{n} - \left(\frac{n+1}{n}\right) \frac{\vartheta(n+1; m, d)}{\vartheta(n; m, d)} \tag{3.45}$$

and

$$\eta = \frac{m}{n-1} - \left(\frac{n}{n-1}\right) \frac{\vartheta(n; m, d)}{\vartheta(n-1; m, d)}.$$

Eliminating η between these two expressions results in a recursion relation for ϑ:

$$m + (n^2 - 1)\frac{\vartheta(n+1; m, d)}{\vartheta(n; m, d)} = n^2\frac{\vartheta(n; m, d)}{\vartheta(n-1; m, d)}. \tag{3.46}$$

Since we supposed η to be a parameter independent of n, not any coefficient $\vartheta(n; m, d)$ will do. Any acceptable value of the ratio of the ϑ's must produce a fixed value of η. There are only two possible cases in addition to one limiting form to which they both tend in a well defined limit. This is the death-knell for intermediate statistics: there is no stationary solution to the master equation, (3.8), that would govern such a form of statistics. There are only three possibilities relating to two forms of particles and one limiting form. No additional types of particles exist.

For $\eta = 1$, condition (3.45) reduces to

$$\frac{m-n}{n+1} = \frac{\vartheta(n+1; m, 1)}{\vartheta(n; m, 1)} = \frac{\binom{m}{n+1}}{\binom{m}{n}},$$

giving the binomial distribution in (3.43). Alternatively, for $\eta = -1$, we have

$$\frac{m+1}{n+1} = \frac{\vartheta(n+1; m, \infty)}{\vartheta(n; m, \infty)} = \frac{\binom{m+n}{n+1}}{\binom{m+n-1}{n}},$$

which yields the negative binomial distribution in (3.43). Between these two, there is a third, limiting, case: $\eta = 0$. Condition (3.45) then becomes

$$\frac{m}{n+1} = \frac{\vartheta(n+1; m, d)}{\vartheta(n; m, d)} = \frac{m^{n+1}/(n+1)!}{m^n/n!},$$

identifying the Poisson distribution,

$$f(n) = \frac{\lambda^n}{n!}e^{-\lambda}, \tag{3.47}$$

as the stationary probability distribution (3.43).

Although we have seen that the Poisson distribution is a limiting form of both the binomial and negative binomial distributions, it—remarkably enough—satisfies the recursion formula exactly. Unlike the binomial and negative binomial distributions, the Poisson distribution does not depend on a given value of d; rather, it requires that m is sufficiently large. To be more precise, it demands $m \to \infty$ and $q \to 0$ such that their product, $qm = \lambda$, is constant.

In this limit, the generating function (3.39) becomes

$$G_P(s) = \left(\frac{(1 - \lambda/m)[1 - (\lambda s/m)^a]}{1 - \lambda s/m}\right)^m. \tag{3.48}$$

Passing to logarithms, we can easily see that the right-hand side tends to $e^{-(1-s)\lambda}$, independently of the magnitude of the occupation number. Hence, (3.48) is the generating function of the Poisson distribution, (3.35), with $\lambda = \bar{n}$.

The transition to classical statistics depends only on the number of cells m, independent of the magnitude of the average number of particles. In the limit as m increases without bound,

$$\sum_{k=0}^{[n/a]} \binom{m}{k} \binom{m+n-ak-1}{m-1} \to \frac{m^n}{n!}$$

if we agree to neglect terms of order m^{-d}. From this we conclude that with the increase of the number of cells, the transition from Bose–Einstein statistics to classical statistics takes place at an earlier stage than Fermi–Dirac statistics. It also shows that the gradual appearance of particle distinguishability is due to the increase of the number of energy cells and not upon how many particles, d, a cell can accommodate.

In §3.4.1 we showed how the logarithm of the generating function can be used to derive the central moments of the distribution. The logarithm of the generating function we are interested in is

$$\ln G_{\mathrm{I}}(s) = m \ln p + m \ln\left[1 - (qs)^a\right] - m \ln(-qs),$$

whose derivatives will be evaluated at $s = 1$. A single differentiation gives the average value

$$\frac{\bar{n}}{m} = \frac{q}{1-q} - a\frac{q^a}{1-q^a}$$

when $s = 1$. This is precisely the expression that Gentile found. For $a = 2$ or $d = 1$, it reduces to

$$\bar{n} = \frac{mq}{1-q},$$

while for $a = \infty$, it becomes

$$\bar{n} = \frac{mq}{1-q}.$$

Thermodynamic considerations have shown that $q = e^{-(\varepsilon-\mu)/T}$ so that these expressions correspond to Bose–Einstein and Fermi–Dirac statistics, respectively. A general expression for the average number of particles can be found from condition (3.45) by multiplying it by q^n and summing over all n. We then obtain $q = \bar{n}/(m - \eta\bar{n})$, which upon rearrangement becomes

$$\bar{n} = \frac{m}{q^{-1} + \eta}.$$

This would correspond to intermediate statistics.

3.5.3   Uniqueness of the Error Laws

We have already had occasion to note that the error laws which satisfy Gauss' principle, (3.24), leading to the average value as the most probable value of the number of particles, all belong to the exponential family of distributions. It is clear from the form of Gauss' principle that in order for there to be maximum probability when $n = \bar{n}$, the stochastic entropy must have the same functional form as the statistical entropy (3.25). This will lead us to the conclusion that only the negative binomial and binomial distributions, in addition to the limiting Poisson distribution, can satisfy Gauss' principle.

Confronting the probability distribution for intermediate statistics, (3.43), with Gauss' error law, (3.24), we see that the stochastic entropy is logarithmically related to the probability coefficient (3.4),

$$S(n) = \ln \vartheta(n; m, d).$$

But the probability coefficient must satisfy the recursion relation, (3.45), which can be written as

$$\ln \left(\frac{m - \eta n}{n} \right) = \ln \left(1 + \frac{1}{n} \right) + \ln \vartheta(n + 1; m, d) - \ln \vartheta(n; m, d).$$

We may approximate this finite difference equation by a differential equation,

$$\frac{d \ln \vartheta}{dn} = \ln \left(\frac{m - \eta n}{n} \right), \qquad (3.49)$$

for sufficiently large n so as to be able to neglect the term $1/n$ in comparison to the other terms. And provided $\eta \neq 0$, we may integrate it to obtain

$$\ln \vartheta(n; m, d) = -n \ln n - \frac{1}{\eta}(m - \eta n) \ln(m - \eta n) + \text{const.} = S(n). \qquad (3.50)$$

When the constant of integration is set equal to $(m/\eta) \ln m$, the stochastic entropy will coincide with the statistical entropy (3.25) when $n = \bar{n}$. This overwhelmingly maximizes the stationary probability distribution so that the average value $\bar{n}$ coincides with the most probable number of particles. The stochastic entropy must be given by the expression (3.50) which, in turn, means that it must satisfy the differential equation (3.49) for sufficiently large values of n. The validity of the differential equation rests upon the recursion relation (3.45), which can only be satisfied for $\eta = \pm 1$ in addition to $\eta = 0$. Only in the first two cases will the stochastic entropy have the form (3.50). The third case is a limiting form of the previous two.

Hence, the mere existence of intermediate statistics would lead to a fatal rupture between statistics and thermodynamics, since averages of extensive variables would no longer be, *overwhelmingly*, the most probable values of the quantities measured. In other words, there would be no connection between the statistical world, where average values prevail, and the thermodynamic world, where no distinction can be made between most probable and actual values.

The requirement that n be sufficiently large is usually made within the context of Stirling's approximation which, if valid, ensures that there will be a single thermodynamics no matter in which ensemble we may be working. We shall now discuss this in greater detail.

3.6   Classical Statistics

The transition from quantum to classical statistics, as the number of energy cells increases without limit, does not lead to the usual results of ensemble theory since fluctuations in the energy are still determined by fluctuations in the particle number. Moreover, in Chapter 6 we will argue that fluctuations in the particle number cannot occur simultaneously with fluctuations in the volume for, in fact, there is a statistical equivalence between the two. So, it appears that as far as the statistics resulting from occupancy problems are concerned, there is no such thing as the microcanonical and canonical ensembles. The chance mechanism is the distribution of the particles among the cells, and the outcomes are the different possible arrangements.

Suppose we did not know that it is the entropy which determines Gauss' law of error (3.24) leading to the average number of particles as the most probable value. Instead we were given this law in terms of a twice-differentiable, but otherwise generic, function φ:

$$f(n) = A \exp\{\varphi'(\bar{n})(n - \bar{n}) + \varphi(\bar{n}) + \psi(n)\}, \qquad (3.51)$$

with no *a priori* connection between the functions φ and ψ. In order to connect (3.51) with a known result of statistical mechanics, we might try to relate the normalization constant

$$e^{\bar{n}\varphi'(\bar{n}) - \varphi(\bar{n})} = \sum_n A e^{n\varphi'(\bar{n}) + \psi(n)}$$

to the statistical mechanical moment generating function, or what we have commonly referred to as the "partition function."[2]

---

[2]The difference between the moment generating function and the partition function is the following. The former is a function of a dummy variable which can be made to take on particular values, depending on how it is defined, when expressions for the moments are desired. This, we have seen, is due to the fact that the generating function is the Laplace transform of a *proper* probability distribution. In thermodynamics, however, the probability distribution is *improper*, having no finite moments. Hence, the Laplace transform variable is no longer a "dummy" variable which can be made to take particular values; rather, it has a well defined range of values with a definite physical significance. Nevertheless, the logarithm of the Laplace transform of this improper distribution, which we will identify with the volume of phase space occupied by the system, still gives the central moments of the distribution that is "conjugate" to this improper distribution. A full discussion will be given in the next chapter.

For fluctuations in the number of particles, the partition function of the "grand-canonical" ensemble would be the most obvious choice so we might set

$$\varphi(\bar{n}) - \bar{n}\varphi'(\bar{n}) = W/T, \qquad (3.52)$$

where $W = -PV$ is referred to as the grand-canonical potential. In doing so, we naturally identify the function φ with the *negative* of the entropy so that the identification (3.52) is none other than the Massieu transform of the entropy with respect to the (average) number of particles,

$$
\begin{aligned}
S\left[\frac{1}{T}, \frac{\mu}{T}, V\right] &= S(\bar{n}) - \left(\frac{\partial S}{\partial \bar{n}}\right)_V \bar{n} \\
&= S(\bar{n}) - \left(\frac{\partial S}{\partial \bar{U}}\right)_{\bar{n},V} \bar{U} - \left(\frac{\partial S}{\partial \bar{n}}\right)_{\bar{U},V} \bar{n} \\
&= S(\bar{n}) - \frac{\bar{U} - \mu\bar{n}}{T} = \frac{PV}{T}.
\end{aligned}
\qquad (3.53)
$$

Gauss' error law (3.51) now takes the form

$$
\begin{aligned}
f(n) &= \exp\left\{-\chi(n - \bar{n}) + S(\bar{n}) - S(n)\right\} \\
&= \exp\left\{-\chi n + S\left[\frac{1}{T}, \frac{\mu}{T}, V\right] - S(n)\right\},
\end{aligned}
\qquad (3.54)
$$

where

$$\chi = \left(\frac{\partial S}{\partial \bar{n}}\right)_V = \left(\frac{\partial S}{\partial \bar{U}}\right)_{\bar{n},V} \varepsilon + \left(\frac{\partial S}{\partial \bar{n}}\right)_{\bar{U},V} = \frac{\varepsilon - \mu}{T}$$

and we have set $\psi(n)$ equal to the stochastic entropy $S(n)$.

Historically, Greene and Callen made this identification based on Einstein's formula for a spontaneous fluctuation. But once having made this identification, it was necessary to assume that, at equilibrium, the stochastic entropy is the same function of its variables as the entropy is of its variables. For if the two did not coincide at equilibrium, it would mean that there would be a separate thermodynamics for the microcanonical and canonical ensembles that would be given by different functional expressions for the stochastic and statistical entropies, respectively. The former is a function of the random variable n while the latter is a function of its average value, $\bar{n}$, but the two would not coincide for $n = \bar{n}$.

The distinction between the ensembles is really academic since the energy is no longer considered to be an independent variable but, rather, it is a function of the average number of particles. It is this concept which is foreign to thermodynamics: thermodynamics assumes that the entropy is a function of all the extensive variables needed to specify the system and the energy is at the head of the list. This is a vestigial factor left over from the occupancy problem of the two forms of quantum statistics.

If we want to retain the functional dependency of the energy on the average number of particles in our classical treatment, we must modify the Greene–Callen principle along the following lines. The numbers of particles and cells are of such magnitude as to warrant Stirling's approximation, and the stochastic entropy is the same function of n that the statistical entropy is of its average value, $\bar{n}$. In view of Gauss' principle, this will guarantee that the average value is also the most probable value of the particle number.

3.6.1 Classical Number Statistics

The Gaussian law of error (3.54) can be written in the (grand-) canonical form[3]

$$f(n) = \frac{e^{\hat{\beta}n(\mu-\varepsilon)}}{\Xi(\hat{\beta},\mu,V)}\Omega(n) \tag{3.55}$$

for a distribution belonging to the exponential family, where $\hat{\beta} = 1/T$. The norming factor,

$$\Xi(\hat{\beta},\mu,V) = \sum_n e^{\hat{\beta}n(\mu-\varepsilon)}\Omega(n) = e^{\hat{\beta}PV}, \tag{3.56}$$

is the partition function of the grand-canonical ensemble. However, we are not studying an open system which is in thermal contact with both a thermostat and a particle reservoir since the energy would be able to fluctuate even when we converted it into a closed system by enclosing it within impermeable walls. Here, the particle fluctuations generate fluctuations in the energy and the energy is no longer an independent variable which can fluctuate on account of the transfer of heat or the performance of work.

In order to derive an expression for the statistical weight, or structure, function, $\Omega(n)$, we consider an urn of m *distinguishable* balls, marked from 1 to m, from which n balls will be drawn. If we agree to draw the n balls sequentially, replacing the ball drawn before the next drawing is made, and to record the numbers on the balls as well as the order of their appearance, then the total number of possibilities is m^n, according to the fundamental rule of probability theory. But when the results are compared with thermodynamics, we find we are off by a factor of $1/n!$ in our expression for the structure function. The ordering then is immaterial, and a bunch of n balls drawn one by one can appear in $n!$ ways so that the number of distinct arrangements is given by (3.38).

With the structure function given by (3.38), the partition function becomes

$$\Xi(\hat{\beta},\mu,V) = \sum_n \frac{1}{n!}\left(e^{\hat{\beta}n(\mu-\varepsilon)}m\right)^n = \exp\left(e^{-\hat{\beta}(\varepsilon-\mu)}m\right). \tag{3.57}$$

This has the same form as the (unnormalized) generating function of the Poisson distribution. In fact, setting

$$\bar{n} = e^{-\hat{\beta}(\varepsilon-\mu)}m, \tag{3.58}$$

---

[3]Statistical mechanics does not make the distinction between any estimate of β and its most likely value $\hat{\beta}$ which is related to the inverse temperature.

the exponential distribution (3.55) is seen to be the Poisson distribution, (3.36). For values of n sufficiently large to warrant Stirling's approximation, the Poisson distribution can be written as

$$f(n) = \left(\frac{\bar{n}}{n}\right)^n e^{n-\bar{n}}.$$

Setting $n = \bar{n}$ is equivalent to placing the distribution at its most probable value since this value overwhelmingly maximizes the probability distribution. In this way the connection is made between statistical mechanics (average values) and thermodynamics (actual values corresponding to "most probable" values).

Illustration: An Ideal Gas

We now show how the foregoing results, based on number fluctuations alone, give the correct results in the case of an ideal gas.

The number of quantum states, or "cells," that a particle can occupy in the momentum range Δp is given by the ratio of the phase volume $4\pi V p^2 \Delta p$ to the elementary unit volume h^3. In terms of the particle energy $\varepsilon = p^2/2M$, it can be written as

$$m(\varepsilon)\,\Delta\varepsilon = 4\pi M \frac{V}{h^3}\sqrt{2M\varepsilon}\,\Delta\varepsilon. \tag{3.59}$$

Solving the expression for the total number of particles,

$$\begin{aligned}
N &= \int_0^\infty \bar{n}(\varepsilon)\,d\varepsilon \\
&= e^{\hat{\beta}\mu}\int_0^\infty e^{-\hat{\beta}\varepsilon} m(\varepsilon)\,d\varepsilon \\
&= e^{\hat{\beta}\mu} V \left(\frac{2\pi M}{\hat{\beta}h^2}\right)^{3/2},
\end{aligned} \tag{3.60}$$

for the chemical potential, μ, we find

$$e^{\hat{\beta}\mu} = \frac{N}{V}\left(\frac{\hat{\beta}h^2}{2\pi M}\right)^{3/2}. \tag{3.61}$$

Now our postulate says that the statistical entropy must be the same function of $\bar{n}$ that the stochastic entropy is of n. And since the latter is given by (3.50), with $\eta = 0$, the former is

$$S(\bar{n}) = \bar{n}\left[1 - \ln\left(\frac{\bar{n}}{m}\right)\right] \tag{3.62}$$

provided $\bar{n}$ is also large enough that Stirling's approximation is valid. Positive entropy implies $\bar{n} < m$ and, in general, $\bar{n} \ll m$. Introducing the expression for the average particle number, (3.58), where the expression for the chemical potential (3.61) has been used, we obtain

$$S(\bar{n}) = \bar{n} \ln \left[\left(\frac{2\pi M}{\hat{\beta} h^2} \right)^{3/2} \frac{V}{N} \right] + \hat{\beta} \bar{n} \varepsilon + \bar{n}. \tag{3.63}$$

This will coincide with the usual expression for the entropy of an ideal gas if we use the equipartition result, $\varepsilon = 3/2\hat{\beta}$.

With the aid of the law of equipartition of energy, $\hat{\beta}$ can be eliminated in (3.63) to give

$$S(\bar{n}) = \bar{n} \ln \left[\left(\frac{4\pi M\bar{U}}{3h^2\bar{n}} \right)^{3/2} \frac{V}{N} \right] + \frac{5}{2}\bar{n}. \tag{3.64}$$

This differs from the fundamental relation of thermodynamics, expressing the entropy as a function of the energy, volume, and number of moles in two very interesting ways. First, the average energy replaces the total energy and second the average number of particles appears instead of the total number of particles. This does not affect the definition of the inverse temperature,

$$\left(\frac{\partial S}{\partial \bar{U}} \right)_{\bar{n},V} = \frac{3}{2} \frac{\bar{n}}{\bar{U}} = \frac{1}{T},$$

which is the equipartition result, but it does effect the expression for the chemical potential,

$$\left(\frac{\partial S}{\partial \bar{n}} \right)_{\bar{U},V} = -\frac{\mu}{T} = \ln \left[\left(\frac{4\pi M\bar{U}}{3h^2\bar{n}} \right)^{3/2} \frac{V}{N} e \right].$$

This expression for the chemical potential differs from the usual one by the term $\ln e = 1$ due to the fact that differentiation has been performed with respect to average number of particles, $\bar{n}$, and not the total number of particles, N, as in the conventional treatment.

Second, our expression for the average number of particles in the energy range $\Delta\varepsilon$,

$$\bar{n}(\varepsilon)\,\Delta\varepsilon = 2N\hat{\beta}\sqrt{\frac{\hat{\beta}\varepsilon}{\pi}} e^{-\beta\varepsilon}\,\Delta\varepsilon, \tag{3.65}$$

gives the exact Maxwell–Boltzmann law for the average number of particles in the velocity range Δv,

$$\bar{n}(v)\,\Delta v \,(= \bar{n}(\varepsilon)\,\Delta\varepsilon) = \left(\frac{2}{\pi} \right)^{1/2} N \left(\frac{M}{T} \right)^{3/2} v^2 e^{-Mv^2/2T}\,\Delta v,$$

when the single-particle energy, ε, is set equal to the kinetic energy of the particle, $\frac{1}{2}Mv^2$. In the first instance, we identified ε with its equipartition value to get the correct expression for the entropy, while in the second instance, we identified it with the kinetic energy of a particle. This is yet another manifestation of what Lorentz called the remarkable insensibility of thermodynamics.

The foregoing results are oridinarily obtained in the canonical, rather than in the grand-canonical, ensemble formalism. The Lagrange multiplier, $\alpha = -\mu/T$, enters through the constraint that the total number of *systems*—not *particles*—remains fixed. But we have seen that the grand-canonical partition function can be used to obtain consistent results which make more physical sense.

Insofar as a problem of statistical estimation is concerned, we must go one step back and work with the probability distribution and not its maximum value which is given by Gauss' error law. The intensive thermodynamic parameters can then be estimated from observations made on their conjugate, extensive variables [cf. §4.3]. Just as the inverse temperature is to be estimated in terms of observations made on the energy, so too, the chemical potential can be estimated from observations made on the number of particles. However, in order for the particle number to be a statistic, we must work in the grand-canonical ensemble which allows the particle number to vary. The constancy of the chemical potential is the criterion for chemical equilibrium. This certainly cannot be achieved in an ensemble comprised of systems whose walls are impermeable to the exchange of particles for there would be no physical mechanism by which such an equilibrium could be established.

3.7   The Error Law as a Moment Generating Function

Gauss' principle offers a very elegant and concise way of determining the moments of the distribution. This was discovered by Greene and Callen but did not refer to Gauss' principle as such. These authors considered only moments of extensive variables. Mixed moments and even moments of intensive variables can also be determined from Gauss' principle. And we shall also see that the moments derived from Einstein's formula for a spontaneous fluctuation [cf. §1.11] are exact to second-order.

Consider any two extensive independent thermodynamic variables, $\bar{X}$ and $\bar{Y}$. Gauss' principle is

$$f_i \equiv f(X_i, Y_i)$$
$$= A \exp\left\{ S(X_i, Y_i) - S(\bar{X}, \bar{Y}) - S_{\bar{X}}(X_i - \bar{X}) - S_{\bar{Y}}(Y_i - \bar{Y}) \right\}, \quad (3.66)$$

where the subscript indicates partial differentiation with respect to that variable and the mean value is defined as the weighted average [cf. §1.8]

$$\bar{X} = \sum_i f_i X_i / \sum_i f_i \qquad (3.67)$$

without necessarily requiring that the distribution (3.66) be normalized.

Now Greene and Callen noticed that

$$\frac{\partial f_i}{\partial S_{\bar{X}}} = -(X_i - \bar{X})f_i. \tag{3.68}$$

This relation can be used to evaluate the covariance in the following way:

$$
\begin{aligned}
\overline{(\Delta X)(\Delta Y)} &= \sum_i f_i (X_i - \bar{X})(Y_i - \bar{Y}) \Big/ \sum_i f_i \\
&= -\sum_i \frac{\partial f_i}{\partial S_{\bar{X}}}(Y_i - \bar{Y}) \Big/ \sum_i f_i \\
&= -\frac{\partial}{\partial S_{\bar{X}}} \sum_i f_i (Y_i - \bar{Y}) \Big/ \sum_i f_i - \frac{\partial \bar{Y}}{\partial S_{\bar{X}}} \\
&= -1/S_{\bar{X}\bar{Y}} \tag{3.69}
\end{aligned}
$$

by definition of the expected value, (3.67).

Instead of following this iterative procedure, it is much neater (and simpler) to observe that the error law (3.66) generates the moments simply by differentiation with respect to the intensive variables, $S_{\bar{X}}$ and $S_{\bar{Y}}$. In other words, we invert the dependencies in (3.66) and consider the error law as a function of the intensive parameters. For instance, the second derivative of the error law with respect to $S_{\bar{X}}$ is

$$\frac{\partial^2 f_i}{\partial S_{\bar{X}}^2} = (X_i - \bar{X})^2 f_i + \left(\frac{\partial \bar{X}}{\partial S_{\bar{X}}}\right)_{S_{\bar{Y}}} f_i.$$

Summing over all i, the left-hand side vanishes and we get the expression for the second moment as

$$\overline{(\Delta X)^2} = -\left(\frac{\partial \bar{X}}{\partial S_{\bar{X}}}\right)_{S_{\bar{Y}}}.$$

Furthermore, the third moment is obtained directly by differentiating three times and summing:

$$\overline{(\Delta X)(\Delta Y)(\Delta Z)} = \left(\frac{\partial^2 \bar{X}}{\partial S_{\bar{Z}} \partial S_{\bar{Y}}}\right)_{S_{\bar{X}}}.$$

3.7.1   Analogy with Parameter Estimation

In order to make the inversion of the functional dependencies in the error function more palatable, we can draw an analogy with a well known method of parameter estimation. By inverting the functional dependency of a probability distribution, in which the parameters upon which the distribution depends are taken as the variables and the realizations of the random variables as the parameters, we can construct a "likelihood" function. Since the random variables are independent and identically distributed, the likelihood function will be the product of the probability distributions for the entire sample. The method of maximum likelihood has already been introduced in §3.3.2 for parameter estimation, and we will return to it in §4.3 in connection with estimating the "state of nature" when a previously isolated system is placed in contact with a reservoir. Since the error law is already evaluated at the maximum likelihood value of the intensive parameters, all we can show is the compatibility of treating the error law as a function of the intensive parameters and the procedure followed in the method of maximum likelihood.

By way of illustration, consider fluctuations in the energy and volume at a fixed value of the particle density. Following the same procedure as in §3.6, the error law,

$$f_i = A \exp \left\{ S(U_i, V_i) - S(\bar{U}, \bar{V}) - S_{\bar{U}}(U_i - \bar{U}) - S_{\bar{V}}(V_i - \bar{V}) \right\},$$

can be rearranged to read

$$f_i = \Omega(U_i, V_i) \exp \left\{ -S_{\bar{U}} U_i - S_{\bar{V}} V_i - S[S_{\bar{U}}, S_{\bar{V}}] \right\}$$

by introducing the Massieu transform,

$$S[S_{\bar{U}}, S_{\bar{V}}] = S(\bar{U}, \bar{V}) - S_{\bar{U}} \bar{U} - S_{\bar{V}} \bar{V},$$

of the entropy with respect to the energy and volume. Since we are interested in the error law as a function of the intensive parameters, we will not need the explicit form of the *a priori* probability distribution, $\Omega(U_i, V_i)$.

Let us recall that we are starting with an isolated system described by $\Omega(U_i, V_i)$ which contains all the information about the mechanical state of the system. We then place the system in contact with reservoirs that are characterized by the parameters $S_{\bar{U}}$ and $S_{\bar{V}}$. These are the parameters which maximize the probability distribution and lead to the error law, as we shall now show.

Taking the logarithm of the error law, we obtain

$$\ln f_i = -S_{\bar{U}} U_i - S_{\bar{V}} V_i - S\left[\frac{1}{\bar{T}}, \frac{\bar{P}}{\bar{T}}\right] + \ln \Omega(U_i, V_i). \tag{3.70}$$

Normally, we would first take the product of the distributions to construct the likelihood function and then differentiate with respect to the parameters to obtain the likelihood equations. Taking the product of the distributions allows us to construct our statistic which will be used in the estimation of the intensive, reservoir parameter. However, we can reverse the procedure and differentiate with respect to $S_{\bar{U}}$, or $S_{\bar{V}}$, and then sum to get

$$\frac{1}{r} \sum_i^r \frac{\partial \ln f_i}{\partial S_{\bar{U}}} = -\frac{1}{r} \sum_i^r U_i - \left(\frac{\partial S[S_{\bar{U}}, S_{\bar{V}}]}{\partial S_{\bar{U}}} \right)_{S_{\bar{P}}} = 0. \tag{3.71}$$

For the exponential family of distributions, the sample mean $(1/r) \sum_i^r U_i$ will always coincide with the mean of the distribution given by the weighted average

(3.67). Thus, the likelihood equation (3.71) can be written as an expression for the first moment,

$$\bar{U} = -\left(\frac{\partial S[S_{\bar{U}}, S_{\bar{V}}]}{\partial S_{\bar{U}}}\right)_{S_{\bar{P}}}. \qquad (3.72)$$

Now the Massieu transform, $S[S_{\bar{U}}, S_{\bar{V}}] = -\bar{n}\bar{\mu}/\bar{T}$, satisfies the Gibbs–Duhem equation

$$\bar{U}\, d(1/\bar{T}) + \bar{V}\, d(\bar{P}/\bar{T}) - \bar{n}\, d(\bar{\mu}/\bar{T}) = 0, \qquad (3.73)$$

when $\bar{n}$ is held constant. Hence, the expected value of the energy (3.72) is none other than

$$\bar{U} = \bar{n}\left(\frac{\partial(\bar{\mu}/\bar{T})}{\partial(1/\bar{T})}\right)_{(\bar{P}/\bar{T})}.$$

An analogous expression is obtained for the expected volume.

At constant $\bar{n}$, the Gibbs–Duhem relation (3.73) becomes the total differential of $G/\bar{T}$, where $G = \bar{n}\bar{\mu}$ is the Gibbs potential, while for $\bar{V}$ constant, it becomes the total differential of $W/\bar{T}$, where $W = -\bar{P}\bar{V}$ is the grand-canonical potential. In both cases, the differential form of the Euler relation is obtained. Hence, the constancy of one of the extensive variables in the Gibbs–Duhem relation reduces it to the total differential of a potential. We will make use of this fact over and over again in the sequel.

Differentiating (3.70) twice and summing over all i give

$$\frac{1}{r}\sum_i^r \frac{\partial^2 \ln f_i}{\partial(1/\bar{T})^2} = \bar{n}\left(\frac{\partial^2(\bar{\mu}/\bar{T})}{\partial(1/\bar{T})^2}\right)_{(\bar{P}/\bar{T})} = -\bar{T}^2 C_{\bar{P}} < 0.$$

This shows that the error law is, indeed, a maximum at $\bar{T}^{-1}$. It is entirely straightforward to show that it is a maximum with respect to all the intensive parameters.

It is important to note that the normalizing terms in the error law introduce a potential, or generating function, which, at a fixed value of the extensive variable, is proportional to the conjugate intensive variable. In §3.6 we saw that for fluctuations in the energy and particle number, it was the grand-canonical potential divided by the temperature, $W/\bar{T} = -\bar{P}\bar{V}/\bar{T}$, which, at constant volume, is a function of $(1/\bar{T})$ and $(\bar{\mu}/\bar{T})$ through the Gibbs–Duhem relation (3.73).

Here, we have seen that for fluctuations in the energy and volume, at a constant number of particles, the generating function is the negative of the Gibbs free energy divided by the temperature, $-G/\bar{T} = -\bar{n}\bar{\mu}/\bar{T}$, which is a function of the other pair of intensive variables, $1/\bar{T}$ and $\bar{P}/\bar{T}$. In fact, from the symmetry of the Gibbs–Duhem relation (3.73) we might be tempted to try to construct an ensemble by considering fluctuations in the volume and particle number at constant energy.[4] We shall now see why this cannot be

---

[4] Recall from §1.8 that the entropy cannot be concave in all three variables simultaneously.

done; we preface our discussion by remarking that the parameter which we would introduce, namely the inverse temperature, by way of the generating function already appears in the intensive parameters of the reservoirs with which the system has been placed in contact. Since it is quite impossible to define a function in terms of itself, no such function exists. In other words, the temperature is a more primitive concept than either the pressure or chemical potential.

3.7.2 A Fifth Ensemble?

If fluctuations in the volume and particle number could occur simultaneously, at a constant value of the energy, it would constitute a fifth ensemble, obtained by placing the system in contact with a particle reservoir that would lead to a state of uniform chemical potential and a movable partition which would equilibrate the pressure. We can express our initial reservations against such an ensemble since there can be no matter transport without energy exchange. However, let us remain open-minded about such a possibility.

A system is fitted with a movable partition which is also permeable to the passage of particles, while restrictive with respect to the energy. At most, such a system can only be realized if there are compensations in changes in volume with those in the particle number. The putative error law would be

$$f_i = A \exp \left\{ S(V_i, n_i) - S(\bar{V}, \bar{n}) - S_{\bar{V}}(V_i - \bar{V}) - S_{\bar{n}}(n_i - \bar{n}) \right\},$$

whose generating function is

$$S \left[\frac{\bar{P}}{\bar{T}}, \frac{\bar{\mu}}{\bar{T}} \right] = S(\bar{V}, \bar{n}) - \frac{\bar{P}}{\bar{T}} \bar{V} + \frac{\bar{\mu}}{\bar{T}} \bar{n} = \frac{\bar{U}}{\bar{T}}$$

at constant $\bar{U}$.

This means that $\bar{T}^{-1} = \bar{T}^{-1}(\bar{P}/\bar{T}, \bar{\mu}/\bar{T})$ and the differential of the generating function is simply the Gibbs–Duhem equation (3.73) at constant energy. In the case of fluctuations in the energy and particle number, the generating function is defined in terms of the pressure as a function of the temperature and chemical potential. At constant volume, its differential gives the Gibbs–Duhem relation. Finally, in the case of fluctuations in the energy and volume, the generating function is defined in terms of the chemical potential which is a function of the temperature and pressure. At constant particle number, its differential is, still again, the Gibbs–Duhem relation.

The apparent symmetry in the three cases stems from the ostensible equivalence of the three intensive variables in the Gibbs–Duhem relation, (3.73). In §1.8 we saw that the entropy cannot be considered as a concave function of all three extensive variables simultaneously. For in such an ensemble, the generating function would vanish identically on account of the Euler relation. Of the three independent extensive variables, it is the energy which has the unique role in the entropy representation. Thermal equilibrium must first be secured before mechanical equilibrium or equilibrium with respect to matter flow.

For a system at constant energy, all that can be surmised is that the ratio $\bar{P}/\bar{T}$, in the case of mechanical equilibrium, or $\bar{\mu}/\bar{T}$, in the case of equilibrium with respect to matter flow, must be constant throughout. But $\bar{P}$ and $\bar{\mu}$ have a physical significance and not their ratios with respect to the temperature. This is borne out when we consider fluctuations in the volume as well as the energy. There the generating function introduces a new intensive variable, the chemical potential, which is made a function of the temperature and pressure via the Gibbs–Duhem relation, (3.73). Likewise, in the situation where the particle number fluctuates as well as the energy, the generating function introduces a new intensive variable, the pressure, which is related to the temperature and chemical potential by means of the Gibbs–Duhem relation. Only in the case where the energy is maintained constant are we forced to reintroduce the temperature that has already appeared in ratios $\bar{P}/\bar{T}$ and $\bar{\mu}/\bar{T}$. This is logically inconsistent since the inverse temperature is made a function of these ratios and hence is used in its own definition!

In the energy representation, this situation would correspond to an adiabatic, movable permeable wall in the form of a piston that separates two subsystems. The adiabatic constraint necessitates our construction of the error law in the energy representation [cf. §1.12]. Such an error law is given by

$$f_i = B \exp\left(-\frac{U(V_i, n_i) - U(\bar{V}, \bar{n}) + \bar{P}(V_i - \bar{V}) - \bar{\mu}(n_i - \bar{n})}{\bar{T}}\right).$$

The generating function of such a putative ensemble would be

$$-\frac{U[\bar{P}, \bar{\mu}]}{\bar{T}} \equiv \left(\frac{\bar{\mu}\bar{n} - U - \bar{P}\bar{V}}{\bar{T}}\right)$$

$$= \ln \sum_i B \exp\left(\frac{-U(V_i, N_i) - \bar{P}V_i + \bar{\mu}n_i}{\bar{T}}\right)$$

$$= -S = \text{const.},$$

whose total derivative is

$$dU[\bar{P}, \bar{\mu}] = -\frac{\bar{T}\,d\bar{S} + \bar{V}\,d\bar{P} - \bar{n}\,d\bar{\mu}}{\bar{T}} + \frac{\bar{S}}{\bar{T}}\,d\bar{T} = -d\bar{S}$$

on account of

$$\bar{S}\,d\bar{T} = \bar{V}\,d\bar{P} - \bar{n}\,d\bar{\mu}, \tag{3.74}$$

which is the Gibbs–Duhem relation in the energy representation. This relation wipes out the dependence of $U[\bar{P}, \bar{\mu}]$ upon $\bar{P}$ and $\bar{\mu}$ leaving it a function only of the entropy which, for an adiabatic process, is constant. The logarithm of the partition function is $-U[\bar{P}, \bar{\mu}]/\bar{T}$ and hence is a pure constant. The total differential is precisely the Gibbs–Duhem relation (3.73), and consequently all moments would vanish. This is still yet another reason why fluctuations in the volume and particle number cannot occur simultaneously.

In other words, one of the independent variables in the generating function must be the inverse temperature itself and not in the ratio $\bar{P}/\bar{T}$ or $\bar{\mu}/\bar{T}$. In order to obtain the moments, every intensive variable must be paired with a factor $1/\bar{T}$, and differentiation is carried out with respect to one ratio, holding all other ratios of intensive variables and temperature constant. Energy exchange comes before all other types of exchanges, and this is what secures thermal equilibrium. Hence, one of the ratios in the generating function must be $1/\bar{T}$ itself. Thermodynamic sense must necessarily make physical sense!

Now if the pressure of the two subsystems are initially unequal and the piston is set free, it will move in the direction of the subsystem with lower pressure. In order to maintain the entropy in each subsystem constant, there must be a flow of particles in the direction in which the piston is moving. Hence, the covariance

$$\overline{(\Delta V)(\Delta n)} = -T\left(\frac{\partial \bar{n}}{\partial \bar{P}}\right)_{\bar{S},\bar{\mu}} < 0,$$

showing that volume and particle number fluctuations are *negatively* correlated. But this implies that $(\partial \bar{V}/\partial \bar{\mu})_{\bar{S},\bar{P}} < 0$, meaning that an increase in the chemical potential will cause a decrease in the volume under adiabatic conditions. However, an increase in the chemical potential is caused by an increase in the number of particles, and hence such a system could never come to equilibrium, let alone be able to predict the final temperatures of the individual subsystems. Temperature cannot ever be defined in an adiabatically isolated system. If the partition was made impermeable, the only way that the volume could change under adiabatic conditions would be to compensate the compressional work by the change in internal energy, $\Delta U = -\bar{P}\,\Delta V$. The changes in the volume would no longer be independent with those in the energy, and consequently, the error law for deviations in the volume would vanish.

It is quite remarkable that the forefathers of thermodynamics were able to exhaust all the physically realizable forms of ensembles! The impossibility of simultaneous fluctuations in the volume and particle number at constant energy or entropy was not even commented upon; even to Landau and Lifshitz, it sounded "obviously true." But no one seemed to think that it was worthwhile to inquire into the statistical repercussions of such an observation. Surely, it can't matter whether we count the number of entrants into a given volume or vary the volume until we have a predetermined number of particles in it. In the former case, the random variable is the particle number, while in the latter, it is the volume. Based on such considerations, we shall develop a statistical equivalence principle in §6.4 between fluctuations in the volume and particle number which, roughly speaking, allows us to interchange a Poisson distribution for the particle number with a continuous gamma density for the volume.

3.7.3   Is the Fundamental Relation Really Fundamental?

What is the advantage of considering the entropy a function of other variables than those given in the fundamental relation? First, let us realize that this involves a Legendre transform on the energy or a Massieu transform on the entropy. The transformed, intensive variable having been liberated could then fluctuate. Second, let us recall that the Massieu transform, as it is used in thermodynamics, converts what was a concave (convex) function in an extensive variable into a convex (concave) function of the conjugate intensive variable. When fluctuations in both extensive and intensive variables are considered, the quadratic form will have no definite sign and hence the concavity property is jeopardized along with the physical realizability of such an ensemble. Third, the generating function for such an ensemble would vanish identically on account of the Euler relation.

Suppose we want to consider the entropy as a function of the enthalpy, $\bar{H} = \bar{U} + \bar{P}\bar{V}$. Then we must also consider it a function of the pressure and particle number. Expanding the entropy in a truncated Taylor series expansion, we have

$$S(H_i, P_i, n_i) - S(\bar{H}, \bar{P}, \bar{n}) - S_{\bar{H}}(H_i - \bar{H}) - S_{\bar{P}}(P_i - \bar{P}) - S_{\bar{n}}(n_i - \bar{n})$$
$$= \tfrac{1}{2}\left[S_{\hat{H}\hat{H}}(H_i - \bar{H})^2 + S_{\hat{P}\hat{P}}(P_i - \bar{P})^2 + S_{\hat{n}\hat{n}}(n_i - \bar{n})^2 + \text{ cross terms}\right],$$

where $\hat{H} \in [H_i, \bar{H}]$, $\hat{P} \in [P_i, \bar{P}]$, and $\hat{n} \in [n_i, \bar{n}]$.

But, as we have mentioned, the entropy is no longer concave in the pressure so that it must be kept constant. At constant pressure, the error law is

$$f_i = A \exp\left\{S(H_i, n_i) - S(\bar{H}, \bar{n}) - \frac{1}{\bar{T}}(H_i - \bar{H}) + \frac{\bar{\mu}}{\bar{T}}(n_i - \bar{n})\right\},$$

which would appear to be reasonable were it not for the fact that the generating function vanishes identically,

$$S - (1/\bar{T})\bar{H} + (\bar{\mu}/\bar{T})\bar{n}$$
$$= \ln \sum_i A \exp\left\{S(H_i, n_i) - (1/\bar{T})H_i + (\bar{\mu}/\bar{T})n_i\right\} = 0,$$

on account of the Euler relation.

This can be seen in another way by calculating the covariance in the enthalpy and particle number. Since the order of differentiation of the error law with respect to $1/\bar{T}$ and $\bar{\mu}/\bar{T}$ is immaterial, we must have

$$\overline{(\Delta H)(\Delta n)} = \bar{T}\left(\frac{\partial \bar{H}}{\partial \bar{\mu}}\right)_{\bar{T}, \bar{P}} = -\left(\frac{\partial \bar{n}}{\partial \bar{T}^{-1}}\right)_{\bar{\mu}, \bar{P}}.$$

This means that there must be a potential, say Φ, whose differential is

$$d\Phi = -\bar{H}\, d(1/\bar{T}) + \bar{n}\, d(\bar{\mu}/\bar{T})$$

at constant pressure. If such a potential existed, it would be the generating function of the ensemble. But, alas, such a potential must be a function of all the intensive variables, and at constant pressure, the equation $d\Phi = 0$ is just the Gibbs–Duhem relation (3.73). We therefore conclude that we cannot tamper with the independent variables in the error law. Nor would it make sense to do so since the stochastic entropy must necessarily describe the purely mechanical features of a microcanonical ensemble. Intensive thermodynamic variables, like the temperature or the chemical potential, enter by placing the previously isolated system in contact with resevoirs, like a heat bath or a particle reservoir.

3.7.4   Moments Involving Intensive Variables

Since the Legendre, or Massieu, transform converts a concave (convex) function of an extensive variable into a convex (concave) function of its conjugate intensive variable, how can we ever hope to obtain information about fluctuations in the intensive variables? Before we can answer this question, we must ask why something which was hitherto considered a mere parameter can suddenly fluctuate? The parameters, which are characteristic of the reservoir, are to be estimated from observations made on their extensive conjugate variables. Since these "estimators"[5] are functions of our observations, they too must fluctuate.

Consider, for instance, the pair of conjugate variables of energy and (inverse) temperature. If we begin with an isolated system, the energy is known precisely while a temperature cannot be defined for it. Once it is placed in a heat bath, the energy of the closed system will begin to fluctuate. The larger the heat bath, the greater the imprecision of the energy while the more precise our knowledge of the temperature will be. What we are alluding to here is the notion of an uncertainty relation that exists between conjugate thermodynamic variables. Although the discussion of the uncertainty relations will be reserved until §4.9.2, what we want to emphasize here is that intensive variables can and do fluctuate. The question is what do the intensive quantities fluctuate about? The way we will develop the fluctuation theory is to transform Gauss' law of error for an extensive variable to one for an intensive variable by a Legendre transform. This does not necessarily mean that the property of the mean value of the extensive set of variables, as the most probable value of the quantity measured, will be extended to the intensive quantities. In §4.11.3 we shall give physical examples where this is not the case; here, we proceed along more formal lines although we will illustrate our method.

For reversible processes, the error law in the energy representation is

$$f_i = B \exp\left(-\frac{U(S_i, X_i) - U(\bar{S}, \bar{X}) - U_{\bar{S}}(S_i - \bar{S}) - U_{\bar{X}}(X_i - \bar{X})}{\bar{T}}\right), \quad (3.75)$$

---

[5]The term estimator in statistics is used to convey the fact that it is a random quantity.

where X can stand for *either* V *or* N. Let us recall that it is the convexity property of the energy which ensures that the probability distribution (3.75) will be a *proper* distribution. The mean square fluctuations in the entropy and volume $(X = V)$ are

$$\overline{(\Delta S)^2} = C_{\bar{P}} \tag{3.76}$$

$$\overline{(\Delta V)^2} = -\bar{T} \left(\frac{\partial \bar{V}}{\partial \bar{P}} \right)_{\bar{T}}. \tag{3.77}$$

Furthermore, the error law (3.75) gives the covariance in the entropy and volume as

$$\overline{(\Delta S)(\Delta V)} = \bar{T} \left(\frac{\partial \bar{V}}{\partial \bar{T}} \right)_{\bar{P}} = C_{\bar{P}} \left(\frac{\partial \bar{T}}{\partial \bar{P}} \right)_{\bar{S}}.$$

This shows that fluctuations in the entropy and volume will be positively correlated if the thermal expansion coefficient $\alpha = (1/\bar{V})(\partial \bar{V}/\partial \bar{T})_{\bar{P}}$ is positive. A positive value of α means that the temperature of a body falls in an adiabatic expansion.

Upon comparing (3.76) and (3.77) with those for the mean square deviations in their intensive conjugate variables, we will be able to relate Le Châtelier's principle to the uncertainty relations between conjugate thermodynamic variables. The difference between the two schemes is the type of processes considered. For instance, in the energy representation we deal with adiabatic volume changes whereas in the entropy representation we have free expansions.

Fluctuations in the Temperature and Pressure

By interchanging the roles of extensive and intensive variables, (3.75) can be converted into an error law for the deviations in the intensive variables from their equilibrium values. The convexity condition of the energy can be transferred to a concavity condition for the Gibbs potential, $G = \bar{\mu}\bar{n}$, as a function of $\bar{T}$ and $\bar{\mu}$ at constant $\bar{n}$ or to a concavity condition for the grand-canonical potential, $W = -\bar{P}\bar{V}$, as a function of $\bar{T}$ and $\bar{\mu}$ at constant $\bar{V}$. We will also be able to appreciate that since the construction of an ensemble depends upon the concavity or convexity of the potential defining that ensemble, these are the only two generalized canonical ensembles which can be constructed.

In other words, we begin with a system in which the temperature is fixed and observe fluctuations in the energy and either the particle number or volume at a constant value of the other quantity. In the former case, the chemical potential is held constant, while in the latter case, the pressure is maintained at a fixed value. In a system which is in contact with a reservoir by means of a rigid diathermal, permeable wall, if we can fix the energy and particle number, we will see fluctuations in the temperature and chemical potential. Or if we have a movable, impermeable diathermal wall, fixing the temperature and pressure will allow us to study fluctuations in the temperature and volume. The transfer from the extensive to the conjugate intensive variable as the independent variable will, of course, be performed by a Legendre or Massieu transform.

A convexity condition is still a convexity condition with the endpoints interchanged so that

$$U(\bar{S}, \bar{V}) - U(S_i, V_i) - T_i(\bar{S} - S_i) + P_i(\bar{V} - V_i) > 0$$

is an equally good candidate for constructing the error law. If we consider a given amount of matter, the number of particles is fixed and the volume can vary. Setting $X = V$ and adding and subtracting $\bar{T}\bar{S}$ and $\bar{P}\bar{V}$, which is tantamount to performing a double Legendre transform on the energy, in the exponent of (3.75), the negative of the convexity condition for U becomes the concavity condition

$$G(T_i, P_i) - G(\bar{T}, \bar{P}) - G_{\bar{T}}(T_i - \bar{T}) - G_{\bar{P}}(P_i - \bar{P}) < 0 \qquad (3.78)$$

for the Gibbs potential G, which we recall is a function of $\bar{T}$ and $\bar{P}$ at constant particle number via the Gibbs–Duhem relation (3.74). The Gibbs potential $G(\bar{T}, \bar{P}) = U(\bar{S}, \bar{V}) - \bar{T}\bar{S} + \bar{P}\bar{V}$ can be viewed as a double Legendre transform on the internal energy with respect to the entropy and volume.

We are thus lead to an error law of the form

$$f_i = B \exp\left(\frac{G(T_i, P_i) - G(\bar{T}, \bar{P}) - G_{\bar{T}}(T_i - \bar{T}) - G_{\bar{P}}(P_i - \bar{P})}{\bar{T}}\right) \qquad (3.79)$$

for deviations in the temperature and pressure from their most probable values which coincide with their thermodynamic values. The generating function is precisely $U/\bar{T}$, where the internal energy is a function of the entropy and volume since

$$\frac{G - \bar{S}\bar{T} - \bar{V}\bar{P}}{\bar{T}}$$
$$= \ln \sum_i B \exp\left(\frac{G(T_i, P_i) + \bar{S}T_i - \bar{V}P_i}{\bar{T}}\right) = \frac{U(\bar{S}, \bar{V})}{\bar{T}}.$$

We are beginning to see a beautiful symmetry emerging between the potential which determines the error law and the generating function of the ensemble.

The mean square fluctuations in the temperature and pressure are

$$\overline{(\Delta T)^2} = \bar{T}\left(\frac{\partial \bar{T}}{\partial \bar{S}}\right)_{\bar{V}} = \bar{T}^2/C_{\bar{V}}, \qquad (3.80)$$

$$\overline{(\Delta P)^2} = -\bar{T}\left(\frac{\partial \bar{P}}{\partial \bar{V}}\right)_{\bar{S}}, \qquad (3.81)$$

respectively. Expression (3.80) explains why measurements in the temperature become increasingly more precise for increasingly large systems. In the limit, there is a negligible difference between the Boltzmann and Gibbs definitions of temperature. The former defines the inverse temperature as the derivative of the logarithm of the density of states,

$$\frac{1}{\bar{T}} = \frac{\partial \ln \Omega(\bar{U})}{\partial \bar{U}},$$

evaluated at its *constrained* maximum at $\bar{U}$, while the latter defines the temperature implicitly through the likelihood equation (3.71).

We may verify the expression for the mean square fluctuation in the pressure by finding the explicit error law (3.79) in the case of an ideal gas. Assume, for simplicity, that the temperature is maintained constant. According to (3.61), the Gibbs potential as a function of the pressure $\bar{P}$ is

$$G = \bar{n}\bar{T}\ln\bar{P} + \text{const.}$$

Using the equation of state of an ideal gas, $\bar{P} = \bar{n}\bar{T}/\bar{V}$, the error law can be written as

$$f_i = B\left(\frac{\bar{V}P_i e}{\bar{n}\bar{T}}\right)^{\bar{n}} e^{-\bar{V}P_i/\bar{T}}. \tag{3.82}$$

Since $\bar{n}$ is certainly a large number, we can use Stirling's approximation and write $\bar{n}! = \bar{n}^{\bar{n}}e^{-\bar{n}}$. The question is what is the normalizing function B?

By the way in which the error law is derived, B can at most be a function of P_i. It is apparent that if we set $B = \bar{n}/P_i$, then (3.82) becomes the gamma density,

$$f_i = \left(\frac{\bar{V}}{\bar{T}}\right)^{\bar{n}} \frac{P_i^{\bar{n}-1}}{\Gamma(\bar{n})} e^{-\bar{V}P_i/\bar{T}}. \tag{3.83}$$

But why this choice? In §4.11.3 we shall appreciate that interchanging the endpoints in the error law with the subsequent application of the Legendre or Massieu transform has the effect of converting the error law into a Bayesian probability density for intensive parameters. According to Bayes' theorem, or the "principle of inverse probability," this probability density is proportional to a product of a likelihood function and a prior density for the intensive quantity. The prior probability density is proportional to B, and if nothing is known about it except that the parameter may take on any value on the right half line, then the most noncommittal choice of the prior probability would be to set it equal to the logarithm of the parameter and hence $B \propto 1/P_i$.

We do not want to enter here into a full discussion of the subjective interpretation of the error law (3.79), as opposed to the frequency interpretation of its extensive counterpart (3.75). We do, however, want to emphasize that (3.83) is a proper probability density and that it gives the correct moments. Using the equation of state of an ideal gas, (3.81) gives the mean square fluctuation in the pressure as

$$\overline{(\Delta P)^2} = \bar{n}\left(\frac{\bar{T}}{\bar{n}}\right)^2,$$

which is precisely the variance of the gamma density (3.83). We cannot use the first moment, $\bar{P} = \bar{n}\bar{T}/\bar{V}$, as a confirmation that we are on the right track since we used this in the derivation of the gamma density (3.83). But we can employ it in an investigation as to what causes the fluctuations in the pressure and what is the most probable value $\bar{P}$. If the fluctuations in the pressure are caused by fluctuations in the volume, then the most probable value of the pressure will be the *harmonic* mean since the most probable value of the volume is the *arithmetic* mean. If, on the other hand, fluctuations in the pressure are due to fluctuations in the particle number, then the most probable value of the pressure will be the harmonic mean since the most probable value of the particle number is the arithmetic mean. As an example of the latter, we can cite Dalton's law of partial pressures.

By introducing an additional piece of information, which is not contained in the error laws themselves when they are written in the energy representation, we can show that the product of the standard deviations in the extensive and intensive conjugate variables satisfies an inequality which has a remarkable resemblance to an uncertainty relation. As we have mentioned, the energy representation applies strictly to reversible processes. With the additional information coming to us in the form of Le Châtelier's principle, which comes very close to making statements about the dynamics of the processes and in which direction the system will proceed if perturbed from equilibrium, we will be able to obtain an inequality that bears a very close resemblance to a thermodynamic uncertainty relation. Thermodynamic uncertainty relations appear much more naturally in the entropy representation and will be discussed in much greater detail in §4.9. However, the second-order derivatives have a more direct connection with thermodynamically measurable quantities in the energy representation, and it is this which enables us to employ Le Châtelier's principle.

Multiplying (3.80) by (3.76), we find

$$\frac{\overline{(\Delta S)^2}\,\overline{(\Delta T)^2}}{\bar{T}^2} = \frac{C_{\bar{P}}}{C_{\bar{V}}},$$

and multiplying (3.81) by (3.77) results in

$$\frac{\overline{(\Delta V)^2}\,\overline{(\Delta P)^2}}{\bar{T}^2} = \left(\frac{\partial \bar{V}}{\partial \bar{P}}\right)_{\bar{T}} \left(\frac{\partial \bar{P}}{\partial \bar{V}}\right)_{\bar{S}}.$$

We now appeal to Le Châtelier's principle, which states that an external interaction which perturbs the system from equilibrium elicits processes within the system that tend to restore it to a state of equilibrium by reducing the effects of the interaction. Stated mathematically,

$$\left|\left(\frac{\partial S_{\bar{X}}}{\partial \bar{X}}\right)_{\bar{Y}}\right| > \left|\left(\frac{\partial S_{\bar{X}}}{\partial \bar{X}}\right)_{S_{\bar{Y}}}\right|, \tag{3.84}$$

where $S_{\bar{\gamma}}$ is the equilibrium value of the force. In other words, the force which perturbs the system from equilibrium is the actual value of the force less its equilibrium value, or the net perturbing force vanishes at equilibrium. Now Le Châtelier's inequality, (3.84), ensures that

$$C_{\bar{P}} > C_{\bar{V}}$$

as well as the fact that the adiabatic compressibility is always smaller in absolute value than the isothermal compressibility,

$$\left| \left(\frac{\partial \bar{V}}{\partial \bar{P}} \right)_{\bar{T}} \right| > \left| \left(\frac{\partial \bar{V}}{\partial \bar{P}} \right)_{\bar{S}} \right|.$$

In turn, these inequalities imply

$$\sqrt{\overline{(\Delta X)^2}} \sqrt{\overline{(\Delta U_X)^2}} > k\bar{T}, \qquad (3.85)$$

where we have reinstated Boltzmann's constant to emphasize that the *product of the standard deviations of the extensive and conjugate intensive variables must always be greater than the thermal energy.*

Were it not for the presence of the temperature, (3.85) would have the appearance of an uncertainty relation between conjugate thermodynamic variables. If we could derive the uncertainty relations in some other way, we would have a proof of Le Châtelier's principle. In the entropy representation, the temperature is absorbed into the definition of the intensive variable as the ratio of the intensive variable and the temperature, and we come out with a genuine uncertainty relation. For instance, the uncertainty between simultaneous measurements of the pressure and volume is

$$\sqrt{\overline{(\Delta V)^2}} \sqrt{\overline{(\Delta P/T)^2}} > k$$

in the entropy representation. However, in the energy representation, (3.85) cannot be reduced to establishing Boltzmann's constant as a lower bound on the uncertainty between fluctuations in conjugate variables. It is entirely comprehensible that we cannot derive the *inequality* in uncertainty relations in the energy representation since, as we shall see in §4.9, the inequality refers strictly to *irreversible* processes.

Recalling our discussion in §1.8, (3.78) is one definition of concavity. An alternative, yet equivalent, definition of strict concavity is given in terms of the quadratic form:

$$G_{\hat{T}\hat{T}}(T_i - \bar{T})^2 + 2G_{\hat{T}\hat{P}}(T_i - \bar{T})(P_i - \bar{P}) + G_{\hat{P}\hat{P}}(P_i - \bar{P})^2 < 0, \qquad (3.86)$$

where $\hat{T} \in [T_i, \bar{T}]$ and $\hat{P} \in [P_i, \bar{P}]$. Since (3.78) and (3.86) are equivalent, we can introduce (3.86) directly into Gauss' error law (3.79).

And the surprising thing is that if we evaluate the differential coefficients at their equilibrium values, the second-order moments calculated from (3.79) and those calculated from the normal approximation to the error law,

$$
f_i = \frac{\sqrt{\|G\|}}{2\pi \bar{T}} \exp \left\{ -\frac{1}{2\bar{T}} \left[\left(\frac{\partial \bar{S}}{\partial \bar{T}} \right)_{\bar{P}} (T_i - \bar{T})^2 - 2 \left(\frac{\partial \bar{V}}{\partial \bar{T}} \right)_{\bar{P}} (P_i - \bar{P})(T_i - \bar{T}) \right. \right.
$$
$$
\left. \left. - \left(\frac{\partial \bar{V}}{\partial \bar{P}} \right)_{\bar{T}} (P_i - \bar{P})^2 \right] \right\}, \tag{3.87}
$$

are identical. In (3.87), the discriminant

$$
\|G\| = - \left(\frac{\partial \bar{S}}{\partial \bar{T}} \right)_{\bar{P}} \left(\frac{\partial \bar{V}}{\partial \bar{P}} \right)_{\bar{T}} - \left(\frac{\partial \bar{S}}{\partial \bar{P}} \right)_{\bar{T}}^2,
$$

and it is understood that $\bar{n}$ is kept constant throughout. Closer inspection shows, in the case of a single fluctuating variable, that the equivalence of the second moments is due to the fact that the first-order derivatives in the error law are evaluated in the reference state.

According to the normal approximation to the error law, (3.87), the mean square fluctuation in the temperature is

$$
\overline{(\Delta T)^2} = -\frac{\bar{T} G_{\bar{P}\bar{P}}}{\|G\|} = -\frac{\bar{T}(\partial \bar{V}/\partial \bar{P})_{\bar{T}}}{\|G\|}.
$$

And since

$$
-\frac{\partial(\bar{V}, \bar{T})}{\partial(\bar{P}, \bar{T})} \frac{\partial(\bar{P}, \bar{T})}{\partial(\bar{V}, -\bar{S})} = \left(\frac{\partial \bar{T}}{\partial \bar{S}} \right)_{\bar{V}},
$$

we can appreciate the fundamental result, first derived by Greene and Callen for second-order moments in the extensive variables, that the normal approximation to Gauss' error law gives the exact expressions for the second-order moments. Here, we have shown it to be valid also for the second-order moments in the intensive variables. Furthermore, we can also appreciate the pivotal role played by the concavity or convexity of the various thermodynamic potentials and the role of the error law, expressed in terms of the concavity or convexity of these potentials, in determining the magnitude of the fluctuations.

Interlude: Second-Order Moments from the Normal Approximation Are Exact

We will now pause to give the Greene and Callen proof of the equivalence of the second-order moments obtained from the exact error law and its normal approximation in the case where several variables fluctuate simultaneously. We will also describe the method employed by Landau and Lifshitz to evaluate pairs of fluctuating variables in the normal approximation.

Greene and Callen have shown that the approximate relation for the error law

$$f_i = \frac{\sqrt{\|S\|}}{2\pi} \exp\left\{ \tfrac{1}{2}\left[S_{\bar{X}\bar{X}}(X_i - \bar{X})^2 + 2S_{\bar{X}\bar{Y}}(X_i - \bar{X})(Y_i - \bar{Y}) \right.\right.$$
$$\left.\left. + S_{\bar{Y}\bar{Y}}(Y_i - \bar{Y})^2 \right]\right\}, \tag{3.88}$$

(which we have referred to as Einstein's formula for a spontaneous fluctuation from equilibrium in §1.11), gives the same second moment as the exact expression obtained from Gauss' law (3.66). According to Einstein's formula (3.88), the second moments will be given by

$$\overline{(\Delta X)(\Delta Y)} = -[S]_{\bar{X},\bar{Y}}^{-1} = -(\text{cofactor } S_{\bar{X},\bar{Y}})/\|S\|, \tag{3.89}$$

where $[S]^{-1}$ is the inverse matrix of the second derivatives of the entropy. We have seen that the general expression for the covariance is (3.69) which, under the assumption that the error law is a function of both extensive variables, can be written explicitly as

$$\overline{(\Delta X)(\Delta Y)} = \left(\frac{\partial \bar{X}}{\partial S_{\bar{Y}}}\right)_{S_{\bar{X}}}.$$

Now,

$$\left(\frac{\partial \bar{X}}{\partial S_{\bar{Y}}}\right)_{S_{\bar{X}}} = \frac{\partial(\bar{X}, S_{\bar{X}})}{\partial(S_{\bar{Y}}, S_{\bar{X}})}$$

$$= -\frac{\partial(\bar{X}, S_{\bar{X}})}{\partial(\bar{X}, \bar{Y})} \frac{\partial(\bar{X}, \bar{Y})}{\partial(S_{\bar{X}}, S_{\bar{Y}})}$$

$$= -\frac{(\partial S_{\bar{X}}/\partial \bar{Y})_{\bar{X}}}{\|S\|},$$

which proves the second-order moments that are obtained from the approximate expression for the error function (3.88) are, in fact, exact. The crucial point is to appreciate that it is the intensive variable which is being held constant in (3.69).

For small fluctuations, Landau and Lifshitz evaluate the coefficients of the quadratic term, in the energy representation, at their equilibrium values and observe that

$$U_{\bar{S}\bar{S}}(\Delta S)^2 + 2U_{\bar{S}\bar{V}}(\Delta S)(\Delta V) + U_{\bar{V}\bar{V}}(\Delta V)^2$$
$$= (\Delta U_{\bar{S}})(\Delta S) + (\Delta U_{\bar{V}})(\Delta V)$$
$$= (\Delta T)(\Delta S) - (\Delta P)(\Delta V).$$

They then obtain the approximate formula (3.88) in the form

$$f_i = \frac{\sqrt{\|U\|}}{2\pi T} \exp\left(\frac{(\Delta P)(\Delta V) - (\Delta T)(\Delta S)}{2T}\right), \tag{3.90}$$

where

$$||U|| = \frac{\partial(U_{\bar{S}}, U_{\bar{V}})}{\partial(\bar{S}, \bar{V})}$$

is the discriminant of the quadratic form.

Then choosing V and T, or S and P, as the pair of independent variables and expressing the deviations of the dependent variables as linear combinations of the deviations in the independent variables, they cast the exponent (3.88) as a sum of squares in the independent variables. The coefficients of these terms are related to their variances on account of the normal form of the distribution.

Since the convexity property of the energy, or the concavity property of the entropy, is involved in establishing the equivalence between the exact and approximate expressions for the second-order moments of extensive variables, the same should be true of the second moments of the intensive variables on the strength of the concavity of the Gibbs potential at constant particle number or the grand-canonical potential at constant volume.

Fluctuations in the Temperature and Chemical Potential

Let us now consider the opposite situation where we focus our attention on a given volume element of the system, thereby allowing the number of particles to vary. Thus $X = n$ will vary at a constant volume $\bar{V}$. At constant temperature and chemical potential, we can study fluctuations in the entropy and particle number, or by performing a double Legendre transform, we can study fluctuations in the temperature and chemical potential at constant values of the conjugate extensive quantities.

Setting $U_{\bar{X}} = \bar{\mu}$ in the error law and adding and subtracting the terms $\bar{T}\bar{S}$ and $\bar{\mu}\bar{n}$ in the exponent of (3.75) result in

$$f_i = B \exp\left(\frac{W(T_i, \mu_i) - W(\bar{T}, \bar{\mu}) - W_{\bar{T}}(T_i - \bar{T}) - W_{\bar{\mu}}(\mu_i - \bar{\mu})}{\bar{T}}\right), \quad (3.91)$$

where $W(\bar{T}, \bar{\mu}) = \bar{U} - \bar{T}\bar{S} - \bar{\mu}\bar{n} = -\bar{P}\bar{V}$, the grand-canonical potential. By now the appearance of a double Legendre transform, this time with respect to $\bar{S}$ and $\bar{n}$, is becoming familiar. Implicit in all the partial derivatives is that $\bar{V}$ is regarded as a constant throughout.

The generating function

$$\frac{W(\bar{T}, \bar{\mu}) + \bar{S}\bar{T} + \bar{n}\bar{\mu}}{\bar{T}} = \ln \sum_i B \exp\left(\frac{W(T_i, \mu_i) + \bar{S}T_i + \bar{n}\mu_i}{\bar{T}}\right)$$

$$= \frac{U(\bar{S}, \bar{n})}{\bar{T}}$$

again displays a symmetry between the potential that determines the error law and the generating function which determines the moments of the distribution. Whereas deviations in the entropy and particle number, from their most probable values, are determined in terms of the energy as a function of entropy and particle number, deviations in the temperature and chemical potential are determined in terms of the grand-canonical potential at constant volume. In the former case, the grand-canonical potential is the generating function, while in the latter case, the energy is the generating function at constant volume.

If we were to begin in the entropy representation and transform the error law for fluctuations in the energy and particle number into one for fluctuations in $1/T$ and μ/T, we would come out with

$$
f_i = A\exp\left\{ S\left[\frac{1}{\bar{T}},\frac{\bar{\mu}}{\bar{T}}\right] - S\left[\frac{1}{T_i},\frac{\mu_i}{T_i}\right] \right.
$$
$$
\left. -\bar{U}\left(\frac{1}{T_i}-\frac{1}{\bar{T}}\right) + \bar{n}\left(\frac{\mu_i}{T_i}-\frac{\bar{\mu}}{\bar{T}}\right) \right\}, \tag{3.92}
$$

where

$$
dS\left[\frac{1}{\bar{T}},\frac{\bar{\mu}}{\bar{T}}\right] = \bar{n}\, d\left(\frac{\bar{\mu}}{\bar{T}}\right) - \bar{U}\, d\left(\frac{1}{\bar{T}}\right) = \bar{V}\, d\left(\frac{\bar{P}}{\bar{T}}\right),
$$

by the Gibbs–Duhem relation (3.73) at constant volume. It is easy to see that $S[1/\bar{T},\bar{\mu}/\bar{T}] = -\bar{P}\bar{V}/\bar{T}$ and the generating function is $-S(\bar{U},\bar{V})$. It is also rather apparent that moment expressions calculated from the error law (3.91) in the energy representation have more of a physical significance than those calculated from the error function (3.92) in the entropy representation. Our aim here is to emphasize the symmetry between the potential which determines the error law and the generating function which determines its moments. This symmetry is shown in Table 3.1.

Table 3.1 Relation between Fluctuating Variables and Potentials

Variables	Potential	Generating Function	Constant
U,V	$S(U,V)$	$S[1/T,P/T]$	$\bar{n}$
U,n	$S(U,n)$	$S[1/T,\bar{\mu}/T]$	V
S,V	$U(S,V)$	$-G(T,P)$	$\bar{n}$
S,n	$U(S,n)$	$-W(T,\bar{\mu})$	V
T,P	$G(T,P)$	$U(\bar{S},\bar{V})$	$\bar{n}$
T,μ	$W(T,\mu)$	$U(\bar{S},\bar{n})$	V
$1/T,P/T$	$-S[1/T,P/T]$	$-S(U,V)$	$\bar{n}$
$1/T,\bar{\mu}/T$	$-S[1/T,\mu/T]$	$-S(U,\bar{n})$	V

Gauss' error law, (3.91), gives the exact expression for the mean square fluctuation in the chemical potential as

$$
\overline{(\Delta\mu)^2} = \bar{T}\left(\frac{\partial\bar{\mu}}{\partial\bar{n}}\right)_{\bar{S},\bar{V}}
$$

by differentiating twice and summing over all i. This must obviously agree with the expression obtained by differentiating the generating function twice with respect to the average number of particles,

$$\overline{(\Delta\mu)^2} = \bar{T}\left(\frac{\partial^2 U}{\partial\bar{n}^2}\right)_{\bar{S},\bar{V}},$$

at constant entropy and volume. And just as the convexity of the energy with respect to entropy and particle number guarantees the existence of an error law for these extensive quantities, the concavity of the grand-canonical potential ensures an error law for fluctuations in the temperature and chemical potential.

In the same way that we have shown that the normal approximation to the error law involving the extensive variables of entropy and particle number gives the exact expressions for the second-order moments, it can be shown that the normal approximation,

$$f_i = \frac{\sqrt{\|W\|}}{2\pi\bar{T}} \exp\left\{\frac{1}{2\bar{T}}\left[-\left(\frac{\partial\bar{S}}{\partial\bar{T}}\right)_{\bar{\mu}}(T_i-\bar{T})^2 - 2\left(\frac{\partial\bar{S}}{\partial\bar{\mu}}\right)_{\bar{T}}(\mu_i-\bar{\mu})(T_i-\bar{T})\right.\right.$$
$$\left.\left. - \left(\frac{\partial\bar{n}}{\partial\bar{\mu}}\right)_{\bar{T}}(\mu_i-\bar{\mu})^2\right]\right\},$$

gives the exact expressions for the mean square fluctuations in the temperature and chemical potential as well as their covariance. We leave this as an exercise for the reader.

Since the second moments obtained from the approximate normal form of the error law are exact, we can check the validity of the error law involving deviations in the temperature and pressure, (3.79), by comparing the predicted second-order moments with the corresponding expressions that are obtained from the well known method of Landau and Lifshitz.

For the mean square fluctuations in temperature and pressure we find

$$\overline{(\Delta T)^2} = \bar{T}^2/C_{\bar{V}},$$

$$\overline{(\Delta P)^2} = -\bar{T}\left(\frac{\partial\bar{P}}{\partial\bar{V}}\right)_{\bar{S}}.$$

Correlations in temperature and pressure fluctuations are given by the covariance,

$$\overline{(\Delta T)(\Delta P)} = -\bar{T}\left(\frac{\partial\bar{T}}{\partial\bar{V}}\right)_{\bar{S}}, \tag{3.93}$$

which could have been obtained in the Landau–Lifshitz scheme by taking P and T as the independent variables and using (3.89). The expression (3.93) shows that fluctuations in the temperature and pressure are positively correlated if the temperature of a body falls in an adiabatic expansion. This is supported by the fact when the independent variables are chosen to be $\bar{V}$ and $\bar{T}$, we have $(\partial\bar{T}/\partial\bar{V})_{\bar{S}} = -(\bar{T}/C_{\bar{V}})(\partial\bar{S}/\partial\bar{V})_{\bar{T}}$. Then using the Maxwell relation $(\partial\bar{S}/\partial\bar{V})_{\bar{T}} = (\partial\bar{P}/\partial\bar{T})_{\bar{V}}$ results in $(\partial\bar{T}/\partial\bar{V})_{\bar{S}} = -(\bar{T}/C_{\bar{V}})(\partial\bar{P}/\partial\bar{T})_{\bar{V}}$.

Since the heat capacity at constant volume is always positive, the covariance (3.93) will have the same sign as $(\partial \bar{P}/\partial \bar{T})_{\bar{V}}$.

Furthermore, we are in no way limited to second-order moments. A typical third-order moment would be

$$\overline{(\Delta T)^3} = \frac{\bar{T}^3}{C_V^2} + \bar{T}^2 \left(\frac{\partial^2 \bar{T}}{\partial \bar{S}^2} \right)_{\bar{V}}.$$

Our analysis also explains why fluctuations in S and P and T and V are statistically independent. Fluctuations in S and V must be simultaneously transformed into those in T and P so that there can be no correlations between S and P and T and V.

3.8   Particle Distinguishability

In §3.6 we contrasted the Poisson description against the canonical ensemble formalism. In relation to black radiation, this would correspond to the Wien and Rayleigh–Jeans limits of the spectrum. In §2.3.2 we showed that the linear term in the variance of the energy was due to the particle nature of the radiation, described by a Poisson distribution for the number of particles, while the second term was related to the wave nature of the radiation, described by a gamma density for the energy. Since there is no trace of particles in the canonical ensemble, how can we talk about particles being distinguishable or indistinguishable?

The argument that particles of an ideal gas are, in fact, indistinguishable is the following. From the expression for the total number of particles, (3.60), we define the *single-particle* partition function as

$$\mu = T(\ln N - \ln \mathcal{Z}_1). \tag{3.94}$$

This is the Gibbs potential, $G = \mu N$, or

$$G = TN(\ln N - \ln \mathcal{Z}_1),$$

per particle. Introducing the Helmholtz potential, $A = G - PV$, and using the equation of state of an ideal gas, $PV = NT$, we have

$$A = TN(\ln N - 1 - \ln \mathcal{Z}_1).$$

If N is large enough so that Stirling's approximation can be used, the expression for the Helmholtz potential becomes

$$A = -T \ln \mathcal{Z}_N,$$

where $\mathcal{Z}_N = \mathcal{Z}_1^N/N!$ is the N-particle partition function.

For "corrected" Boltzmann counting the product of the single-particle partition functions has to be divided by $N!$, the factorial of the number of *identical* particles. It is hard to believe that by writing $\mathcal{Z}_N$ as the product $\mathcal{Z}_1^N$, the number of states of the system has been overcounted $N!$ because there are $N!$ ways of permuting the particles among themselves which are physically indistinguishable. Permutations of single-particle partition functions have neither a physical nor a statistical sense since they are a function only of the temperature and volume and are not objects to be permuted.

Using (3.94) to eliminate the chemical potential in (3.65) and rearranging, we have

$$f(\varepsilon) = \frac{\bar{n}(\varepsilon)}{N} = \beta \frac{(\beta\varepsilon)^{1/2}}{\Gamma(\frac{3}{2})} e^{-\beta\varepsilon} = \frac{e^{-\beta\varepsilon}}{\mathcal{Z}_1(\beta)} m(\varepsilon), \qquad (3.95)$$

where we have identified $\bar{n}(\varepsilon)/N$ with the probability density of finding a system with energy ε in the canonical ensemble. The role of the structure function is now played by the density of states, $m(\varepsilon) = C\varepsilon^{1/2}/\Gamma(\frac{3}{2})$, for a system with three degrees-of-freedom, and the single-particle partition function is $\mathcal{Z}(\beta) = C\beta^{-3/2}$, where C is a constant that depends on the specific nature of the mechanical system.

The probability distribution (3.95) is a gamma density. It was introduced into mathematical statistics by the astronomer and geodesist Friedrich Robert Helmert in 1876 and was rediscovered by Karl Pearson in 1900. Ironically enough, J. Willard Gibbs, who never dealt with statistical problems in astronomy or the social sciences, which were the primary objects of study of the nineteenth-century statisticians, could have claimed priority had not mathematical and physical statistics developed independently of one another. It is also the probability distribution in the Rayleigh–Jeans limit of black radiation, as we have shown in §2.3.2, where the corpuscular nature of the photons has all but disappeared. We have also had to reinterpret what we mean by the structure function, since the random variable is now the energy rather than the number of particles. From the analogy of black radiation we would expect this to be the least likely place to decide whether the particles are really distinguishable or indistinguishable. Yet if we can tag the particles by their energies, we will be able to decide in favor of the distinguishability of the particles which is in contradiction with the "corrected" method of Boltzmann counting.

Suppose we have an ideal gas of N particles enclosed in a box of volume V. By boring a small hole in our box and allowing r particles to escape while recording their energies, we obtain a random sample, $\varepsilon = \{\varepsilon_1, \varepsilon_2, \ldots, \varepsilon_r\}$, of size r. Here we are sampling without replacement. Since the particles are independent and identically distributed, the joint distribution for r observations is

$$dP(\varepsilon) = dF(\varepsilon_1)\, dF(\varepsilon_2) \cdots dF(\varepsilon_r), \qquad (3.96)$$

where the probability distribution is equipped with a density,

$$dF(\varepsilon_i) = f(\varepsilon_i)\, d\varepsilon_i.$$

Setting

$$\varphi_i = \int_0^{\varepsilon_i} dF(x) = F(\varepsilon_i),$$

we have that the φ's and the ε's are in the same order, since the distribution F is a nondecreasing function of ε. The joint distribution (3.96) can now be written as

$$dP = d\varphi_1 \, d\varphi_2 \cdots d\varphi_r. \tag{3.97}$$

Suppose now we order the particle energies so that $\varepsilon_i < \varepsilon_{i+1}$ for every i. These inequalities imply that the distribution functions are ordered such that

$$0 \le \varphi_1 \le \varphi_2 \le \cdots \le \varphi_r \le 1.$$

Since the probability distribution $P(\varepsilon)$ is continuous, we can assign a zero probability to the event that any two energies will be exactly equal. Integrating (3.97) over this domain,

$$\int_0^1 \cdots \left[\int_0^{\varphi_3} \left(\int_0^{\varphi_2} d\varphi_1 \right) d\varphi_2 \right] \cdots d\varphi_r,$$

gives the factor $1/r!$. Then for *ordered* statistics, the joint distribution of the φ's is

$$dP_{\text{ord}} = r! \, dF(\varepsilon_1) \, dF(\varepsilon_2) \cdots dF(\varepsilon_r).$$

Since the particles can be distinguished by their different energies, the joint probability density will be given by

$$
\begin{aligned}
r! \prod_{i=1}^r f(\varepsilon_i) &= \frac{r!}{Z_1^r} \prod_{i=1}^r e^{-\beta \varepsilon_i} m(\varepsilon_i) \\
&= \frac{r!}{Z_1^r} \left(\frac{C}{\Gamma(\frac{3}{2})} \right)^r \exp\left\{ -\beta r \bar{\varepsilon} + \frac{1}{2} \sum_{i=1}^r \ln \varepsilon_i \right\},
\end{aligned}
$$

where $\bar{\varepsilon} = (1/r) \sum_{i=1}^r \varepsilon_i$ is the sample mean energy. The factorial, $r!$, which has hitherto been introduced to take into account indistinguishability of the particles, actually arises because the particles can be distinguished according to their energies!

Bibliographic Notes

§3.1

In spite of its age,

- W.A. Whitworth, *Choice and Chance* (Hafner, New York, 1965),

which is a reprint of the 1901 5th edition, is still the classic textbook of elementary combinatorial analysis. Time has not faded its lucidity and completeness.

The combinatorial basis for Bose–Einsten and Fermi–Dirac statistics is discussed in §II.5a of

- W. Feller, *An Introduction to Probability Theory and Its Applications*, Vol. 1, 3rd ed. (Wiley, New York, 1968), which is a classic text in elementary probability theory. Feller, being a mathematician, observes the peculiar use of the term "statistics." He also alludes to the possibility of there being a third model for other types of particles which are not described by Bose–Einstein or Fermi–Dirac statistics, since neither can claim universality.

Such an intermediary model had been given in

- G. Gentile, junior, "Osservazioni sopra le statistiche intermedie," *Nuovo Cimento* **19**, 493-497 (1940).

The derivation Bose–Einstein statistics was given in

- S.N. Bose, *Z. Phys.* **26**, 178 (1924).

The incident regarding its publication (i.e., the rejection from *Philosophical Magazine* and its subsequent translation into German by Einstein) is well documented in

- A. Pais, *Subtle is the Lord...The Science and Life of Albert Einstein* (Oxford Univ. Press, New York, 1982), Ch. 23.

Even in those days, good work was rejected by narrow-minded referees, and it took the stature of an Einstein to grasp the importance of it.

The Ehrenfest–Oppenheimer paper referred to in the text is

- P. Ehrenfest and J.R. Oppenheimer, "Note on the statistics of nuclei," *Phys. Rev.* **37**, 338 (1931).

Pais points out in

- A. Pais, *Inward Bound* (Oxford Univ. Press, New York, 1986), p. 285

that Ehrenfest and Oppenheimer rediscovered the theorem on the statistics of nuclei that was first proved by Wigner in

- E.P. Wigner, *Math. Naturwiss. Anzeiger Ungar. Ak. Wiss.* **46**, 576 (1929).

The discovery of Fermi–Dirac statistics was made by Enrico Fermi who, several weeks after reading Pauli's work on his exclusion principle, asked Orso M. Corbino to present his note on a new form of statistics to the *Accademia Nazionale dei Lincei*. It was published in

- E. Fermi, "Sulla quantizzazione del gas perfetto monoatomico," *Rend. Lincei* **3**, 145-149 (1926),

which was followed by a longer paper in

- F. Fermi, "Zur Quantelung des idealen einatomigen Gases," *Z. Phys.* **36**, 902-912 (1926).

Gentile's application of his intermediate statistics was published posthumously in

- G. Gentile, junior, "Le statistiche intermedie e le proprietà dell'elio liquido," *Nuovo Cimento* **20**, 9 (1942).

The new identity of intermediate statistics under the pseudonym of "parastatistics" had its beginnings in

- O.W. Greenberg, *Phys. Rev. Lett.* **13**, 598 (1964);

- O.W. Greenberg and D. Zwanziger, *Phys. Rev.* **150**, 1177 (1966).

Ter Haar's negative result for intermediate statistics is given in a letter to the editor,

- D. ter Haar, "Gentile's intermediate statistics," *Physica* **17**, 199-200 (1952).

§3.2

Markov processes, master equations, and the like are discussed in almost every book on stochastic processes and probability theory. In our opinion, two of the clearest presentations are

- D.R. Cox and H.D. Miller, *The Theory of Stochastic Processes* (Chapman and Hall, London, 1965), Ch. 4;

- N.G. van Kampen, *Stochastic Processes in Physics and Chemistry* (North-Holland, Amsterdam, 1981), Chs. IV, V, and VI.

Our presentation follows rather closely the latter reference which is noted for its critical appraisal of the different stochastic formulations.

§3.3

The derivation of quantum statistics from genuine stationary probability distributions follows

- B.H. Lavenda, "Derivation of quantum statistics from Gauss' principle and the second law," *Int. J. Theoret. Phys.* **27**, 1371-1381 (1988).

In

- J. Güémez and S. Velasco, "A probabilistic method to calculate the probability distribution of a gas," *Eur. J. Phys.* **8**, 75-80 (1987)

the hypergeometric distribution was used in the description of a hard sphere gas. Using the hypergeometric distribution to account for particle repulsion in configuration space and the Pólya distribution to account for attractive interactions in momentum space, we derived van der Waals' equation in

- B.H. Lavenda, "A probabilistic derivation of van der Waals' equation," *Found. Phys. Lett.* **3**, 285-290 (1990).

For more details, see §7.7. The hypergeometric and Pólya distributions and their approximations are discussed in Feller's book in §II.6 and §V.8, respectively.

Einstein's derivation of Bose–Einstein statistics, in which he refers to the degeneracy parameter to take into account the conservation of particles, is given in

- A. Einstein, "Quantentheorie des einatomigen idealen Gases," *Sitzungsberichte, Akademie der Wissenschaften* 261-267 (1924).

§3.4

We again refer to Feller's *An Introduction to Probability Theory and Its Applications* Vol. 1, Ch. XI, for a more detailed discussion of generating functions.

§3.5

This section follows our article,

- B.H. Lavenda and J. Dunning-Davies, "The case against intermediate statistics," *J. Math. Phys.* **30**, 1117-1121 (1989).

§3.6

The discussion of this section follows

- B.H. Lavenda and J. Dunning-Davies, "Statistical thermodynamics without ensemble theory," *Ann. d. Physik* (in press).

§3.7

The reference to the Callen and Greene principle is from

- R.F. Greene and H.B. Callen, "On the formalism of thermodynamic fluctuation theory," *Phys. Rev.* **83**, 1231-1235 (1951).

They emphasize that this principle "is a fundamental theorem of statistical mechanics, rooted in the enormously high dimensionality of the phase spaces (speaking classically) of thermodynamic systems, and responsible for the fact that there is a single general thermodynamics, rather than a 'microcanonical thermodynamics' and a separate 'canonical thermodynamics.'"

The major result of the paper is that the second moments of the normal approximation to the error law for extensive variables, or Einstein's formula, are exact. Moreover, the higherorder moments calculated from Einstein's formula contain different powers of Boltzmann's constant so that they cannot be correct. They arrive at this conclusion by observing that "transformations among purely thermodynamic quantities cannot involve Boltzmann's constant," which belongs to the microscopic world of fluctuations and not to the macroscopic world of thermodynamics.

The quote taken from Landau and Lifshitz is found in

- L.D. Landau and E.M. Lifshitz, *Statistical Physics*, 2nd ed. (Pergamon, Oxford, 1969), §114.

In this section, they derive moment expressions based on the approximate expression (3.88) in the energy representation.

Every so often we come across in the literature attempts to construct new ensembles. One recent attempt,

- J.R. Ray and H.W. Graben, "Fourth Adiabatic Ensemble," *J. Chem. Phys.* **93** 4296-4298 (1990),

purports to have constructed an *adiabatic* ensemble that admits simultaneous fluctuations in the volume and particle number. We have shown in §3.7.2 that simultaneous fluctuations in the volume and particle number cannot occur at constant energy and there is even greater reason to discredit their occurrence under adiabatic conditions. The idea is to consider an ensemble with the chemical potential, pressure, and a "new" potential $R \equiv TS$ as independent variables. The potential R can be maintained constant only when the entropy varies inversely to the temperature, and this definitely goes against the grain of the second law. Alternatively, if the entropy is held constant, the differential of the entropy reduces to the Gibb–Duhem equation, while if the temperature is held constant, then a mere identity results. Therefore, we are led to the conclusion that the forefathers of statistical thermodynamics exhausted all the possibilities!

§3.8

This section follows our article

- B.H. Lavenda and J. Dunning-Davies, "Classical particles and order statistics," *Phys. Lett.* **140A**, 90-93 (1989).

For more details on order statistics, see

- H.A. David, *Order Statistics*, 2nd ed. (Wiley, New York, 1981).

Chapter 4

Processing Information

4.1 Introduction

In courses in statistical mechanics, one usually introduces the canonical distribution in the following way. Consider a space Γ of all microstates which are compatible with the given macrostate of the system. Define a measure $d\Gamma$ which is usually taken to be the classical phase space volume element

$$d\Gamma = dp_1 \cdots dq_m, \tag{4.1}$$

where p_i and q_i are, respectively, the momenta and coordinates of the individual molecules. Now, obviously we can't follow the individual molecules, nor would there be any point in doing so even if we could, for the question would arise of how the individual motions would combine to give us information on the drastically reduced number of macroscopic variables $\{E, \mathbf{X}\}$ which are the internal energy and the external parameters that are necessary to specify the state of the system.

The relation between the microscopic and macroscopic measures is

$$d\mathcal{V} = \Omega(E, \mathbf{X}) \, dE \, d\mathbf{X} = \int_{\mathcal{R}} d\Gamma, \tag{4.2}$$

where the region $\mathcal{R}$ contains all those microstates for which E is in dE and for which the extensive variables $\mathbf{X}$ are in $d\mathbf{X}$. Frequently, the number of particles as well as the volume are included in the set $\mathbf{X}$.[1] To simplify matters, let us suppose that we hold constant the set of extensive parameters $\mathbf{X}$. The individual energies of the molecules will usually turn out to be a quadratic function of the momenta and coordinates. Thus, the distance from the origin to any point in the phase volume will be proportional to $\sqrt{E}$. Just as the area of a circle is proportional to the square of its radius and the volume of a sphere

---

[1]It is the customary procedure to make the measure density a function of the energy, volume, *and* particle number. This, however, is incorrect as will be shown in §6.4. As far as fluctuations are concerned, the volume and particle number representations offer equivalent, though different, descriptions of the same phenomenon.

is proportional to the cube of its radius, so the volume of a hypersphere of radius $\sqrt{E}$ is proportional to the $2m$th power of its radius,

$$\mathcal{V} = \frac{(C \cdot E)^m}{\Gamma(m+1)}, \tag{4.3}$$

where C is a constant which depends upon the properties of the specific system under consideration and $2m$ is the number of degrees-of-freedom. Hence, the measure Ω is the "surface area" of the volume $\mathcal{V}$,

$$\Omega = \frac{C^m E^{m-1}}{\Gamma(m)}. \tag{4.4}$$

Given this measure, we seek to determine the equilibrium properties of the system.

Now, Ω has its roots in two worlds: the macroscopic one in which it is specified as a function of the energy and the set of external parameters and the microscopic one in which it is proportional to the number of complexions or the number of ways a given macroscopic state can be realized. As a procedure for inference, we want to predict that behavior which can happen in the greatest number of ways, consistent with what we know about the system (i.e., its macroscopically measurable variables). But in what sense is Ω a probability density for the occurrence of a macroscopic state for a given energy? Certainly, the measure (4.4) is not normalizable; that is, its integral over all values of the energy is not unity. In fact, it is clear from (4.4) that Ω is a positive and monotonically increasing function of the energy and it increases without bound as $E \to \infty$.

Now, in order to make the integral converge, we might try to multiply Ω by a factor $\exp(-\beta E)$, where β is a positive parameter, which can overpower it at some point $\hat{E}$. The integrand in

$$\mathcal{Z}(\beta) = \int_0^\infty e^{-\beta E} \Omega(E) \, dE \tag{4.5}$$

has an enormously sharp peak at E^*, and most of the contribution to the integral (4.5) comes from the immediate neighborhood about the peak. And since the integral (4.5) converges for some positive value of the parameter β, we may look upon

$$f(E; \beta) = \frac{e^{-\beta E}}{\mathcal{Z}(\beta)} \Omega(E) \tag{4.6}$$

as a *proper* probability density whose integral over all values of E is indeed unity. But what is the physics behind the mathematical trick of multiplying the rapidly increasing factor $\Omega(E)$ by an even more rapidly decreasing factor $\exp(-\beta E)$ and what is the physical significance of the parameter β which is introduced?

The crucial point is (4.4): "$\Omega(E)$ is an analytic function which does not increase faster than a certain power of E as $E \to \infty$." This hypothesis is due to the Russian mathematician A. I. Khinchin, and that physical systems satisfy such a hypothesis is a consequence of the central limit theorem. Provided that (4.4) holds, $e^{-\beta E}\Omega(E)$ can be expanded about its mean value. The mean value is finite and so too is the variance. As we have mentioned, there is a compromise between the increase in $\Omega(E)$ and the decrease in $\exp(-\beta E)$ so that one can approximate the conjugate distribution by

$$\frac{e^{-\beta E}}{\mathcal{Z}(\beta)}\Omega(E) = \left\{2\pi\overline{(\Delta E)^2}\right\}^{-1/2}\exp\left\{-(E-\bar{E})^2/2\overline{(\Delta E)^2}\right\}$$
$$+ \text{small order term}, \qquad (4.7)$$

which is the normal distribution. Hence, behind the hypothesis (4.4) lies the (local) law of large numbers. We will return to this point in the next section, but first we have to obtain expressions for the first, $\bar{E}$, and second, $\overline{(\Delta E)^2}$, central moments.

The probability density $f(E;\beta)$ is the *canonical* or *posterior* density as opposed to the *structure* function or *prior* density. The latter summarizes all the relevant information concerning the mechanical structure of the system *prior* to making an observation. In the simplest case, the observation is made on the energy, and the exponential function in (4.6) provides the coupling between the heat reservoir and our previously isolated system. In order to make a measurement, the system must be transformed in a closed one by placing it in contact with a heat bath. The reservoir parameter, β, is then to be estimated by observations made on the energy. Finally, to ensure that $f(E;\beta)$ is a *proper* probability density, there is the normalizing factor $\mathcal{Z}(\beta)$ which depends upon the type of system under investigation and the reservoir parameter β.

Multiplying (4.6) by E and integrating over all values of the energy give

$$\bar{E} = \int_0^\infty Ef(E;\beta)\,dE = \frac{1}{\mathcal{Z}(\beta)}\int_0^\infty Ee^{-\beta E}\Omega(E)\,dE. \qquad (4.8)$$

The average value of the energy $\bar{E}$, apart from the sign, can also be determined by differentiating the logarithm of (4.5) with respect to β. We then obtain

$$\bar{E} = -\frac{\partial}{\partial\beta}\ln\mathcal{Z}. \qquad (4.9)$$

This shows that the effect of multiplying Ω by the exponential decreasing factor is to construct a *generating* function whose derivatives with respect to the parameter β give the moments of the distribution.

We can easily see that the derivative of (4.9) gives

$$-\frac{\partial}{\partial\beta}\bar{E} = \frac{\partial^2}{\partial\beta^2}\ln\mathcal{Z}$$
$$= -\frac{1}{\mathcal{Z}^2}\left(\frac{\partial\mathcal{Z}}{\partial\beta}\right)^2 + \frac{1}{\mathcal{Z}}\frac{\partial^2\mathcal{Z}}{\partial\beta^2}$$

$$= -\bar{E}^2 + \frac{1}{\mathcal{Z}} \int_0^\infty E^2 f(E; \beta) \, dE$$

$$= -\bar{E}^2 + \overline{E^2} = \overline{E^2} - 2\bar{E}^2 + \bar{E}^2$$

$$= \overline{\left(E - \bar{E}\right)^2} \equiv \overline{(\Delta E)^2}. \tag{4.10}$$

This is the second moment about the mean value or what is usually referred to as the *dispersion*, or variance, of the distribution. The infinite number of higher moments can be calculated in a completely analogous way by continually differentiating $\ln \mathcal{Z}$ with a sign change for an odd number of derivatives. These moments will give more detailed information about the skewness of the distribution. Normally, we will not be concerned about such fine details since we will be interested in the limit of a large number of degrees-of-freedom where the canonical distribution becomes the normal distribution.

4.2   The Central Limit Theorem

Equipped with the expressions for the first and second central moments, (4.8) and (4.10), we may now show that the form of the structure function, (4.4), implies the central limit theorem. What we need is an explicit expression for the generating function $\mathcal{Z}(\beta)$. In doing so we will anticipate many of the results of later sections.

In order to bring out the symmetry between the phase space volume element (4.3) and the generating function (4.5), we rewrite the latter in terms of the former as

$$\mathcal{Z}(\beta) = \int_0^\infty e^{-\beta E} \, \mathcal{V}\{dE\}, \tag{4.11}$$

which we suppose exists for $\beta > 0$. The measure $\mathcal{V}$ is an improper distribution function (i.e., it is not normalizable), which is defined for $E \geq 0$. When inversion of (4.5) cannot be done explicitly, it is natural to look for asymptotic relations which are valid as $E \to \infty$. We will see that the behavior of $\mathcal{Z}(\beta)$ as $\beta \to 0$ determines the asymptotic form of $\mathcal{V}(E)$ as $E \to \infty$. This is qualitatively clear from the form of the Laplace transform (4.11). If β is very large, $e^{-\beta E}$ is negligible except when E is very small so that $\mathcal{Z}(\beta)$ will depend upon the behavior of $\mathcal{V}(E)$ near zero and vice versa.

Historically, the deduction of the limiting behavior of $\mathcal{V}(E)$ from that of $\mathcal{Z}(\beta)$ is called a Tauberian theorem in probability theory. Except in very simple cases, Tauberian theorems are difficult to handle since such theorems are valid only under restrictions on $\Omega(E)$ that prevent it from oscillating rapidly as $E \to \infty$. Fortunately enough, we have no such problem in thermodynamics since it will turn out that Ω is a monotonically increasing function of E in all systems of interest.

Let us introduce the reciprocal pair of variables t and τ such that $t\tau = 1$. The scaling $\beta \to \tau\beta$ in the Laplace transform (4.11) implies $E \to tE$. Hence, $\mathcal{Z}(\tau\beta)$ is the Laplace transform of the improper distribution function, $\mathcal{V}(tE)$.

This implies that $\mathcal{Z}(\beta)$ is some power of β while $\mathcal{V}(E)$ is some power of E. Since $\mathcal{Z}(\beta)$ decreases while $\mathcal{V}(E)$ increases as their respective variables increase, the power must be negative in the former and positive in the latter. It is also easy to see that they must be of the same absolute value.

Thus, for $m > 0$, $\mathcal{Z}(\beta) \sim \beta^{-m}$ as $\beta \to 0$ while $\mathcal{V}(E) \sim E^m$ as $E \to \infty$. We see that the behavior of $\mathcal{Z}(\beta)$ near the origin determines the asymptotic behavior of $\mathcal{V}(E)$. We will now show that the form of $\mathcal{V}$ implies the central limit theorem; that is, the canonical density, (4.6), is adequately described by the normal density.

Rearranging (4.7), we have

$$\Omega(E) = \frac{e^{\beta E}\,\mathcal{Z}(\beta)}{\left\{2\pi\overline{(\Delta E)^2}\right\}^{1/2}} \exp\left\{-\frac{(E - \bar{E})^2}{2\overline{(\Delta E)^2}}\right\},$$

where we have neglected the small order term since our intention is to evaluate the expression at $E = \bar{E}$. We then obtain[2]

$$\Omega(\bar{E}) = \frac{e^{\hat{\beta}\bar{E}}\,\mathcal{Z}(\hat{\beta})}{\sqrt{2\pi\partial^2 \ln \mathcal{Z}(\hat{\beta})/\partial\beta^2}}, \tag{4.12}$$

where $\hat{\beta}$ is that value of β determined from the implicit relation (4.9). Since we have determined the functional form of the generating function $\mathcal{Z}(\beta)$ to be β^{-m} as $\beta \to 0$, it follows that $\hat{\beta} = m/\bar{E}$ from (4.9). Then from (4.10) we find $\partial^2 \ln \mathcal{Z}(\hat{\beta})/\partial\beta^2 = \bar{E}^2/m$. Introducing these relations into (4.12) leads to

$$\Omega(\bar{E}) = \bar{E}^{m-1} m / \sqrt{2\pi m}\, m^m e^{-m}.$$

It will be appreciated that the denominator is *precisely* Stirling's expression for $m!$ so that, in fact, we recover the functional form of the structure function in (4.4). Thus, we may conclude that the form of the structure function implicitly implies that the "conjugate" density, (4.6), is amenable to the local law of large numbers, which says that the distribution function of a sequence of independent and identically distributed random variables tends to the normal law as the sequence increases without limit. This means that by our choice of the improper distribution, $\mathcal{V}(E)$, we are limiting ourselves to a single limit law—albeit a very important one.

The canonical probability density,

$$f(E; \beta) = \beta\frac{(\beta E)^{m-1}}{\Gamma(m)}e^{-\beta E}, \tag{4.13}$$

belongs to the family of gamma densities. When $m = 1$, the density is exponential. For $m < 1$, the probability density has no mode while for $m > 1$ it has a single mode at $E = (m-1)/\beta$. If we take the limit of (4.13) as $m, \beta \to \infty$, with

---

[2] $\partial \ln \mathcal{Z}(\hat{\beta})/\partial\beta$ denotes the derivative with respect to β evaluated at $\hat{\beta}$.

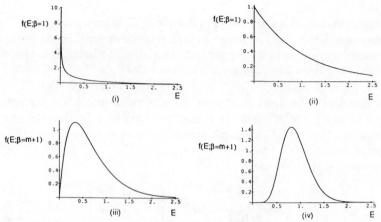

Figure 4.1: Four special cases of the gamma probability density of mean 1 for increasing values of m: (i) $m = \frac{1}{2}$; (ii) $m = 1$; (iii) $m = 2$; (iv) $m = 10$. Notice how the density becomes concentrated about the mean value as m increases.

m/β fixed and equal to the mean value $\bar{E}$, the canonical density will asymptotically approach the normal density centered at $\bar{E}$. Ultimately, as $m \to \infty$, all the probability will tend to concentrate at $E = \bar{E}$. This is shown in Fig. 4.1 for four special cases with increasing values of m.

The parameter m is usually of the order of Avogadro's number, but it is sufficient that it be at least about 10 so that we are able to apply Stirling's approximation to (4.13). Taking logarithms, we have

$$\ln f(E; \beta)$$
$$= m \ln(\beta E) - \left(m + \tfrac{1}{2}\right) \ln m - \tfrac{1}{2} \ln 2\pi - \beta E + O\left(m^{-1}\right), \quad (4.14)$$

since we are certainly allowed to neglect 1 with respect to m. Now introducing

$$E = \frac{1}{\beta}(m + x)$$

into (4.14) and assuming that $|x|/m \ll 1$, we expand the logarithm of $\ln(1 + x/m)$ as a power series and retain the first two terms in the series since they will be multiplied by m. We then obtain

$$f(E; \beta) = \frac{\beta}{\sqrt{2\pi m}} \exp\left\{ -\frac{\beta^2}{2m} \left(E - \frac{m}{\beta} \right)^2 \right\} \qquad (4.15)$$

upon exponentiating both sides, where the factor β is the Jacobian of the transformation. This is the normal density with precisely the same first and second moments of the original distribution.

It is interesting to observe that if we were not told what was the random variable in (4.14), we would certainly confuse the canonical probability density for E with the Poisson distribution for m in the case $m \gg 1$. The *continuous* random variable, E, has expectation m/β while the average of *discrete* random variable, m, is βE. We will discuss the relation between these two nonconjugate variables in §5.2; our purpose here is simply to show that we again obtain the normal form as $m \to \infty$. No matter how we look at it, the normal distribution seems to crop up everywhere in the asymptotic limit of a large number of degrees-of-freedom.

We now introduce

$$m = \beta E + z$$

into (4.14) to obtain

$$\ln f(m) = -\left(\beta E + z + \tfrac{1}{2}\right) \ln\left(1 + \frac{z}{\beta E}\right) + z - \tfrac{1}{2}\ln(2\pi\beta E) + O\left(m^{-1}\right).$$

Again supposing that the deviations for expected behavior are small, $|z|/\beta E \ll 1$, we expand the logarithm and keep only the dominant term. In this way we get

$$f(m) = \frac{1}{\sqrt{2\pi\beta E}} \exp\left\{-\frac{(m - \beta E)^2}{2\beta E}\right\},$$

in the limits as $m \to \infty$ and $|z|/\beta E \to 0$. This is again a normal probability density, but in contrast to (4.15), m is now the random variable instead of the energy, E. In either case, we have shown that with the increase in m, the probability becomes increasingly concentrated about its mean value. This is responsible for the success of thermodynamics, which deals only with mean values, and the predictions of statistical mechanics since the modes and means coincide. As we have seen in §1.4, the connection between thermodynamics and statistical mechanics is through the mean values.

Independent and identically distributed random variables approach other kinds of limit laws as the sample size increases without limit. These are known as the Lévy "stable" distributions. Like the prior probability density Ω, the strictly stable laws for positive random variables do not possess finite integral moments. For such limit laws it is impossible to speak about fluctuations about a mean value since it does not exist. We would not see any clustering about the mean value, and rare events would play a more prominent role than in the normal case. Thus, the expression for the phase space volume element, (4.3), *a priori* selects out the normal distribution as the limit law for "normal" thermodynamic systems. However, since other limit laws are known to exist in nature, it would be too restrictive to suppose that thermodynamics be limited to the normal case. Certainly, we would have to modify our notions of what phase space looks like, together with Boltzmann's principle, in order to achieve such an application of thermodynamics to systems which tend to stable laws other than the normal one. But such considerations would take us too far afield, and we will limit ourselves in this book to a classical expression of the form (4.3).

4.3 Maximum Likelihood

We remarked in §4.1 that the peak of the probability density (4.6) is so sharp that it becomes extremely difficult to distinguish the most probable value, or the mode, $\hat{E}$ from the mean or average value $\bar{E}$. Now, if we make a lot of measurements on the energy, the sum of them divided by the number of measurements (i.e., the "sample" mean) will be essentially indistinguishable from the mean of the distribution (4.8). The more measurements we make, the better the coincidence should be. But it is a peculiarity of densities of the form (4.6), which belong to a family known as exponential distributions, that the coincidence will occur for any finite number of observations—even for a single observation or a sample of size 1! We can thus choose the value of the parameter β so that the experimental value of the energy coincides with its expected value, given by (4.8), and use (4.9) to solve for β. This equation can be inverted to obtain β so long as (4.10) does not vanish.

This is our first encounter with a problem of "point" estimation in which we want to obtain a reasonable "guess" of the parameter β on the basis of observations made on the internal energy. A statistician would call β a "state of nature," and mathematical statistics provides us with a framework for determining the value of β which has the largest "likelihood" of occurrence [cf. §1.13 and §3.3.2].

Mathematically, the generating function, (4.5), is the Laplace transform of the prior measure, Ω. In statistical thermodynamics, these two quantities are given special names: the generating function, $\mathcal{Z}(\beta)$, is referred to as the partition function, and the improper prior density, $\Omega(E)$, is known as the structure function. As we have mentioned in the previous section, the structure function in (4.2) contains all the relevant information about the mechanical structure of the corresponding physical system. Usually, it is a monotonically increasing function from 0 to ∞ as E varies between these limits. A well known property of the Laplace transform is that if the integral (4.5) converges for one value of $\beta = \beta'$, it will do so for all values which are larger than β'. In actual physical systems, it is found that the integral converges for any $\beta > 0$.

Now, the importance in transferring our attention to the canonical density, $f(E; \beta)$ in (4.6), lies in the fact that we have taken an *improper* density, $\Omega(E)$, with infinite moments and transformed it into a *proper* distribution, $f(E; \beta)$, with finite moments. In order to accomplish this, we had to introduce a parameter β which, as we have mentioned, characterizes the heat reservoir in which our system has been placed in thermal contact. Once we identify β thermodynamically, this physical transformation, which is summarized in (4.6), will become perfectly transparent. In order to do so, we have to introduce the notion of what we mean by a "likelihood" function.

Parenthetically, a word is in order concerning the type of information that can be conveyed between the prior density, $\Omega(E)$, and the posterior density, $f(E; \beta)$. For the unphysical situation in which the energy has the value $E' < 0$, the prior density vanishes and so too will the posterior density. This is just a commonsense principle stating that if some event (the value observed for the internal energy in this case) is virtually impossible, then no evidence whatsoever can lend it credibility. Now back to the problem of the thermodynamic identification of β.

The logarithm of the ratio of the posterior to the prior densities defines what statisticians call the likelihood function, whose logarithm is

$$\mathcal{L}(\beta; E) \equiv \ln\left(\frac{f(E; \beta)}{\Omega(E)}\right) = -\beta E - \ln \mathcal{Z}(\beta). \tag{4.16}$$

In contrast to the posterior density, $f(E; \beta)$, the log-likelihood function, $\mathcal{L}(\beta; E)$, inverts the dependence between E and β. From the point of view of the likelihood function, E is the *parameter* and β is the *variable*! Of course, $e^{\mathcal{L}(\beta; E)}$ is not a probability distribution in the sense that $f(E; \beta)$ is, since the integral of the likelihood function over all values of β is not unity. Moreover, the value of the integral would depend upon the arbitrary scale of β. So we cannot talk about the likelihood of a value of β, attributing a greater probability for some values of β than others. Therefore, we must content ourselves with comparing the likelihoods of different values of β.

The method we are discussing is attributed to Sir Ronald Fisher and is known as the "theory of maximum likelihood." In essense, it is a takeoff of a much earlier method due to the Reverend Thomas Bayes which we will discuss in §4.8.1. Later on in this chapter, we will argue that since the energy is a fluctuating variable, so too will be its intensive, or conjugate, variable β, but first we must find out what the most likely value of β is and attribute to it a physical significance.

The true value of β will be the one with the largest likelihood. So we want to maximize the likelihood function with respect to β. Since the maximum of a positive function and its logarithm coincide, it will prove more convenient to maximize the logarithm of the likelihood function (4.16). The stationary condition,

$$\frac{\partial}{\partial \beta}\mathcal{L}(\beta; E) = -E - \frac{\partial}{\partial \beta}\ln \mathcal{Z}(\beta) = 0, \tag{4.17}$$

is known as the "likelihood" equation. But this is none other than the equation for the mean value (4.9) so that we can identify E with the mean $\bar{E}$ of the distribution.

Here we have touched on a subtle, but very important, property of probability densities of the form (4.6). Since the energies, E_i, are independent and identically distributed random variables, the sequence of observations $E_1, E_2, \ldots, E_r$ will give rise to a likelihood function whose logarithm is given by

$$\mathcal{L}(E_1, E_2, \ldots, E_r; \beta) = \sum_{i=1}^{r} \ln f(E_i; \beta),$$

dependent upon a single parameter β whose true, but unknown, value is $\hat{\beta}$. But because (4.6) is an *exponential* density, the data is summarized by the single statistic, $\bar{E}$, which is the sample average so that

$$\mathcal{L}(\bar{E}; \beta) = \sum_{i=1}^{r} \ln f(E_i; \beta) = -r[\beta \bar{E} + \ln \mathcal{Z}(\beta)]$$

is proportional to the sample size r.

In general, the sample mean will coincide with the mean of the distribution only in the limit as $r \to \infty$. But we have seen that the two coincide even in the case $r = 1$ for the exponential density (4.6). In fact, only exponential distributions possess the property of a "sufficient" statistic for any sample size—even a sample of size 1. We will discuss the property of sufficiency in greater detail in §4.9.2 and §4.10. Let it suffice here to note that the mean energy is a sufficient statistic for estimating the temperature since the conditional probability for finding the energy of any subsystem, given the total energy, $r\bar{E}$, is independent of their common temperature.

As we have already mentioned, it will be possible to solve the implicit relation (4.17) for the value of β as a function of the average energy, $\hat{\beta} = \beta(\bar{E})$, which in statistician's jargon is known as the *estimator*, by inverting it provided $\hat{\beta}$ is indeed the maximum likelihood estimate since, in this case,

$$\left(\frac{\partial^2 \mathcal{L}}{\partial \beta^2} \right)_{\beta = \beta(\bar{E})} = - \left(\frac{\partial^2 \ln \mathcal{Z}}{\partial \beta^2} \right)_{\beta = \beta(\bar{E})} < 0. \qquad (4.18)$$

This condition is indeed familiar to us since it is none other than the negative of the mean square fluctuation in energy [cf. Eq. (4.10)]. And because

$$\overline{(\Delta E)^2} = \overline{E^2} - \bar{E}^2 = \overline{(E - \bar{E})^2} > 0, \qquad (4.19)$$

the condition for a maximum, (4.18), will be satisfied. Strictly speaking, the inequality sign in this important convexity relation should be $\geq$, but since the inequality in (4.19) can be related to the thermodynamic stability criterion that the heat capacity should be positive, we will exclude the limiting case where an equality sign applies.

How many values of β can we expect to determine from the likelihood equation (4.17)? The answer lies in the properties of the partition function, (4.5). According to its definition,

1. $\mathcal{Z}(\beta)$ is positive,

2. $\mathcal{Z}(\beta)$ is a monotonically decreasing function of β,

3. $\mathcal{Z}(\beta) \to \infty$ as $\beta \to 0$, and

4. $\partial^2 \ln \mathcal{Z} / \partial \beta^2 > 0$ for $\beta > 0$.

From these properties of the generating function, we can draw conclusions about the properties of the log-likelihood function. According to properties 2 and 3, $\mathcal{L}(\beta; E) \to -\infty$ as $\beta \to 0$, and since

$$e^{-\mathcal{L}(\beta; E)} > e^{\beta E} \int_0^{E/2} e^{-\beta x} \Omega(x) \, dx > e^{\beta E/2} \int_0^{E/2} \Omega(x) \, dx,$$

it is also apparent that $\mathcal{L} \to -\infty$ as $\beta \to \infty$. Because of property 4, the log-likelihood function is a *concave* function of β since its second derivative coincides with that of $-\ln \mathcal{Z}$, as shown in (4.18). Hence, it must necessarily possess a *single* maximum at $\beta(\bar{E})$, which is the unique solution to the likelihood equation (4.17).

4.4   Entropy and Maximum Likelihood

Therefore, whatever $\beta(\bar{E})$ may turn out to be, it has a unique value determined from the likelihood equation (4.17). We must now make an appeal to thermodynamics, and in particular to the second law, to discover its thermodynamic significance. In order to appreciate the full content of the second law, we reinstate the dependence on the volume by making the energy a function of it. This can be criticized in view of our interpretation of the structure function as containing all the relevant information about the mechanical structure of the isolated system. We shall leave to §4.8 the task of showing the equivalence of this interpretation with the one we are about to present.

In order to arrive at the *desired* result, we write the partition function (4.5) as the phase average

$$\mathcal{Z}(\beta) = \int \cdots \int e^{-\beta E} \, dp_1 \cdots dq_m, \qquad (4.20)$$

where we have used the microscopic and macroscopic measure equivalence (4.2). The integration in (4.20) is extended over all "phases" or values of the momenta and coordinates. We assume that *the volume dependence is contained implicitly in the energy*. Differentiating both sides of (4.20) with respect to the volume, we obtain

$$\frac{\partial \mathcal{Z}}{\partial V} = -\beta \int \frac{\partial E}{\partial V} e^{-\beta E} \, dp_1 \cdots dq_m.$$

Dividing both sides by $\mathcal{Z}$ and β then gives

$$\frac{1}{\beta} \frac{\partial \ln \mathcal{Z}}{\partial V} = -\overline{\frac{\partial E}{\partial V}} = P, \qquad (4.21)$$

which defines the pressure, P.

We are now ready to identify the optimal value, $\hat{\beta} = \beta(\bar{E})$, of the parameter β. The total differential of the maximum log-likelihood function (4.16) is

$$d\mathcal{L}(\hat{\beta}; V) = -\left(\bar{E} + \frac{\partial}{\partial \beta} \ln \mathcal{Z}(\hat{\beta}, V) \right) d\hat{\beta} - \hat{\beta} \, d\bar{E} - \frac{\partial}{\partial V} \ln \mathcal{Z}(\hat{\beta}, V) \, dV.$$

But on account of the expressions for the mean value of the energy, (4.9), and the pressure, (4.21), this reduces to

$$d\mathcal{L}(\hat{\beta}; V) = -\hat{\beta} \left\{ d\bar{E} + P \, dV \right\}. \qquad (4.22)$$

We can now appreciate that since the left-hand side is a total differential, so too must be the right-hand side of (4.22).

By making appeal to:

- the first law of thermodynamics,

$$d\bar{E} = \delta Q + \delta W,$$

  where δQ is the quantity of heat added to the system in an infinitesimal transition between two neighboring states of equilibrium and $\delta W = -P\,dV$ is the mechanical work done on the system in effectuating the transition, and

- the second law,

$$\frac{\delta Q}{T} = dS,$$

  where T is the absolute temperature,

it becomes apparent that the optimal value of the parameter β must be a function of the absolute temperature. In fact, thermodynamics *defines* the absolute temperature by the relation

$$\hat{\beta} = \beta(\bar{E}) = 1/T, \qquad (4.23)$$

where T is measured in energy units.

We thus come out with the very intuitive result of equating the entropy with the negative of the maximum log-likelihood function

$$S(\bar{E}, V) = \ln \mathcal{Z}(\hat{\beta}, V) - \hat{\beta}\frac{\partial}{\partial\beta}\ln \mathcal{Z}(\hat{\beta}, V) = -\mathcal{L}(\hat{\beta}, V). \qquad (4.24)$$

Since the expression for the mean value of the energy, (4.9), enables us to express $\bar{E}$ and $\beta(\bar{E})$ in terms of one another, we have the option of considering the entropy as a function of $\bar{E}$, corresponding to a fundamental relation of thermodynamics, or as a function of $\beta(\bar{E})$, which is more natural in the canonical ensemble. In the way we have expressed the entropy in (4.24), as a Legendre transform of the logarithm of the generating function, the first option seems the more natural. Yet, there is something to be gained by considering a function with the same form as (4.24) for arbitrary β and E. Such a function will be shown, in the next section, to obey a minimax principle, being convex in β and weaking concave in E.

4.5   The Minimax Principle

In analogy with the way the mean "discrimination information" is defined,

$$\mathcal{I}(f) = \int f(E; \beta)\ln\left(\frac{f(E; \beta)}{\Omega(E)}\right)\,dE, \qquad (4.25)$$

as the expectation of the log-likelihood ratio $\ln[f(E;\beta)/\Omega(E)]$, we may define an entropy function by

$$
\begin{aligned}
\mathcal{S}(f) &= -\int f(E;\beta)\ln\left(\frac{f(E;\beta)}{\Omega(E)}\right)dE \\
&= -\int g(E)\ln g(E)\, d\mathcal{V}(E),
\end{aligned}
\tag{4.26}
$$

where g stands for the ratio, f/Ω, of the posterior to the prior distribution. Owing to the fact that Ω is a defective probability density, in that $\int \Omega(E)\,dE = \infty$ in the usual case where the system has more than a single degree of freedom, we cannot employ the simple inequality $x\ln(x/y) - x + y \geq 0$ to establish that (4.26) is negative or equivalently that (4.25) is positive. In fact, since $\Omega \gg 1$, which is the maximal value of f, (4.26) is nonnegative.[3] A perfect "resemblance" between the prior and posterior densities would imply a vanishing entropy. We now want to show that the optimal posterior density maximizes the entropy (4.26) in the sense of "largest value."

Developing the convex function $\varphi(g) = g\ln g$ in a truncated Taylor series expansion about

$$
g^\star = \frac{f^\star(E;\beta)}{\Omega(E)} = \frac{e^{-\beta E}}{\mathcal{Z}(\beta)}
\tag{4.27}
$$

results in

$$
\begin{aligned}
\varphi(g) &= \varphi(g^\star) + (g - g^\star)\varphi'(g^\star) + \tfrac{1}{2}(g - g^\star)^2\varphi''(\hat{g}) \\
&= g - g^\star - [\beta E + \ln\mathcal{Z}(\beta)]g + \tfrac{1}{2}(g - g^\star)^2\hat{g}^{-1},
\end{aligned}
$$

where $\hat{g}$ lies between g and $g^\star$. Integrating with respect to the "measure" $\mathcal{V}(E)$, we obtain

$$
S(\bar{E},\beta) - \mathcal{S}(f) \geq 0,
$$

where

$$
\mathcal{S}(f^\star) = S(\bar{E},\beta) = \beta\bar{E} + \ln\mathcal{Z}(\beta)
\tag{4.28}
$$

since

$$
\int (g - g^\star)^2\varphi''(\hat{g})\, d\mathcal{V}(E) = \int (f - f^\star)^2\hat{f}^{-1}\, dE \geq 0
$$

and

$$
\int f\, dE = \int f^\star\, dE = 1.
$$

It is important to bear in mind that we would have obtained (4.28) had we maximized (4.26) subject to the conditions (4.8) and that f is normalized. Viewed in these terms, β and $\ln\mathcal{Z}$ are the Lagrange multipliers for these constraints, respectively. In other words, the exponential distribution $f^\star$ maximizes the entropy subject to the two constraints. However, maximizing the

---

[3]If Ω were a *proper* probability density, one could employ Jensen's inequality $\mathcal{Z}(\beta) \geq e^{-\beta\bar{E}}$, for the convex function $e^{-\beta E}$, to establish the positive semidefiniteness of the entropy, $\beta\bar{E} + \ln\mathcal{Z}(\beta) \geq 0$.

entropy with respect to the constraints is not equivalent to Gauss' principle, contrary to what is commonly believed.

In the exponential distribution (4.27), β is an arbitrary, positive parameter, whereas in Gauss' principle, it is a solution to the stationary condition

$$\frac{\partial}{\partial \beta} S(\bar{E}, \beta) = 0. \tag{4.29}$$

Denoting this solution by $\beta(\bar{E})$ and introducing it into (4.27) give

$$f^\star[E; \beta(\bar{E})] = \frac{e^{-\beta(\bar{E})E}}{\mathcal{Z}[\beta(\bar{E})]} \Omega(E), \tag{4.30}$$

which is Gauss' principle for the following reason.

The Young–Fenchel transform,

$$S(\bar{E}) = \inf_{\beta} [\beta \bar{E} + \ln \mathcal{Z}(\beta)],$$

reduces to the Legendre transform

$$S(\bar{E}) = \beta(\bar{E})\bar{E} + \ln \mathcal{Z}[\beta(\bar{E})] \tag{4.31}$$

because $\ln \mathcal{Z}(\beta)$ is a smooth convex function whose norm grows slower than linearly. Introducing (4.31) into (4.30) gives the usual form of Gauss' error law,

$$f^\star(E, \bar{E}) = A \exp \left\{ S(E) - S(\bar{E}) - S'(\bar{E})(E - \bar{E}) \right\}, \tag{4.32}$$

where we have used Boltzmann's principle,

$$S(E) = \ln \Omega(E) + \text{const.}, \tag{4.33}$$

to introduce the stochastic entropy.

Alternatively, we may write the Young–Fenchel transform for the dual function as

$$\ln \mathcal{Z}(\beta) = \sup_{\bar{E}} [S(\bar{E}) - \beta \bar{E}]$$

which, since $S(\bar{E})$ grows slower than linearly, reduces to the Legendre transform

$$\ln \mathcal{Z}(\beta) = S[\bar{E}(\beta)] - \beta \bar{E}(\beta).$$

The crucial point is that the thermodynamic entropy must be a concave function of the energy in order to get the Legendre transform. In addition, the Legendre transform defines the optimal value of the parameter β corresponding to inverse temperature.

The function (4.28) is a saddle function which is strictly convex in β and weakly concave in $\bar{E}$ since it satisfies the saddle inequality

$$S(\bar{E}_+, \beta_+) - S(\bar{E}_-, \beta_-) - (\bar{E}_+ - \bar{E}_-) \left. \frac{\partial S}{\partial \bar{E}} \right|_{\bar{E}_+, \beta_+} - (\beta_+ - \beta_-) \left. \frac{\partial S}{\partial \beta} \right|_{\bar{E}_-, \beta_-}$$

$$= \ln \mathcal{Z}(\beta_+) - \ln \mathcal{Z}(\beta_-) - (\beta_+ - \beta_-) \frac{\partial}{\partial \beta} \ln \mathcal{Z}(\beta_-) > 0$$

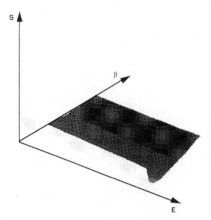

Figure 4.2: Saddle property of the entropy.

for any two pairs of endpoints E_+, β_+ and E_-, β_- due to the strict convexity of $\ln \mathcal{Z}(\beta)$. Such a saddle function is shown in Fig. 4.2. Therefore, for fixed $\bar{E}$, (4.28) is a convex function of β with minimum value (4.31). This is precisely the thermodynamic entropy.

The thermodynamic entropy is also given by

$$S(\bar{E}) = \ln \Omega(\bar{E}) + \text{const.},$$

showing it to be the same function of $\bar{E}$ that the stochastic entropy, (4.33), is of E. Since the thermodynamic entropy is a concave function of the energy, it follows that

$$\ln \Omega(\bar{E}) \geq \ln \overline{\Omega(E)}.$$

And since $- \int f \ln f \, dE$ is nonnegative, we have the hierarchy of inequalities

$$S(\bar{E}, \beta) \geq S(\bar{E}) \geq \overline{S(E)},$$

where

$$S(\bar{E}) = \min_\beta S(\bar{E}, \beta).$$

As an illustration of Gauss' error law (4.32), consider the case of an ideal gas. According to Boltzmann's principle (4.33), the stochastic entropy is proportional to the logarithm of the phase volume[4]:

$$S(E) = \ln \left[\frac{(C \cdot E)^m}{m!} \right].$$

---

[4]Since for a large number of degrees-of-freedom, "essentially all the volume lies very near to the surface," we could equally as well define the entropy in terms of the surface area Ω. The entropy defined as $\ln \Omega(E)$ differs from the value obtained above only by the additive terms $\ln m - \ln E$ which are negligible in comparision to $m \ln E$ and $m \ln m$. However, such a definition of entropy would not conform to Gauss's principle (4.32) since the normalization function A must be independent of the mean value $\bar{E}$. Only in the case where the entropy is defined as $\ln \mathcal{V}$ is this true.

If there is not to be a separate thermodynamics for the microcanonical and canonical ensemble (i.e., the Greene–Callen principle that we discussed in §3.6), the thermodynamic entropy, $S(\bar{E})$, must be the same function of $\bar{E}$ that the stochastic entropy is of the random energy E. Hence, Gauss' principle (4.32) for a perfect gas is

$$f[E; \beta(\bar{E})] = \beta(\bar{E}) \frac{[\beta(\bar{E})E]^{m-1}}{\Gamma(m)} e^{-\beta(\bar{E})E}, \qquad (4.34)$$

where we have set $A = m/E$ and used the second law in the form

$$\frac{\partial}{\partial \bar{E}} S(\bar{E}) = \frac{m}{\bar{E}} = \beta(\bar{E}). \qquad (4.35)$$

As Gibbs originally noted, this distribution depends only on the number of degrees-of-freedom and is independent of the particular nature of the system.

The gamma density, (4.34), can easily be transformed into a chi-square distribution for a system with $2m$ degrees-of-freedom. The square of a random variable with a Gaussian distribution is said to have a chi-square distribution. As we have shown in §4.2, this points to the fact that asymptotic theorems, like the law of large numbers and the central limit theorem, are actually behind the success of thermodynamics. In other words, thermodynamics has captured the elegance of these asymptotic theorems which is reflected in the generality of the results, such as the law of equipartition of energy, (4.35). It also warns us not to attempt to formulate more fundamental, microscopic theory of thermodynamics since these asymptotic theorems of probability are actually at the heart of thermodynamics.

4.6   The Second Law

Consider a "composite" system formed from two subsystems whose thermodynamic functions are labeled by the indices 1 and 2 while those pertaining to the composite system are unindexed. The conservation of energy requires

$$E = E_1 + E_2. \qquad (4.36)$$

We want to compare the sum of the entropies of the two subsystems to the entropy of the composite system. Since the entropy is given by (4.31), we also need to know how the logarithms of the generating functions of the subsystems combine.

The phase volume of the composite system can be expressed as the product of the phase volumes of the subsystems,

$$\begin{aligned}
\mathcal{V}(E) &= \int_{\mathcal{V}_E} d\mathcal{V} = \int_{\mathcal{V}_{E_1}} d\mathcal{V}_1 \int_{\mathcal{V}_{E_2}} d\mathcal{V}_2 \\
&= \int_{\mathcal{V}_{E_1}} \mathcal{V}_2 \left(E - E_1 \right) d\mathcal{V}_1,
\end{aligned}$$

where $\mathcal{V}_{E_1}$ denotes that part of the phase space volume which contains microstates whose energies are less that E_1 while $\mathcal{V}_{E_2}$ contains those states with energies less than $E_2 = E - E_1$. Introducing the definition of the structure function of subsystem 1, according to $\Omega_1 = d\mathcal{V}_1/dE_1$, we have

$$\mathcal{V}(E) = \int_0^E \mathcal{V}_2\,(E - E_1)\,\Omega_1\,(E_1)\,dE_1.$$

Observing that $\mathcal{V}_2(E - E_1) = 0$ for $E_1 > E_2$, the upper limit of integration can be extended to infinity:

$$\mathcal{V}(E) = \int_0^\infty \mathcal{V}_2\,(E - E_1)\,\Omega_1\,(E_1)\,dE_1.$$

Then differentiating with respect to the total energy E gives

$$\Omega(E) = \int_0^\infty \Omega_2\,(E - E_1)\,\Omega_1\,(E_1)\,dE_1. \tag{4.37}$$

This is the fundamental relation between structure functions of the subsystems and the composite system. We shall call this the *composition* law. Its importance lies in the fact that if Ω_1 is the (improper) density of E_1 and Ω_2 is the density of E_2, then the sum $E_1 + E_2$ will have the density $\Omega_1 * \Omega_2$, where $*$ denotes the convolution [cf. §3.4.1].

Given a partitioning of the system into two subsystems, we may ask how is the energy shared between them? The distribution of E_1, given that the total energy is E, is

$$\frac{\Omega_1(E_1)\,\Omega_2(E - E_1)}{\Omega(E)} \propto x_1^\alpha (1 - x_1)^\beta,$$

where $x_1 = E_1/E$. If the system is partitioned into two subsystems with equal number of degrees-of-freedom, $\alpha = \beta$ and Fig. 4.3 shows that the energy will be equally divided between the two. The average energy is the most probable value of the energy of each of the subsystems. If the first system is smaller than the second, $\alpha < \beta$, and the maximum will be shifted toward the left, meaning that it acquires a smaller fraction of the total energy. This is what we would expect would happen, and it can be seen to be a consequence of the central limit theorem.

The composition law for the structure functions (4.37) enables us to determine a product relation for the partition functions when the subsystems have a common value of β. Based on the composition law (4.37), the generating function of the composite system (4.5) can be decomposed into

$$
\begin{aligned}
\mathcal{Z}(\beta) &= \int e^{-\beta E}\Omega(E)\,dE = \int e^{-\beta E}\,dE \int \Omega_2\,(E - E_1)\,\Omega_1\,(E_1)\,dE_1 \\
&= \int \Omega_1\,(E_1)\,e^{-\beta E_1}\,dE_1 \int e^{-\beta(E - E_1)}\Omega_2\,(E - E_1)\,dE \\
&= \int e^{-\beta X}\Omega_1(X)\,dX \int e^{-\beta Y}\Omega_2(Y)\,dY \\
&= \mathcal{Z}_1(\beta)\,\mathcal{Z}_2(\beta),
\end{aligned}
\tag{4.38}
$$

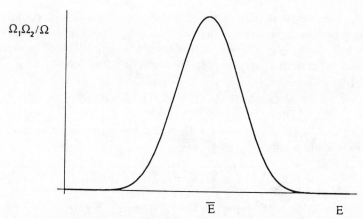

Figure 4.3: Probability density of the energy resulting from the partitioning of a system described by the microcanonical ensemble when both subsystems contain the same number of degrees-of-freedom.

where we have used the energy conservation law (4.36).

Thus, we have derived the important property that *the partition function of the composite system is the product of the partition functions of the individual subsystems, provided they have a common temperature.* Obviously, this will hold true no matter how many subsystems the composite system is broken up into, provided, of course, that all are at the same temperature.

Equipped with the product law for the partition functions, (4.38), we return to the original problem of determining the behavior of the entropy when two subsystems are combined to form a composite system. The law of energy conservation, (4.36), and the product law, (4.38), of the partition functions imply

$$\beta E + \ln \mathcal{Z}(\beta) = \beta E_1 + \ln \mathcal{Z}_1(\beta) + \beta E_2 + \ln \mathcal{Z}_2(\beta),$$

so that the entropy of the composite system is

$$
\begin{aligned}
S(E, \beta) &= \beta E + \ln \mathcal{Z}(\beta) \\
&= \beta E_1 + \ln \mathcal{Z}_1(\beta) + \beta E_2 + \ln \mathcal{Z}_2(\beta) \\
&= S_1(E_1, \beta) + S_2(E_2, \beta),
\end{aligned}
$$

the sum of the entropies of the subsystems when they have a common value of β. The functions $\beta E_i + \ln \mathcal{Z}_i(\beta)$ reach a minimum when $\beta = \beta_i(\bar{E}_i) \equiv \hat{\beta}_i$ for $i = 1, 2$, so it follows that

$$S(E, \beta) \geq S_1(\hat{\beta}_1) + S_2(\hat{\beta}_2). \tag{4.39}$$

This inequality asserts that

> *thermal interactions of bodies at different temperatures cause the entropy to increase. Only in the case that the bodies were originally at the same temperature will the entropy of the composite system show no tendency to increase.*

A lower limit to the increase in entropy, due to thermal interaction, can be established with the aid of the concavity criterion

$$S(\bar{E}) - S(E) \geq \beta(\bar{E})(\bar{E} - E). \tag{4.40}$$

By interchanging endpoints of the interval in the above expression and adding the two results, we get

$$(E - \bar{E})[\beta - \beta(\bar{E})] \leq 0. \tag{4.41}$$

This says that the slope of a strictly concave function is monotonically decreasing. We recall from §1.8 that herein lies the essence of the second law. The inequality ensures that energy passes from higher to lower temperature.

Inequalities (4.40) and (4.41) can be employed to prove the following important result. Suppose that we allow two subsystems 1 and 2, at temperatures T_1 and T_2 to interact thermally and give them a sufficient amount of time so as to secure a state of thermal equilibrium at some intermediary temperature T corresponding to an inverse temperature $\beta(\bar{E})$. With $E = E_i$ in (4.41) we have

$$(\bar{E} - E_i)(\beta(\bar{E}) - \beta_i) \leq 0. \tag{4.42}$$

For subsystem 2 we use the fact that $E_2 = 2\bar{E} - E_1$ and thus obtain

$$(E_1 - \bar{E})(\beta(\bar{E}) - \beta_2) \leq 0. \tag{4.43}$$

By letting inequalities (4.42) and (4.43) hold for subsystems 1 and 2, respectively, we get the string of inequalities

$$\beta_2(\bar{E} - E_1) \leq \beta(\bar{E})(\bar{E} - E_1) \leq \beta_1(\bar{E} - E_1). \tag{4.44}$$

Introducing the first inequality into (4.40) results in

$$S(\bar{E}) - S(E_1) \geq \frac{\bar{E} - E_1}{T_2}. \tag{4.45}$$

Instead of using the fact that $\bar{E}$ is the average energy, we could have used the property that $-\ln \mathcal{Z}(\beta)$ is a monotonic increasing function of β. We could also have replaced the average energy of the composite system by the average energy of the first component since $\beta(\bar{E}_1) = \beta(\bar{E}_2) = \beta(\bar{E})$, which says that the composite system must come to a common final temperature. Then, we would have obtained Khinchin's theorem,

$$S(\bar{E}_1) - S(E_1) \geq \frac{\bar{E}_1 - E_1}{T_2},$$

asserting that

> the entropy change of any component resulting from thermal interaction cannot be inferior to the energy difference of that component divided by the initial temperature of the subsystem which it interacts with.

This is another illustration of the importance of the concavity property of the entropy in thermodynamics and, in particular, in establishing a lower bound on the entropy change.

4.7   A Measure of Irreversibility

Consider a body at temperature T which interacts with a medium at temperature T_0 and suppose for concreteness that $T_0 > T$. All quantities referring to the medium will have an index 0 while the unindexed quantities will pertain to the body.

In addition to heat transfer from the medium to the body, an amount of work, $\sum_i F_i^0 \, dq_i^0$, is done by the medium on the body. A set of external parameters q_i will be necessary to specify the state of the body. On account of conservation, the changes in the external parameters of the body and medium are related by $dq_i = -dq_i^0$ for each i. The forces F_i^0 represent the applied, or *observed*, forces acting on the body while the forces that the body is *expected* to exert on the medium are given by

$$\bar{F}_i = \frac{1}{\beta(\bar{E})} \frac{\partial}{\partial q_i} \ln \mathscr{Z} = \begin{cases} (1/\beta(\bar{E}))\overline{\partial \ln \Omega / \partial q_i}, \\[2ex] -\overline{\partial E / \partial q_i}. \end{cases} \tag{4.46}$$

The reason for the two definitions of the forces is due to the apparent choice we have for designating which term in the partition function (4.5) shall carry the functional dependency of the external parameters. We have already come across this in the definition of the pressure (4.21) and dedicate the next section to a comparison of the two definitions. Here, we shall again opt for the second definition in (4.46).

The total derivative of the maximum log-likelihood function (4.22) is now given by

$$\begin{aligned} d\mathscr{L}(\beta(\bar{E}), \{q_i\}) &= -\beta(\bar{E}) \, d\bar{E} - \sum_i \frac{\partial}{\partial q_i} \ln \mathscr{Z} \, dq_i \\ &= -\beta(\bar{E}) \left\{ d\bar{E} - \sum_i \frac{\overline{\partial E}}{\partial q_i} \, dq_i \right\} \\ &= -\beta(\bar{E}) \left\{ d\bar{E} + \sum_i \bar{F}_i \, dq_i \right\}. \end{aligned} \tag{4.47}$$

When the first law is applied to the body,

$$d\bar{E} = \delta Q + \sum_i F_i^0 \, dq_i^0,$$

it determines the change in its average energy for an infinitesimal transition from one equilibrium state to a neighboring one. Introducing this expression into (4.47) results in

$$d\mathscr{L} = -\beta(\bar{E}) \left\{ \delta Q + \sum_i \left(\bar{F}_i - F_i^0 \right) dq_i \right\} = -\beta(\bar{E}) T \, dS,$$

where dS is the increase in the entropy of the body. Since the left-hand side is a total differential, so, too, must be the right-hand side. Hence, $\beta(\bar{E}) \propto 1/T$, where T is the temperature of the body.

We now apply the global conservation of energy, $\bar{E} + \bar{E}_0 = \text{const.}$, or equivalently $d\bar{E} = -d\bar{E}_0$, where the change in the average energy of the medium is

$$d\bar{E}_0 = \delta Q_0 - \sum_i F_i^0 \, dq_i^0.$$

Inserting this expression into (4.47) gives

$$
\begin{aligned}
d\mathcal{L}(\beta(\bar{E}), \{q_i\}) &= \beta(\bar{E}) \left\{ \delta Q_0 + \sum_i (\bar{F}_i - F_i^0) \, dq_i \right\} \\
&= \beta(\bar{E}) T_0 \, dS_0,
\end{aligned}
\tag{4.48}
$$

where dS_0 is the change in entropy of the medium. Since the entire system (body + medium) is isolated, the second law predicts that

$$dS + dS_0 \geq 0.$$

This transforms Eq. (4.48) into the inequality

$$d\mathcal{L}(\beta(\bar{E}), \{q_i\}) \geq -\frac{T_0}{T} \, dS.$$

Using expression (4.47) for the left-hand side, the inequality becomes

$$d\bar{E} - T_0 \, dS + \sum_i \bar{F}_i \, dq_i \leq 0. \tag{4.49}$$

We have already come across this well known inequality in §1.10. We recall that it expresses the fact that spontaneous irreversible processes occurring in the body will cause the quantity given on the left-hand side of (4.49) to decrease until it attains its minimum value at equilibium. Under the conditions of constant temperature, where the body's temperature equals the temperature of the medium T_0, and constant volume, which we identify with the only external coordinate q, it is the Helmholtz potential of the body which will decrease until a state of equilibrium is attained between the body and medium.

4.8   The Principle of Adiabatic Invariance

We now return to the comparison of the two definitions of the force in (4.46). Using the second definition of the force in (4.46), we have

$$\frac{\partial}{\partial q_i} \int e^{-\hat{\beta} E - \ln \mathcal{Z}} \Omega(E) \, dE = -\hat{\beta} \frac{\overline{\partial E}}{\partial q_i} - \frac{\partial}{\partial q_i} \ln \mathcal{Z} = 0, \tag{4.50}$$

implying that the structure function, Ω, is independent of the external coordinates q_i. This choice of the definition of the force is often referred to as the principle of "adiabatic invariance" of the structure function since in the words of Paul Ehrenfest, it "must always depend on those quantities which remain invariant under adiabatic influencing."

An *adiabatic* process is one in which external forces acting on a *thermally isolated* system cause sufficiently slow variations in the external coordinates. In such a process, the entropy of the body does not change, and hence the process is *reversible*. This is entirely reasonable because a purely mechanical action on the system—provided it occurs sufficiently slowly—should have no influence on the entropy. The conventional argument is the following.

Consider the variation of a single external coordinate q. Its rate of change in time, $\dot{q}$, will cause a change in the entropy. If $\dot{q}$ is sufficiently small, the rate of change of the entropy, $\dot{S}$, can be expanded in powers of $\dot{q}$. Since $\dot{S}$ will be zero if $\dot{q}$ is zero, the zero-order term in the expansion is zero. In other words, the entropy does not change if there are no mechanical changes imposed upon the system. The first-order term, too, will vanish because $\dot{q}$ can be of either sign while $\dot{S}$ must always be positive. Consequently, the series starts off with a quadratic term, $\dot{S} \propto \dot{q}^2$, so that the thermodynamic force $dS/dq \propto \dot{q}$. As the velocity $\dot{q}$ tends to zero, so too does dS/dq. Consequently, if the mechanical changes on the system are carried out sufficiently slowly, the entropy will be independent of the external coordinate. According to Boltzmann's principle, (4.33), the structure function will also be independent of the external coordinate.

The adiabatic condition, (4.50), has been claimed to be responsible for the fact that the quantity $\hat{\beta}\, dQ$, when calculated over a cycle, is path independent,

$$\oint \hat{\beta}\, dQ = -\left\{ \mathcal{L}(\hat{\beta}, \{q_i\})_{\text{final state}} - \mathcal{L}(\hat{\beta}, \{q_i\})_{\text{initial state}} \right\}. \qquad (4.51)$$

This may be demonstrated in the following way.

Multiplying the first law,

$$dQ = d\bar{E} + \sum_i \bar{F}_i\, dq_i,$$

through by $\hat{\beta}$ and using the second definition of the forces in (4.46) give

$$\hat{\beta}\, dQ = \hat{\beta}\, d\bar{E} - \hat{\beta} \sum_i \overline{\frac{\partial E}{\partial q_i}}\, dq_i. \qquad (4.52)$$

The total change in energy of the system, in its infinitesimal transition from one equilibrium state to a neighboring one, is

$$d\bar{E}(\hat{\beta}, \{q_i\}) = \frac{\partial \bar{E}}{\partial \beta}\, d\hat{\beta} + \sum_i \frac{\partial \bar{E}}{\partial q_i}\, dq_i.$$

Introducing this into expression (4.52) gives

$$\hat{\beta}\, dQ = \hat{\beta} \left\{ \frac{\partial \bar{E}}{\partial \beta}\, d\hat{\beta} + \sum_i \left(\frac{\partial \bar{E}}{\partial q_i} - \overline{\frac{\partial E}{\partial q_i}} \right) dq_i \right\}.$$

In the presence of fluctuations, the average of a derivative is not the same as the derivative of the average so that the last two terms do not cancel one another.

Now the crucial step comes when we introduce the second equality in (4.46). We then obtain

$$\hat{\beta} \, dQ = \hat{\beta} \left\{ \frac{\partial \bar{E}}{\partial \beta} \, d\hat{\beta} + \sum_i \frac{\partial \bar{E}}{\partial q_i} \, dq_i \right\} + \sum_i \frac{\partial}{\partial q_i} \ln \mathcal{Z} \, dq_i.$$

Furthermore, by writing

$$\sum_i \frac{\partial}{\partial q_i} \ln \mathcal{Z} \, dq_i = d \ln \mathcal{Z}(\hat{\beta}, \{q_i\}) - \frac{\partial}{\partial \beta} \ln \mathcal{Z} \, d\hat{\beta}$$

and using the expression for the average energy (4.9), we finally get

$$\begin{aligned}
\hat{\beta} \, dQ &= \frac{\partial}{\partial \beta} (\hat{\beta} \bar{E}) \, d\hat{\beta} + \sum_i \frac{\partial}{\partial q_i} (\hat{\beta} \bar{E}) \, dq_i + d \ln \mathcal{Z}(\hat{\beta}, \{q_i\}) \\
&= d \left\{ (\hat{\beta} \bar{E}) + \ln \mathcal{Z}(\hat{\beta}, \{q_i\}) \right\}.
\end{aligned}$$

This proves that (4.51) is, indeed, path independent.

However, the proof we have given necessarily implies that

$$\overline{\frac{\partial \ln \Omega}{\partial q_i}} \equiv 0,$$

which, according to the first definition in (4.46), would imply a vanishing force. Yet, there are arguments in favor of the first definition.

First, the adiabatic condition, (4.50), implies a certain "conditioning" of the random variable, E, upon the external coordinates. Although this is certainly true for its average, $\bar{E}$, it can hardly apply to the random variable itself. Second, according to our definition (4.2), the structure function, Ω, contains all the relevant information concerning the mechanical structure of the thermally isolated system. Hence, we would conclude that the structure function must depend on the external coordinates. In particular, if the only external coordinate corresponds to the volume of the system, then

$$\frac{1}{\hat{\beta}} \overline{\frac{\partial \ln \Omega(E, V)}{\partial V}} = P \tag{4.53}$$

is the definition of the pressure.

The resolution of this paradox can be found in what Lorentz called the "remarquable insensibilité de la formule de Boltzmann" insofar as the Boltzmann and Gibbs definitions of entropy coincide at equilibrium. What we have defended so far [i.e., the second definition in (4.46)] is essentially Gibbs' definition while the first definition in (4.46) belongs to Boltzmann. Let us look at Boltzmann's viewpoint more closely.

According to Boltzmann, the (inverse) temperature is defined by

$$\frac{\partial \ln \Omega(\bar{E}, \{q_i\})}{\partial \bar{E}} = \beta(\bar{E}) \equiv \hat{\beta}, \qquad (4.54)$$

and the expected forces are

$$\bar{F}_i = \frac{1}{\hat{\beta}} \frac{\partial \ln \Omega(\bar{E}, \{q_i\})}{\partial q_i}. \qquad (4.55)$$

This definition of the force differs from the first equality (4.46) in that the maximum likelihood estimate replaces the average. In general, an average of a function of a random variable will not be equal to the function of its average. However, it is a particularly significant property of the exponential family of distributions, like the canonical distribution (4.6), that the maximum likelihood estimates and their average values do coincide. This is responsible for the fact that Boltzmann's definition of (inverse) temperature (4.54) is none other than the second law, $1/T = (\partial S/\partial \bar{E})_V$.

We can show that (4.55) is the most probable value of the force inasmuch as it is a condition for maximum probability. As a function of β, it does not make any difference if we consider the log-likelihood function or the logarithm of the posterior probability, $f(E; \beta)$, since the structure function is independent of β. This is no longer the case if we intend to vary E, so that we must consider the posterior probability density, (4.6), whose total differential is

$$d \ln f(E, \{q_i\}; \beta) = -E \, d\beta - \beta \, dE - \frac{\partial \ln \mathcal{Z}}{\partial \beta} \, d\beta$$
$$- \sum_i \frac{\partial \ln \mathcal{Z}}{\partial q_i} \, dq_i + \frac{\partial \ln \Omega}{\partial E} \, dE + \sum_i \frac{\partial \ln \Omega}{\partial q_i} \, dq_i, \qquad (4.56)$$

where we have reinstated the dependencies on the external coordinates. Evaluating (4.56) at the maximum value (4.9) gives

$$d \ln f(\bar{E}, \{q_i\}; \hat{\beta})$$
$$= -\left[\hat{\beta} - \frac{\partial \ln \Omega(\bar{E}, \{q_i\})}{\partial E} \right] d\bar{E} - \sum_i \left(\hat{\beta} \bar{F}_i - \frac{\partial \ln \Omega(\bar{E}, \{q_i\})}{\partial q_i} \right) dq_i.$$

The necessary condition for an extremum, $d \ln f = 0$, is satisfied by requiring the coefficients of the differentials to vanish. But this is nothing more than Boltzmann's definition of temperature, (4.54), and the *most probable* values of the forces, (4.55).

Gibbs called $\bar{\eta} \equiv -\overline{\ln \Omega}$ the average "index of probability" and attributed to it all the properties of an entropy with its sign reversed. In his treatise *Elementary Principles in Statistical Mechanics*, which is in fact not "elementary" by any standards, Gibbs employed interchangeably both definitions of the force (4.46). In the first instance, Gibbs wrote the partition function as the phase average of $\exp(-\beta E)$, as in (4.20), where he assumed that the dependency on the external coordinates is implicitly contained in the energy, E. He then compared the differential of the logarithm of (4.20) with the thermodynamic relation

$$dA = -\frac{\bar{\eta}}{\hat{\beta}^2} d\hat{\beta} + \sum_i \overline{\frac{\partial E}{\partial q_i}} dq_i,$$

where $A = -(1/\hat{\beta}) \ln \mathcal{Z}$ is the Helmholtz potential, and obtained the second definition of the forces in (4.46), since the thermodynamic relation (in present-day notation) is

$$dA = -S\,dT - \sum_i \bar{F}_i\,dq_i. \tag{4.57}$$

This gives the well known thermodynamic relation for the forces

$$\bar{F}_i = -\frac{\partial A}{\partial q_i},$$

which is valid only at equilibrium.

In the second instance, Gibbs replaced the phase space volume element, $d\Gamma$, in (4.1) by $e^{-\eta}\,dE$ or, equivalently, $\Omega\,dE$, which reduces the $2m$-fold integral in (4.20) to a simple integral whenever the limits can be expressed by the energy alone. In place of (4.20) we now have

$$e^{-\beta A} = \int e^{-\beta E} \Omega(E, \{q_i\})\,dE,$$

where the structure function now carries the dependency on the external coordinates. Comparing the total differential of the logarithm of this expression with the thermodynamic relation (4.57) gives the first definition in (4.46). We should point out that although the two definitions of the force given in Eq. (4.46) are equivalent at equilibrium, this will not be true out of equilibrium. The coincidence of the two definitions at equilibrium is a consequence of the remarkable insensitivity of Boltzmann's definition of entropy.

From an aesthetic point of view, the first definition is to be favored over the second since a random variable, like the energy, cannot be conditioned by the external parameters, although these can and do modify its realizations. Moreover, since the structure function contains all the relevant information about the mechanical nature of the physical system, it should be a function of the external coordinates. The method of maximum likelihood bears this out.

4.9 Thermodynamic Uncertainty Relations

In §3.7.4 we came close to obtaining an uncertainty relation between conjugate thermodynamic variables when we appealed to Le Châtelier's principle in evaluating the product of the standard deviations in the energy representation. The uncertainty relations will appear much more naturally in the entropy representation, especially when we recognize that the inequality is related to the irreversible nature of the process.

In more mundane terms, the uncertainty relations are seen to be the result of applying Schwarz's inequality to conjugate variables. The conjugate variables involved may be Fourier or Laplace duals. The former gives rise to Heisenberg's uncertainty relations in quantum mechanics, or Garbor's relations in communication theory, while the latter gives nascence to the "thermodynamic uncertainty relations." In both instances, the more precisely one variable is measured, the fuzzier its conjugate becomes.

How does the uncertainty in measurement come about in thermodynamics? Consider the simplest case of a system placed in contact with a heat reservoir. The energy ceases to be a thermodynamic function and becomes a random quantity because a knowledge of all the external parameters which specify the system is insufficient to determine the energy. The energy of the system is said to fluctuate. Observations made on the energy can be used to estimate its conjugate intensive variable. As the thermostat shrinks in size, the measurements of the energy become increasingly more precise at the expense of being able to assign a definite energy to the system. Alternatively, as the thermostat grows in size, the energy measurements become less precise and a definite temperature can be assigned to the system in the limit of an infinite thermostat. An infinite heat reservoir means an infinite heat capacity. Since the latter is inversely proportional to the dispersion in the temperature fluctuations, it tends to zero in the limit.

To derive the uncertainty relations between energy and inverse temperature, let us consider the Legendre transform (4.31). Differentiating twice, we obtain

$$\frac{\partial^2 S}{\partial \bar{E}^2} = \frac{\partial \beta}{\partial \bar{E}}.$$

Since the dispersion in the energy is given by (4.10), we have the well known result of Legendre dual functions:

$$\frac{\partial^2 \ln \mathcal{Z}}{\partial \beta^2} = - \left(\frac{\partial^2 S}{\partial \bar{E}^2} \right)^{-1}. \tag{4.58}$$

Although this has no apparent resemblance to an uncertainty relation, it will turn out that this is a statement of minimum uncertainty in energy and inverse temperature.

4.9.1   The Bayesian Approach to Thermostatics

Any estimate of the (inverse) temperature is itself a random variable since it is a function of the sample energies. And in order to study fluctuations in β, we employ Bayes' theorem:

$$\boxed{\text{Posterior probability} \propto \text{likelihood} \times \text{prior probability}}$$

or in symbols

$$\tilde{f}(\beta; \bar{E}) \propto e^{\mathcal{L}(\beta; \bar{E})} \tilde{f}(\beta), \tag{4.59}$$

where $\tilde{f}(\beta)$ is the prior density. We have already mentioned that β cannot be interpreted as a random variable in the frequency sense but, rather, in the sense of degree-of-belief that some values of β are more probable than others. Expressed in words, Bayes' theorem states that

> the probability that the unknown parameter has the value β given the statistic $\bar{E}$ is proportional to the likelihood of β given $\bar{E}$ multiplied by the initial probability of β.

The stumbling block in putting Bayes' theorem into practice has always been how to choose the prior probability density, $\tilde{f}(\beta)$. In the absence of any information about β, it has been common practice to use the uniform prior density which has been formalized into the so-called Bayes–Laplace principle of insufficient reason. This rings a familiar tone with Boltzmann's approach of setting the *a priori* probabilities equal to one another in the multinomial distribution since we have no reason to consider the occupancy of one cell as being more probable than the occupancy of any other cell [cf. §1.3].

The Bayes–Laplace principle of setting $\tilde{f}(\beta)\, d\beta \propto d\beta$, or uniform assessment, was criticized by Harold Jeffreys in the case where there is only a finite range of possible values for β. As an example, Jeffreys cited the law of continuity connecting mass and volume of a substance. If the uniform rule is adopted for one of the variables, Jeffreys contended that it will be incorrect for the other variable because mass and specific volume are reciprocals of one another. Rather, Jeffreys argued that improper prior densities should be used to represent "knowing little or nothing" based on certain invariance properties of the prior densities.

According to Jeffreys, since the parameter β can take on values between 0 and ∞, we should take the logarithm of its prior density as being uniformly distributed over the whole line. In other words, if we set $\vartheta = \ln \beta$, then the prior probability density of ϑ is $\tilde{f}(\vartheta)\, d\vartheta \propto d\vartheta$ over the entire range $-\infty < \vartheta < \infty$. And because $d\vartheta = d\beta/\beta$, it implies that

$$\tilde{f}(\beta)\, d\beta \propto d\beta/\beta \tag{4.60}$$

is an *improper* probability density, where β varies between 0 and ∞. Jeffreys then queried why should we take unity to imply certainty? Isn't it rather only a convention? Why shouldn't $\int d\beta/\beta$ make a better definition of certainty when improper probability densities are used?

Since both $\int_0^a d\beta/\beta$ and $\int_a^\infty d\beta/\beta$ are infinite, the ratio $\Pr\{0 < \beta < a\}/\Pr\{a < \beta < \infty\}$ is indeterminate, meaning that nothing can be said about which interval is more probable to contain the random variable. This indeterminacy can be taken as a formal statement of our ignorance. The invariancy of the improper prior, $\tilde{f}(\beta) \propto 1/\beta$, to transformations of the form $\vartheta = \beta^n$, since

$d\beta/\beta$ and $d\beta^n/\beta^n$, is further support in favor of using an improper density to represent ignorance.

Observe that the proportionality term in Bayes' theorem is the inverse of

$$\tilde{Z}(\bar{E}) = \int e^{\mathcal{L}(\beta;\bar{E})} \tilde{f}(\beta) \, d\beta. \tag{4.61}$$

Differentiating its logarithm gives

$$\frac{\partial \ln \tilde{Z}}{\partial \bar{E}} = -\bar{\beta}. \tag{4.62}$$

How is the average value of β related to the most probable value $\hat{\beta} = \beta(\bar{E})$ found from the likelihood equation (4.17)?

We have seen that the likelihood function factors into a product of r factors for a sample of size r because our observations, E_i, are independent and identically distributed random variables. We may write the integrand of (4.61) as

$$\exp\left\{\mathcal{L}(\beta; \bar{E}) + \ln \tilde{f}(\beta)\right\}$$

and note that since $\mathcal{L}(\beta; \bar{E})$ increases like r, it will eventually dwarf $\ln \tilde{f}(\beta)$ as $r \to \infty$. Hence, the integrand of the generating function can be expressed as

$$\begin{aligned}
&\exp\{\mathcal{L}(\beta; \bar{E})\}\tilde{f}(\beta) \\
&= \exp\left\{\mathcal{L}(\hat{\beta}; \bar{E}) + \tfrac{1}{2}(\beta - \hat{\beta})^2 \frac{\partial^2}{\partial \beta^2}\mathcal{L}(\hat{\beta}; \bar{E}) + \cdots\right\} \tilde{f}(\hat{\beta}),
\end{aligned} \tag{4.63}$$

on expanding $\mathcal{L}(\beta; \bar{E})$ by Taylor's theorem about $\hat{\beta}$. The remainder term in the series expansion can be neglected since $\sigma_r^{-2} = -\partial^2\mathcal{L}/\partial\beta^2$ is of order r and the term $\exp\{-(\beta - \hat{\beta})^2/2\sigma_r^2\}$ will be appreciably different from zero for $\beta - \hat{\beta}$ of order $1/\sqrt{r}$. The cubic term will be $r^{-3/2}$ times its coefficient, which is of order r so that the product will be of order $1/\sqrt{r}$ and hence negligible. And since the main contribution to the integral of (4.63) comes in the neighborhood of the global maximum at $\hat{\beta}$, the lower limit of integration can be extended to $-\infty$. We then come out with the asymptotic normal approximation

$$e^{\mathcal{L}(\hat{\beta};\bar{E})} \tilde{f}(\hat{\beta}) e^{-(\beta-\hat{\beta})^2/2\sigma_r^2},$$

which upon integrating over β yields

$$\tilde{Z}(\bar{E}) = \left\{2\pi \left(\frac{\partial^2 \ln \mathcal{Z}}{\partial \beta^2}\right)^{-1}\right\}^{1/2} \exp[-\hat{\beta}\bar{E} - \ln \mathcal{Z}(\hat{\beta})]\tilde{f}(\hat{\beta}).$$

This is the asymptotic expression for the generating function as the sample size increases without bound. Hence, for large sample sizes, the average value (4.62) coincides with the maximum likelihood value $\hat{\beta} = \beta(\bar{E})$.

This allows us to identify $-\ln \tilde{\mathcal{Z}}(\bar{E})$ with the entropy $S(\bar{E})$ since, according to the second law,

$$\frac{\partial S(\bar{E})}{\partial \bar{E}} = \hat{\beta} = -\frac{\partial \ln \tilde{\mathcal{Z}}(\bar{E})}{\partial \bar{E}}.$$

A second differentiation yields

$$\begin{aligned}
\frac{\partial^2 S}{\partial \bar{E}^2} &= \frac{\partial \hat{\beta}}{\partial \bar{E}} \\
&= -\frac{\partial^2 \ln \tilde{\mathcal{Z}}}{\partial \bar{E}^2} = -\overline{(\Delta \beta)^2}.
\end{aligned}$$

Replacing the second derivative of the entropy by the negative of the variance of β in the Legendre dual relation (4.58) leads immediately to the uncertainty relation

$$\overline{(\Delta E)^2}\,\overline{(\Delta \beta)^2} = 1, \tag{4.64}$$

which displays the physical content of (4.58).

Restoring Boltzmann's constant in (4.64) gives

$$\sqrt{\overline{(\Delta E)^2}}\,\sqrt{\overline{(\Delta T^{-1})^2}} = k,$$

which shows that k represents the minimum uncertainty in the simultaneous measurements of thermodynamic conjugate variables, just as Planck's constant h represents the minimum uncertainty in quantum mechanics. The product of the errors of simultaneous measurements made on conjugate thermodynamic variables can never be smaller than the order of Boltzmann's constant. Boltzmann's constant places a lower limit on the precision of measurements in thermodynamics, like Planck's constant does in quantum mechanics. In summary, we may say that if we could imagine a world with a much greater value of k, then such a world would show fluctuations on a macroscopic scale.

4.9.2   Uncertainty and the Second Law

All inequalities in thermodynamics stem, in one form or another, from the second law. And from the thermodynamic viewpoint, it would be gratifying to know that the second law, rather than a geometric inequality, is at the core of the thermodynamic uncertainty relations. With the aid of the mean discrimination information, we will now show that this is, indeed, the case.

We define the mean discrimination information according to (4.25) and suppose that β differs slightly from its optimal value, $\hat{\beta} = \beta(\bar{E})$, by a small amount $\Delta \beta$. The corresponding mean discrimination information is

$$\mathcal{I}(\hat{\beta}, \hat{\beta} + \Delta\beta) = \int f(E; \hat{\beta}) \ln \frac{f(E; \hat{\beta})}{f(E; \hat{\beta} + \Delta\beta)}\, dE. \tag{4.65}$$

We may approximate the logarithm using the series expansion

$$\ln(1 + x) = x - \tfrac{1}{2}x^2 + o(x^2),$$

where $o(x^2)$ denotes terms of powers greater than x^2. We can thus write

$$
\ln \frac{f(E; \hat{\beta} + \Delta\beta)}{f(E; \hat{\beta})}
$$

$$
= \ln\left\{1 + \frac{f(E; \hat{\beta} + \Delta\beta) - f(E; \hat{\beta})}{f(E; \hat{\beta})}\right\} = \frac{f(E; \hat{\beta} + \Delta\beta) - f(E; \hat{\beta})}{f(E; \hat{\beta})}
$$

$$
- \frac{1}{2}\left[\frac{f(E; \hat{\beta} + \Delta\beta) - f(E; \hat{\beta})}{f(E; \hat{\beta})}\right]^2 + o\left[(\Delta\beta)^2\right].
$$

Replacing the logarithm in (4.65) by this expression and expanding $f(E; \hat{\beta} + \Delta\beta)$ to lowest order in $\Delta\beta$ give

$$
\mathcal{I}(\hat{\beta}, \hat{\beta} + \Delta\beta)
$$

$$
= -\frac{(\Delta\beta)^2}{2}\int\left[\frac{\partial^2 f(E; \hat{\beta})}{\partial\beta^2} - \left(\frac{\partial}{\partial\beta}\ln f(E; \hat{\beta})\right)^2 f(E; \hat{\beta})\right]dE + o\left[(\Delta\beta)^2\right]
$$

$$
= \frac{(\Delta\beta)^2}{2}\int\frac{\partial^2}{\partial\beta^2}\ln \mathcal{Z}(\hat{\beta})f(E; \hat{\beta})\,dE + o\left[(\Delta\beta)^2\right]
$$

$$
= \frac{(\Delta\beta)^2}{2}\mathcal{I}_F + o\left[(\Delta\beta)^2\right], \tag{4.66}
$$

where $\mathcal{I}_F = \overline{(\Delta E)^2}$ is the "Fisher information," defined here in terms of the dispersion in the energy, (4.10).

For small values of $\Delta\beta$, we can develop the energy about its average value in a series in $\Delta\beta$,

$$
E = -\frac{\partial}{\partial\beta}\ln \mathcal{Z}(\hat{\beta}) - \frac{\partial^2}{\partial\beta^2}\ln \mathcal{Z}(\hat{\beta})\Delta\beta + o\left[(\Delta\beta)^2\right],
$$

or equivalently,

$$
\Delta E = -\frac{\partial^2}{\partial\beta^2}\ln \mathcal{Z}(\hat{\beta})\,\Delta\beta + o\left[(\Delta\beta)^2\right]
$$

$$
= -\overline{(\Delta E)^2}(\Delta\beta) + o\left[(\Delta\beta)\right]. \tag{4.67}
$$

We can thus express the mean discrimination information as the product of deviations

$$
\mathcal{I}(\hat{\beta}, \hat{\beta} + \Delta\beta) = -\tfrac{1}{2}\Delta E\,\Delta\beta + o\left[(\Delta\beta)^2\right], \tag{4.68}
$$

showing more clearly the inverse relationship between the deviations in energy and inverse temperature.

In contrast to the second-order expression for the information discrimination, the Fisher information is

$$
\mathcal{I}_F = -\frac{\Delta E}{\Delta\beta} = C_V T_f^2, \tag{4.69}
$$

where C_V is the heat capacity at constant volume and T_f is the final equilibrium temperature. This provides us with the relation

$$\mathcal{I}(\hat{\beta}, \hat{\beta} + \Delta\beta) = \tfrac{1}{2}\mathcal{I}_F(\Delta\beta)^2 + o\left[(\Delta\beta)^2\right]$$

between the discrimination information of Kullback and Leibler and the Fisher information. With the aid of the second law we will now show that the Fisher information is the lower bound on the dispersion in β which characterizes *reversible* processes.

Consider two systems, labeled 1 and 2, which are placed in thermal contact. If their temperatures T_1 and T_2 are unequal and no work is delivered, there will be an increase in entropy by an amount

$$\Delta S_1 + \Delta S_2 = \left(\frac{1}{T_1} - \frac{1}{T_2}\right)\Delta E_1 \geq 0, \qquad (4.70)$$

on account of the conservation of energy, $\Delta E_1 = -\Delta E_2$. This means that if $T_2 > T_1$, heat will flow into the first subsystem, leading to an increase in its energy, $\Delta E_1 > 0$. We now want to translate this into an inequality which sets a lower limit in the dispersion in the inverse temperature. The only restriction we will impose is that the temperature difference be small, and consequently, the energy change of either subsystem will also be small. This is on account of the fact that the Fisher information is the coefficient of a second-order term in the series expansion of the discrimination information, (4.68).

In the case that $T_2 > T_1$, the entropy of the second subsystem will decrease by an amount

$$\Delta S_2 = -\frac{\Delta E_1}{T_2}, \qquad (4.71)$$

while the entropy of the first subsystem will increase. In order to determine this change in the entropy, we consider that both systems comprise an ideal gas with the same constant heat capacity. The increase in the entropy of the first subsystem will then be

$$\begin{aligned}
\Delta S_1 &= C_V \ln\left(\frac{T_2}{T_1}\right) = C_V \ln\left(\frac{T_1 + \Delta T_1}{T_1}\right) \\
&\simeq C_V \frac{\Delta T_1}{T_1} = -C_V T_1 \Delta\beta_1,
\end{aligned} \qquad (4.72)$$

since the difference in the temperatures of the two subsystems is assumed to be small. Let $\beta_1(E_1)$ be the estimator of $\hat{\beta}_1$ based on the observation E_1. To relate the variation in the inverse temperature of the first subsystem to the variation in its energy, we invert (4.67) to obtain

$$\Delta\beta_1 = -\overline{(\Delta\beta_1)^2}\,\Delta E_1 \qquad (4.73)$$

for small energy differences. Introducing (4.73) into (4.72) leads to

$$\Delta S_1 = \mathcal{I}_F\overline{(\Delta\beta_1)^2}\,\Delta E_1/T_2, \qquad (4.74)$$

where $T_f = \sqrt{T_1 T_2}$ in the Fisher information, (4.69), which is the *minimum attainable final temperature of the composite system.*

The sum of the entropy differences, (4.71) and (4.74), then satisfies

$$(\Delta S_1 + \Delta S_2) = \left(\mathcal{I}_F \overline{(\Delta \beta_1)^2} - 1 \right) \Delta E_1 / T_2 \geq 0,$$

according to the second law, (4.70). And because $\Delta E_1 \geq 0$, it implies that

$$\overline{(\Delta \beta_1)^2} \geq 1/\mathcal{I}_F. \tag{4.75}$$

Hence, the inverse of the Fisher information is the lower bound on the dispersion in the inverse temperature. The equality sign will apply only in the case where the temperatures of the two systems are equal.

In the opposite situation, where $T_1 > T_2$, there will be a positive variation in the energy of the second subsystem with a corresponding increase in the entropy given by $\Delta S_2 = \mathcal{I}_F \overline{(\Delta \beta_2)^2} \Delta E_2 / T_1$. Adding this to the decrease in the entropy of the first subsystem, $\Delta S_1 = -\Delta E_2 / T_1$ gives

$$(\Delta S_1 + \Delta S_2) = \left(\mathcal{I}_F \overline{(\Delta \beta_2)^2} - 1 \right) \Delta E_2 / T_1 \geq 0.$$

Now since $\Delta E_2 \geq 0$, we again find (4.75).

Inequality (4.75) provides a reference for measuring the quality of estimators in a wide range of situations. It tells us what "best" is in terms of the accuracy of an estimator. Introducing (4.69) into (4.75) leads immediately to the thermodynamic uncertainty relation in its most general form:

$$\overline{(\Delta \beta)^2} \geq 1/\overline{(\Delta E)^2}. \tag{4.76}$$

The Fisher information lives up to its name: it is the "amount of information" on β contained in a single observation, E. As the heat capacity increases, the lower limit on the dispersion in the inverse temperature decreases. The more information about β provided by an observation, E, the smaller we expect the variance of a good estimator to be. For a sample of r observations, the Fisher information is r times the information provided by each observation.

However, in conventional thermodynamics, r is not the sample size but, rather, the number of subdivisions of the original system. In the next section we shall see that this is tied up with the condition that any subdivision must leave intact the additivity of the thermodynamic extensive variables and that nothing can beat the macroscopic estimation of the inverse temperature. The latter is ensured by the property that the statistic used in the estimation of the inverse temperature be *sufficient*. The statement that "E is sufficient for the estimation of β" basically means that the variation of E does not depend on β. In thermodynamics this is interpreted to mean that the probability that the first system has an energy E_1, given a fixed total energy, is independent of the common temperature of the two subsystems at thermal equilibrium. In other words, the equality in (4.76) applies to subsystems which are in thermal equilibrium. And *only sufficient estimators have a minimum variance given by the inverse of the Fisher information*. Surely, if we could find some way to violate the uncertainty relation between energy and inverse temperature, this would no longer be true. Both the uncertainty relation and the property of sufficiency are characteristic to the exponential family of distributions so it is no wonder why thermodynamics has selected out this family of distributions. Violations of the uncertainty relations would be analogous to the destruction of the probabilistic interpretation of quantum mechanics which could then be formulated as a "hidden variable" theory in which a causal intepretation of matter would be restored.

Relation (4.75) also brings out another important point. Thus far, we have considered β as a mere parameter whose optimal value $\beta(\bar{E})$ is equal to the inverse temperature. We now see it to be something more than a "mere" parameter: energy fluctuations are related to fluctuations in the temperature. In terms of the absolute temperature, measured in degrees, this can be expressed through the suggestive inequality

$$\sqrt{\overline{(\Delta E)^2}}\sqrt{\overline{(\Delta T^{-1})^2}} \geq k, \tag{4.77}$$

identifying Boltzmann's constant as the elementary unit in statistical thermodynamics, just as Planck's constant is the elementary unit of a quantum of energy.

The minimum uncertainty in measurements of conjugate thermodynamic variables occurs for *reversible* processes in which the total entropy change vanishes. If the system is isolated, the internal energy is known with unlimited precision. Alternatively, if the system is in contact with an infinite thermostat, having infinite heat capacity, the dispersion in the energy would be infinite, and since E is a "sufficient" statistic for estimating β, the variance in the temperature vanishes. This means that the temperature can be determined with unlimited precision.

4.10 Statistical Correlations and Irreversibility

The uncertainty relations between conjugate thermodynamic variables, like energy and inverse temperature, come under the heading of Cramér–Rao inequalities in mathematical statistics. Since the Fisher information is usually proportional to the number of observations, the lower bound of the variance tends to zero as the number of observations becomes indefinitely large. This has an intuitive appeal in that, if we are willing to make more and more observations, what was only probable should become almost certain.

However, thermodynamics does not purport to make a statistical sampling of the energy of a system. What then plays the role of the sample number? We could, of course, divide our system into smaller and smaller subsystems with energies E_i. In the limit, if we had a Maxwell demon at our disposal who could determine how the energy is partitioned among the different degrees-of-freedom of the molecules, uncertainty would give way to certainty. But can we actually beat the macroscopic estimation of β which is itself a macroscopically defined quantity?

Let us consider a simple illustration. Suppose we have a set of r independent trials $x_1, \ldots, x_r$, where each x_i is either 1 ("success") or 0 ("failure"). If ϑ is an unknown parameter which represents the probability of success at each trial, we might want to use the sum $\sum_{i=1}^{n} x_i = s$, representing the number of successes, to obtain information about ϑ. The conditional probability of ϑ given s is $\binom{r}{s}^{-1}$. This does not depend upon the probability of success at each trial. It means that once the total number of successes s is given, additional information concerning the order of occurrence of successes and failures tells us no more about ϑ than what we already know. It also tells us that, at each trial, there is the same probability of success or failure and the successive trials are uncorrelated. Any statistic for which the conditional probability is independent of the parameter to be estimated is a sufficient statistic.

Whereas there is nothing to prohibit us from making more and more trials, we cannot endlessly breakup our system into smaller and smaller subsystems and expect that the energies, E_i, of the different subsystems will be additive and statistically independent of one another. At any level, where the additivity of the extensive thermodynamic variables holds, the sample mean energy will be a sufficient statistic for estimating the parameter β. But we know that as the sample size grows without limit, the sample mean approaches the mean of the distribution so that increasing the sample size seems to be beneficial. This would be true were we not dealing with the exponential family of distributions which give sufficient statistics *for any sample size*.

The fact that there is a lower bound to the number of possible subdivisions that can be made, in such a way that the individual subsystems retain their thermodynamic character, is related to the concept of entropy. In contrast to the usual statistical situation in which high probability becomes almost certainty as the sample size increases without limit, we cannot imagine an infinitely fine subdivision of our original thermodynamic system, since it would no longer be a statistical object and, consequently, its entropy would shrink to zero.

In the language of mathematical statistics, we would say that $\beta(E)$ of β is not a *consistent* estimator. A consistent estimator would approach the true value as the number of observations increases without bound. The inconsistency of the estimator is due to the fact that its variance is independent of the sample size. The estimator $\beta(E)$ is, however, *unbiased* since its average value coincides with the true value of the parameter. We will now use the property of *unbiasedness* to establish a lower bound on the variance of the energy.

The lower bound on the variance provides some important information on the type of correlations that exist between conjugate thermodynamic variables.

Since $\beta(E)$ is an unbiased estimator of β,

$$\int_0^\infty [\beta - \beta(E)] \frac{e^{-\beta E}}{\mathcal{Z}(\beta)} \Omega(E) \, dE = 0.$$

Differentiating under the integral sign with respect to β gives

$$1 - \int [\beta - \beta(E)] \left(E + \frac{\partial \ln \mathcal{Z}}{\partial \beta}\right) f(E; \beta) \, dE$$

$$= 1 + \overline{(\beta(E) - \beta)(E - \bar{E})} = 1 + \overline{(\Delta\beta)(\Delta E)} = 0. \qquad (4.78)$$

The covariance, $\overline{(\Delta\beta)(\Delta E)}$, can be used to define the *correlation* coefficient which measures the "degree of association" between the two random variables. Since we want the correlation coefficient to assume values between -1 and 1 inclusive, we define it as

$$\varrho \equiv \overline{(\Delta E)(\Delta \beta)} \Big/ \sqrt{\overline{(\Delta E)^2} \, \overline{(\Delta \beta)^2}}. \qquad (4.79)$$

If $\varrho = 0$, the variables are uncorrelated, while if $\varrho = \pm 1$, the variables are correlated linearly. Positive values of ϱ mean that one variable tends to increase with the other (*positive* correlation) while in the opposite case where one increases at the expense of the other, we would say there is *negative* correlation.

Equation (4.78) tells us that β and E are correlated linearly and negatively. The latter attests to the fact that the energy increases with the temperature or E decreases with β. Since $\rho^2 \leq 1$, it follows that

$$\overline{(\Delta\beta)(\Delta E)}^2 \leq \overline{(\Delta\beta)^2} \, \overline{(\Delta E)^2}. \qquad (4.80)$$

According to (4.78) the square of the covariance is unity, and consequently, (4.80) gives precisely the uncertainty relation (4.77) when the square root of both sides is taken and the temperature is measured in degrees. Inequality (4.80) is Schwarz's inequality stated in terms of expected values.

The values of ρ on the closed interval $[-1, 0]$ can be used as a measure of the degree of correlation between the conjugate variables. Moreover, it provides some valuable insight into the modification that the equations of state must incur as the system moves away from equilibrium. A value of ρ near 0 indicates a lack of linearity between E and β. It does not preclude the possibility of some *nonlinear* relationship. Therefore, the value $\rho = 0$ does not necessarily mean that the variables are statistically independent, but the converse is true.

In any *reversible* process, the variance $\overline{(\Delta\beta)^2}$ will always achieve its lower bound given by $1/\mathcal{I}_F$ in (4.75). For *irreversible* processes, the variance will be larger, meaning there is less information about β at our disposal from any observation made on E. Energy E will no longer be a sufficient statistic for estimating β, which in turn implies that it is no longer governed by a distribution belonging to the exponential family.

The very fact that we can speak about a variance for the inverse temperature seems to imply that β is something more than a mere parameter which characterizes the heat reservoir. Perhaps it should be considered as a random variable in its own right. We have also seen that there exists a certain "duality" between fluctuations in E and those in β which has permitted us to express deviations in one in terms of deviations in the other through (4.67) and (4.73). This possibility that β actually fluctuates brings us to the dual interpretation of thermostatistics.

4.11   Duality in Thermodynamics and Statistics

Thermodynamics and mathematical statistics classify variables into two categories according to whether they are *extensive* or *intensive* and *observable* or *estimable*. Bayesian statistics makes the further distinction of whether the variables are to be interpreted in the *limit-of-frequency* sense as opposed to the *degree-of-belief* sense.

Furthermore, there exists an inherent asymmetry in the two categories where prominence is given to the primary set of variables which are extensive and observable. The primary set defines the state of the system in thermodynamics, while in statistics, it comprises the sample data. The secondary set links the system to the outside world; it consists of the intensive variables in thermodynamics and those parameters which define the "state of nature" in statistics. These variables are *derived* quantities: in thermodynamics, they are obtained by partially differentiating the fundamental relation that describes the thermodynamic properties of the system, while in statistics, they are estimated in terms of the observables. This correspondence is shown in Table 4.1.

Table 4.1 Thermodynamical and Statistical Correspondence between Extensive and Intensive Variables

Extensive Variables	Intensive Variables
observable	estimable
frequency	degree-of-belief
frequency distribution	Bayes' distribution
sample data	state of nature
fundamental relation	equation of state
additive	uniform

The estimators of the intensive variables, which are random quantities since they are functions of the sample data, are to be interpreted in the sense of limit-of-frequency as opposed to estimation which elicits a degree-of-belief interpretation. Therefore, the secondary set must also possess well defined probability distributions, and we will now derive these distributions from the Gaussian law of error for their conjugate extensive variables.

Observable variables are measurable, and since a measurement can never be performed with unlimited precision, a distribution in the possible values that the extensive variables can assume will be set up about their most probable values. Since these variables are *additive*, the most probable value will be the sample mean which, we will assume, coincides with the mean of the distribution. This, in general, will be rigorously so when the sample size is allowed to increase without bound. As we have already mentioned in the last section, the sample mean will coincide with the mean of the distribution for any sample size because we are dealing with the exponential family of distributions.

Estimable variables are parameters upon which the probability distributions of the extensive variables depend; they define the state of nature of the system by relating the probability distributions to the properties of the reservoirs with which the system has been placed in contact. If the state of nature was known with certainty, an observation would not be informative and there would be no need to deal with probability distributions at all. The estimators of the parameters defining the state of nature are random variables since they are functions of the sample data.

The natural choice for the sufficient statistic for the energy is the sample mean,

$$\bar{E} = \frac{1}{r}\sum_{i=1}^{r} E_i, \qquad (4.81)$$

owing to the additivity property of the energy. We recall that the assertion that (4.81) is a sufficient statistic for the estimation of the inverse temperature simply means that the conditional probability of observing the value E_i of any subsystem, given the total energy $r\bar{E}$ of the composite system, is independent of the common value of the temperature when the subsystems are brought into thermal contact and left alone for a sufficiently long period of time so as to secure thermal equilibrium.

Any estimate of the temperature is itself a random variable since it is a function of the sample energies. A sufficient statistic requires that the converse is not true. This is in complete harmony with the second law: the concavity of the entropy with respect to the energy ensures that there is a unique solution, $\hat{\beta} = \beta(\bar{E})$, to

$$S'(\bar{E}) = \hat{\beta}, \qquad (4.82)$$

while the convexity of the entropy with respect to the inverse temperature is independent of the energy [cf. §4.5].

There are two different types of probability distributions for the inverse temperature, and both can be derived from the fundamental error law for the energy. In one case, we want a rational degree-of-belief of the inverse temperature given the sample mean energy. The randomness of β is in the sense of degree-of-belief that certain values are more probable than others. In the next section we show how the likelihood function can be obtained simply by switching the endpoints of the interval in the error law for the energy. This function is a measure of the "likelihood" of β, and it is only natural to choose that value whose likelihood is a maximum. Since we are dealing with the Legendre transform of the entropy, the stationary condition is entirely equivalent to the second law, which defines the inverse temperature. We inquire into the probability of making an error by "guessing" a certain value of β whose true value is $\hat{\beta}$. This probability distribution is derived by performing a Legendre transform on the entropy in the fundamental error law for the energy. In exactly the same way that the error law for an extensive variable is expressed in terms of the concavity criterion of the entropy, the error law governing its conjugate intensive variable is expressed in terms of the condition of convexity of the Legendre transform of the entropy.

4.11.1   The Bayes Distribution

The second law, (4.82), defines the inverse temperature which provides knowledge of the "state of nature" of the system, since it relates the probability distribution to the physical state in which the system is found. However, in the same way that it is essential to determine the probability of deviations in an extensive variable from its most probable value, so too, it is necessary to consider the statistical character of the knowledge of the state of nature. For if the state of nature were known exactly, no observation would be informative.

We can use the second law, (4.82), to convert the error law for the energy (4.32) into

$$f(E_i; \hat{\beta}) = A \exp\left\{-\hat{\beta}E_i + S(E_i) - L(\hat{\beta})\right\}, \qquad (4.83)$$

for which the best estimator of β, namely $\hat{\beta}$, has the highest probability. The function

$$L(\hat{\beta}) = S(\bar{E}) - \hat{\beta}\bar{E} \qquad (4.84)$$

is the Legendre transform of the entropy (or more correctly, the Massieu transform) with respect to the energy. Consulting the canonical expression for the entropy, (4.31), the Legendre transform is simply

$$L(\hat{\beta}) = \ln \mathcal{Z}(\hat{\beta}).$$

If we want to find a measure of rational belief in a value of β, we would start with a probability distribution

$$f(E_i; \beta) = A \exp\left\{S(E_i) - L(\beta) - \beta E_i\right\} \qquad (4.85)$$

instead of its maximum value given by (4.83). The Legendre transform $L(\beta)$ in (4.85) is related to the moment generating function by

$$\exp\{L(\beta)\} = \sum_{i=1}^{r} A \exp\left\{-\beta E_i + S(E_i)\right\}.$$

The distribution (4.85) depends upon a parameter β which is to be estimated from a sample of r observations on the energy. To this end, we form the joint, or "sample," distribution whose logarithm is

$$\sum_{i=1}^{r} \ln f(E_i; \bar{E}) = \mathcal{L}(\beta; \bar{E}) \propto -r \left[\beta \bar{E} + L(\beta) \right] \tag{4.86}$$

since the energies of the r subsystems are independent and identically distributed. Owing to the additivity of the energy, the log-likelihood function (4.86) depends on the observations of the energy only through the sample mean, (4.81).

Taken as a function of β, the log-likelihood function, (4.86), for a fixed sample mean $\bar{E}$ provides a subjective probability measure for different values of β. As discussed in §4.3, the maximum likelihood method inverts the functional dependency in (4.86), where β is now considered as the variable and $\bar{E}$ the parameter. Only in the case (i.e., the Bayes case) where β is equipped with a prior probability density can a probabilistic intepretation be given to (4.86); otherwise, we must be content with comparing the likelihoods of different values of β.

The maximum likelihood estimate is the solution of [cf. (4.17)]

$$\frac{\partial}{\partial \beta} \mathcal{L}(\beta; \bar{E}) = -r \left[\bar{E} + \frac{\partial L(\beta)}{\partial \beta} \right] = 0 \tag{4.87}$$

since

$$\frac{\partial^2}{\partial \beta^2} \mathcal{L}(\hat{\beta}; \beta) = -r \frac{\partial^2 L(\hat{\beta})}{\partial \beta^2} < 0.$$

This is a consequence of the fact that, in thermodynamics, the strict concavity property of the entropy is transformed into the stricty convexity of its Legendre dual,

$$L''(\hat{\beta}) = -1/S''(\bar{E}) > 0. \tag{4.88}$$

By the implicit function theorem, the likelihood equation (4.87) may be solved for the maximum likelihood estimate, $\hat{\beta} = \beta(\bar{E})$. Since S and L are Legendre transforms of one another, the maximum likelihood estimate, $\hat{\beta}$, will coincide with the thermodynamic definition of the inverse temperature, (4.82).

We have observed in §3.7.4 that a convexity criterion is invariant under the interchange of the endpoints. In terms of the entropy this means that

$$S(\bar{E}) - S(E_i) - S'(E_i)(\bar{E} - E_i) < 0 \tag{4.89}$$

is an equally valid definition of concavity. But how are we to interpret $S'(E_i)$? If E_i denotes the energy of the ith subsystem, each of these subsystems is characterized by a parameter β_i. Moreover, if nothing is known about these parameters, we may, invoking the principle of insufficient reason, expect them all to have a common value,

$$\beta = S'(E_i). \tag{4.90}$$

Certainly, when the subsystems are allowed to interact thermally and left alone for a sufficiently long period of time, we should expect that all the β's will be the same. This also implies that the Legendre transform

$$L(\beta) = S(E_i) - \beta E_i \tag{4.91}$$

will be the same for each subsystem.

Interchanging the endpoints in Gauss' principle (4.32) results in

$$\tilde{f}(\bar{E}; E_i) = A \exp\left\{ S(\bar{E}) - S(E_i) + S'(E_i)(E_i - \bar{E}) \right\}. \tag{4.92}$$

Owing to the concavity condition (4.89), it will again be a proper density and normalized if we let A, the integrating function in Gauss' principle, depend upon the E_i. Moreover, when the principle of insufficient reason (4.90) is introduced into (4.92), there results

$$\tilde{f}(\beta; \bar{E}) = A \exp\left\{ -\beta \bar{E} - L(\beta) + S(\bar{E}) \right\}, \tag{4.93}$$

where A can be a function of β. Expression (4.93) is the distribution of β given the sample mean $\bar{E}$ and, as such, it should be subject to Bayes' theorem, (4.59).

Following our lead in §3.7.4 we set $A = \hat{f}(\beta)$ and introduce the likelihood function in (4.86) into (4.93). It can then be written as

$$\tilde{f}(\beta; \bar{E}) = \frac{e^{\mathcal{L}(\hat{\beta};\bar{E})}}{\tilde{\mathcal{Z}}(\bar{E})} \hat{f}(\beta), \tag{4.94}$$

which is precisely Bayes' theorem, (4.59), equipped with the correct normalization factor, (4.61). Viewed in these terms, the ratio $e^{\mathcal{L}(\hat{\beta};\bar{E})}/\tilde{\mathcal{Z}}(\bar{E})$ is to be interpreted as the conditional probability density of β given the mean value of the energy $\bar{E}$. Hence, by switching the endpoints of the interval in Gauss' law of error and invoking the principle of insufficient reason, we have interchanged "cause" and "effect," resulting in Bayes theorem (4.94).

As an illustration of Bayes' theorem and a corroboration of Jeffreys' rule for selecting the prior density $\hat{f}(\beta)$, let us consider an ideal gas whose fundamental relation is

$$S(E, V, N) = N \ln\left(\frac{E^{3/2} V}{N^{5/2}} \right). \tag{4.95}$$

The Legendre transform of the entropy (4.95) with respect to the energy is

$$L(\beta) = -N \ln\left[\frac{N}{V} \left(\frac{2\beta e}{3} \right)^{3/2} \right], \tag{4.96}$$

where e is the base of the Napierian logarithm.

Introducing (4.95) and (4.96) into Bayes' theorem, (4.59),

$$\begin{aligned}
\tilde{f}(\beta; \bar{E}) &= \exp\left\{ -\beta \bar{E} + S(\bar{E}) - L(\beta) \right\} \hat{f}(\beta) \\
&= \frac{1}{\Gamma(\frac{3N}{2} + 1)} (\beta \bar{E})^{3N/2} e^{-\beta \bar{E}} \hat{f}(\beta).
\end{aligned}$$

This will turn out to be a gamma density,

$$\tilde{f}(\beta; \bar{E}) = \bar{E} \frac{(\beta\bar{E})^{3N/2-1}}{\Gamma(\frac{3N}{2})} e^{-\beta\bar{E}} \tag{4.97}$$

if we choose

$$\tilde{f}(\beta) = \frac{3N/2}{\beta} \tag{4.98}$$

as the prior density. The choice of the prior density as (4.98) is Jeffreys' rule, (4.60), for a variable that can take on all values on the right half interval. In fact, it is Jeffreys' prior (4.98) that gives the correct expression for the entropy when it is defined as

$$e^{-S(\bar{E})} = \int_0^\infty e^{-\beta\bar{E}-L(\beta)} \tilde{f}(\beta) \, d\beta$$
$$= \left(\frac{N}{V}\right)^N \left(\frac{2e}{3\bar{E}}\right)^{3N/2} \Gamma\left(\frac{3N}{2} + 1\right).$$

Using Stirling's approximation, it is readily seen that this gives the entropy of an ideal gas, (4.95).

The mean of the gamma density (4.97) is

$$\bar{\beta} = \frac{3N/2}{\bar{E}}, \tag{4.99}$$

which is simply equipartition of energy in a system with $3N$ degrees-of-freedom. But now it is the expression for the average of what was formally a parameter specifying the state of nature of the reservoir. In comparison with the gamma density (4.34) the variate E and parameter β have now changed roles. So the law of equipartition of energy may be interpreted as the mean value of either E or β, depending upon which is allowed to fluctuate.[5]

Moreover, the second central moments of the gamma densities, (4.34) and (4.97), satisfy the minimum uncertainty relation. The variance of (4.97) is $\sigma_\beta^2 = 3N/2\bar{E}^2$, while the variance of (4.34) is $\sigma_E^2 = 3N/2\bar{\beta}$, where $\bar{\beta} = \beta(\bar{E})$. Their product $\sigma_\beta^2 \sigma_E^2 = 1$, in conformity with the equality sign in the uncertainty relation (4.77).

It is quite remarkable that the same probability density governs both fluctuations in the conjugate thermodynamic variables; one variable fluctuates at the expense of the other, which is maintained at its expected value. The expected values are given by the law of equipartition of energy, (4.35) or (4.99), depending upon whether E or β fluctuates. The law of large numbers is surely at work here!

---

[5]There is still another possibility which will be discussed in §5.2.1 in which the number of degrees-of-freedom is allowed to fluctuate. It is, indeed, noteworthy that a single law should give rise to three different interpretations depending upon the physical situation.

4.11.2   Moments in the Dual Representation

We have seen that fluctuations in the energy cause fluctuations in its conjugate intensive variable. In order to display the symmetry that exists between the two conjugate variables, we consider the energy to be a continuous random variable and compare the variances of the probability densities (4.83) and (4.93).

The logarithm of the moment generating function of the probability density (4.83),

$$L(\hat{\beta}) = \ln \int_0^\infty A \exp\left\{-\hat{\beta}E + S(E)\right\} dE, \qquad (4.100)$$

is a completely monotone function since its derivatives alternate in sign. The first moment is just the stationarity condition of the likelihood function, (4.87), while the second central moment is

$$L''(\hat{\beta}) = \overline{(E - \bar{E})^2} \equiv \overline{(\Delta E)^2} > 0, \qquad (4.101)$$

which is equivalent to (4.10). Expression (4.101) clearly brings out the relation between the convexity of the Legendre transform of the entropy and the stability properties of thermodynamic systems: since $\overline{(\Delta E)^2} = -\bar{E}'(\hat{\beta})$, the heat capacity $C_V = -\hat{\beta}^2 \bar{E}'(\hat{\beta})$ is necessarily positive.

The logarithm of the moment generating function of the conditional probability density (4.93),

$$S(\bar{E}) = -\ln \int_0^\infty A \exp\left\{-\beta\bar{E} - L(\beta)\right\} d\beta, \qquad (4.102)$$

is also a completely monotone function. A comparison of (4.100) and (4.102) shows that the roles of the Legendre dual functions have been interchanged, although their physical meanings must be kept quite distinct. The "structure" function, $\Omega(E) = \exp[S(E)]$, represents the density of states in the energy interval between E and $E + dE$. Although we may formally consider $\Omega(\beta) = \exp[-L(\beta)]$ as a "density of states" for the parameter to lie in the interval between β and $\beta + d\beta$, it cannot be given a physical interpretation analogous to the structure function which contains all the mechanical information about the state of the *isolated* system. This is certainly a manifestation of the fact that the random variable, whose probability density is (4.93), cannot be interpreted in the limit-of-frequency sense but, rather, it must be interpreted in the sense of degree-of-belief.

The moments of the probability density (4.93) can be determined by differentiating the logarithm of the generating function (4.102). The first moment is none other than the second law, (4.82), provided the average and most probable values coincide. The second central moment of (4.93) is obtained by differentiating twice the negative of the logarithm of the generating function (4.102). We then obtain

$$-S''(\bar{E}) = \overline{(\beta - \hat{\beta})^2} \equiv \overline{(\Delta\beta)^2} > 0. \qquad (4.103)$$

Then introducing (4.101) and (4.103) into the Legendre dual relation, (4.88), we obtain the minimum uncertainty relation. In other words, (4.103) coincides with the inverse of the Fisher information, defined in (4.69); this represents the smallest value of the variance of any unbiased estimator of the inverse temperature. In fact, it represents the lower limit of the variance among *all* unbiased estimators.

Here we see a certain asymmetry creeping into the dual representations. Expression (4.93) is the probability density of β given $\bar{E}$ while (4.83) is the probability density of E given $\hat{\beta}$. But $\hat{\beta}$ is a function of $\bar{E}$, so it does not depend on any arbitrary value of the parameter β. This is as it should be since β is to be estimated by some statistic of the observations E and not the other way around!

4.11.3 Dual Probability Distributions

In this section, we derive the probability distributions for intensive variables that can be interpreted in the limit-of-frequency sense. Since these will turn out to be error laws, it is essential that we specify their most likely values.

The error laws for extensive variables will all have a structure similar to (4.32) which identifies the sample mean as the most probable value of the quantity measured. The most probable value is determined by the property of *additivity*. There is no reason to suppose that the error laws for intensive variables will identify the mean value with the most probable one, and consequently, their laws of error may have an entirely different structure than (4.32). But just as the intensive variables are derived quantities in thermodynamics, so too will be their error laws. It is precisely the Legendre transform which allows the error laws for intensive variables to be derived from the error laws for their conjugate variables. The specification of the most likely values of the intensive variables must come from the physical situation under consideration.

Introducing the Legendre transform (4.84) and

$$L(\beta_i) = S(E_i) - \beta_i E_i \qquad (4.104)$$

into the error law for the energy (4.32) gives

$$f(\beta_i; \hat{\beta}) = A \exp\left\{-L(\hat{\beta}) + L(\beta_i) + L'(\beta_i)(\hat{\beta} - \beta_i)\right\}, \qquad (4.105)$$

which is a proper probability distribution on account of the concavity criterion of the Legendre transform of the entropy. A comparison of the Legendre transforms, (4.91) and (4.104), shows that we are no longer claiming ignorance about the initial values of the parameters β_i, which determine the state of nature of the r subsystems. We will now show that (4.105) is a law of error which identifies mean values, other than the sample mean, as the most probable values of the quantity measured.

As an illustration, consider an ideal gas with the fundamental relation given by (4.95). By combining our observations $\beta_1, \beta_2, \ldots, \beta_r$ in some as yet unknown way, we obtain some average value (4.99) which we suppose is also its most probable value. For this value, the Legendre transform (4.96) will be given by

$$L(\hat{\beta}) = -N \ln \left[\frac{N}{V} \left(\frac{2\hat{\beta}e}{3} \right)^{3/2} \right],$$

while the Legendre transforms for the individual observations will be given by

$$L(\beta_i) = -N \ln \left[\frac{N}{V} \left(\frac{2\beta_i e}{3} \right)^{3/2} \right].$$

Introducing these expressions into the error law (4.105) results in

$$f(\beta_i; \hat{\beta}) = \frac{3N}{2\beta_i} \left(\frac{\hat{\beta}}{\beta_i} \right)^{3N/2} \exp \left\{ -\frac{3N}{2} \hat{\beta} \left(\frac{1}{\beta_i} - \frac{1}{\hat{\beta}} \right) \right\}, \qquad (4.106)$$

where we have set $A = 3N/2\beta_i$.

Expression (4.106) is recognized as the inverted gamma density. It is obtained from the gamma density for the energy (4.34) by setting $\beta_i = 3N/2E_i$ and recalling that the Jacobian of the change of variables is $3N/2E_i^2$. Since (4.106) is obtained from a proper normalized probability density by a simple one-to-one differentiable change of variable, it too is a proper normalized density.

The relationship between the error law for the extensive variable and the other derivable forms, including the error law for the intensive variable is shown in Fig. 4.4.

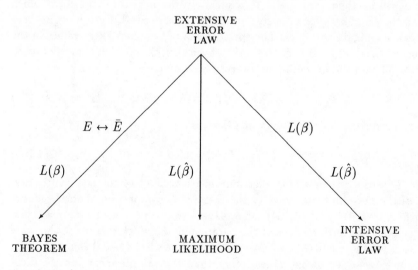

Figure 4.4: Derivation of three other densities from the error law where $E \leftrightarrow \bar{E}$ indicates to switch the endpoints of the interval and L means to introduce the Legendre transform.

The *random* variable β is equipped with a proper *posterior* probability density, (4.106), which is the inverse of the canonical distribution (4.6). It is truly astonishing that statistical thermodynamics should fit so perfectly into a probabilistic framework. We shall now show that the inverted gamma density, (4.106), is the error law leading to the *harmonic* mean as the most probable value of the inverse temperature.

In order that

$$\sum_{i=1}^{r} \frac{f_i'}{f_i} = 0, \tag{4.107}$$

where the prime now stands for differentiation with respect to $\hat{\beta}$, be equivalent to the definition of the harmonic mean,

$$\sum_{i=1}^{r} \left(\frac{1}{\beta_i} - \frac{1}{\hat{\beta}} \right) = 0, \tag{4.108}$$

we require

$$\frac{f'(\beta_i; \hat{\beta})}{f(\beta_i; \hat{\beta})} + \vartheta''(\hat{\beta}) \left(\frac{1}{\beta_i} - \frac{1}{\hat{\beta}} \right) = 0,$$

where $\vartheta''(\hat{\beta})$ may be interpreted as a Lagrange multiplier for the constraint (4.108). Integrating by parts then gives

$$f(\beta_i; \hat{\beta}) = A \exp \left\{ -\vartheta'(\hat{\beta}) \left(\frac{1}{\beta_i} - \frac{1}{\hat{\beta}} \right) + \psi(\ln \hat{\beta}) - \psi(\ln \beta_i) \right\}, \tag{4.109}$$

where

$$\psi(\ln x) = \int \frac{\vartheta'(x)}{x^2} \, dx.$$

We have used the Greene–Callen principle to guide us in the choice of the integration function $\psi(\ln \beta_i)$.

The error law (4.109) identifies the harmonic mean as the most probable value of the inverse temperature. This law of error coincides with (4.106) for a Lagrange multiplier given by $\vartheta'' = 2N/2$.

We have run the gamut: we started with the gamma density for the energies of the various subsystems. We then transformed from the E_i to the β_i, claiming that each subsystem should be characterized by a different value of β. Then we showed that the resulting probability density, the inverted gamma density, is an error law leading to the harmonic mean as the most probable value. But this identifies the arithmetic mean of the temperature as the most probable value of the temperature measured. Herein lies the physical significance of the error law.

Consider two identical bodies that are initially at the temperatures T_1 and T_2, where $T_1 > T_2$. A quantity of heat, $-\delta Q_1$, is withdrawn from the hotter body and an amount, δQ_2, is introduced into the colder one. In a completely *irreversible* process, no work is delivered and the final temperature of the composite system is the arithmetic mean of the two initial temperatures, $\hat{T} = \frac{1}{2}(T_1 + T_2)$. Owing to the fact that no work is done, this is the highest attainable temperature that the composite system can achieve. Since $\hat{T}$ is the arithmetic mean, $\hat{\beta}$ is the harmonic mean.

According to the deterministic viewpoint of thermodynamics, the composite system will certainly reach the final temperature $\hat{T}$ after thermal equilibrium has been established. However, the outcome of bringing two bodies initially at different temperatures into thermal contact is less than certain since the quantity of heat which is transferred from one body to the other is the *uncontrollable* part of the energy. The controllable part of the energy is the work, and given that this is zero, all we can surmise is that $\hat{T}$ is the most probable value of the temperature of the composite system. In other words, the act of thermal interaction makes the predicted outcome less than certain.

At the other extreme, we have a *reversible* process. Consider two identical bodies that have initial temperatures T_1 and T_2 and which are subsequently brought into thermal contact. In a reversible process, the maximum amount of work is performed and the final temperature of the composite system will be the *lowest* possible temperature that can be attained. The total entropy change is

$$\frac{\delta Q_1}{T_1} + \frac{\delta Q_2}{T_2} = 0. \tag{4.110}$$

Assuming the heat capacity $C = \delta Q/dT$ to be constant, (4.110) becomes

$$\frac{dT_1}{T_1} + \frac{dT_2}{T_2} = 0.$$

Integrating over the entire process, where the temperature of the first body varies from T_1 to $\hat{T}$ while that of the second changes from T_2 to $\hat{T}$, the final temperature $\hat{T}$ is found to be

$$\hat{T} = \sqrt{T_1 T_2}.$$

Therefore, for r systems which are brought into thermal contact, with the exchange of heat taking place reversibly, the minimum attainable final temperature is given by the *geometric* mean

$$\hat{T} = \sqrt[r]{\prod_{i=1}^{r} T_i}, \tag{4.111}$$

or equivalently,

$$\sum_{i=1}^{r} \ln\left(\frac{T_i}{\hat{T}}\right) = 0. \tag{4.112}$$

Requiring the likelihood equation (4.107), in which the prime now stands for differentiation with respect to $\hat{T}$, to be equivalent to (4.112), we set terms with the same index in the sums proportional to one another. Upon integrating, we obtain

$$f(T_i; \hat{T}) = A \exp\left\{ \varphi'(\hat{T}) \ln\left(\frac{T_i}{\hat{T}}\right) + \chi(\ln \hat{T}) - \chi(\ln T_i) \right\} \qquad (4.113)$$

as the law of error for which the geometric mean is the most probable value of the temperature measured, where

$$\chi(\ln x) = \int \frac{\varphi'(x)}{x}\, dx.$$

If the geometric mean is the most probable value of the quantity measured, the arithmetic mean of the logarithms of the observations is the most probable value of the logarithms of the quantity. Since the entropy of an ideal gas is related logarithmically to the temperature, we set $\varphi' = r \ln \hat{T} = \bar{S}$, and $2\chi(\hat{T}) = r(\ln \hat{T})^2 = \bar{S}^2/r$. This has the effect of transforming the error law (4.113) for the temperature into the normal error law

$$f(S_i; \bar{S}) = \frac{1}{\sqrt{2\pi r}} \exp\left\{ -\frac{(S_i - \bar{S})^2}{2r} \right\} \qquad (4.114)$$

for the entropy.

In a reversible process, the entropy is the independent random variable and the error law is expressed in terms of the convexity property of the energy, as we have seen in §1.12. There we have shown that normal density (4.114) is the small fluctuation limit, which is achieved in the limit as the number of degrees-of-freedom increases without bound. We have always maintained that it is immaterial whether we make observations on a single system in time or on many subsystems at any given instant. The parameter r may be identified as half the number of degrees-of-freedom that was introduced in §1.12 or the size of the sample.

Large sample theory in statistics has developed primarily because of the existence of certain theorems in the theory of probability that give excellent approximate results when the sample size is large. Here, we have seen how the central limit theorem is obtained in the large sample limit [cf. §4.2].

The mean square fluctuation of the entropy $\overline{(\Delta S)^2} = r$ corresponds to that of an ideal gas, and the relative fluctuation is proportional to $r^{-1/2}$, tending to zero as the number of degrees-of-freedom becomes indefinitely large. The small fluctuation limit is incorporated into the error law

$$f(T_i; \hat{T}) = A \exp\left\{ -\frac{r}{2}\left(\ln \frac{T_i}{\hat{T}} \right)^2 \right\} \qquad (4.115)$$

by the fact that the average entropy is proportional to the logarithm of the geometric mean value of the temperature which is the lowest attainable temperature. We therefore expect fluctuations in reversible processes to be substantially smaller than in irreversible processes. However, the distribution in the intensive variable, (4.115), differs from that of the corresponding extensive variable, (4.114), in one fundamental aspect: whereas the normal law for the entropy, (4.114), satisfies the condition that negative and positive errors of the same absolute magnitude are equally probable, the same is not true for the distribution in the temperature.

This is a good illustration of how the asymptotic limit theorems of probability, like the law of large number and the central limit theorem, lie at the heart of thermodynamics. It also dissuades us from searching for a more fundamental, microscopic theory since the general results, like the law of equipartition of energy, are none other than a manifestation of the extreme elegance of these asymptotic limit laws of probability theory.

Bibliographic Notes

§4.1

The general properties of phase space can be found in any one of the following excellent references, listed in chronological order:

- J.W. Gibbs, *Elementary Principles in Statistical Mechanics* (Yale Univ. Press, New Haven, 1902), Ch. VIII;

- H.A. Lorentz, *Lectures on Theoretical Physics*, Vol II, transl. by L. Silberstein and A.P.H. Trivelli (Macmillan, London, 1927), Ch. IV of "Entropy and Probability";

- F. Perrin, *Mécanique Statistique Quantique* (Gauthier-Villars, Paris, 1939), Ch. III;

- A.I. Khinchin, *Mathematical Foundations of Statistical Mechanics*, transl. by G. Gamow (Dover, New York, 1949), Ch. IV;

- A. Blanc-Lapierre and A. Tortrat, "Statistical mechanics and probability theory," in *Proc. Third Berkeley Symposium on Mathematical Statistics and Probability Theory*, Vol. III (Univ. Calif. Press, Berkeley, 1956), pp. 145-170;

- A. Blanc-Lapierre, *Mécanique Statistique* (Masson, Paris, 1967), Première Partie.

§4.2

Our discussion of the Tauberian and the central limit theorems follow

- W. Feller, *An Introduction to Probability Theory and Its Applications*, Vol. II, 2nd ed. (Wiley, New York, 1971), §XIII.5 and §XVI.7.

In the pair of papers,

- B.H. Lavenda and C. Scherer, "The statistical inference approach to generalized thermodynamics: I. Statistics" and "II. Thermodynamics," *Nuovo Cimento* **100**, 199-213, 215-227 (1988),

we studied the thermodynamic implications of dispersion in small thermodynamic systems within the realm of validity of the normal approximation to the canonical distribution.

§4.3

The maximum likelihood method is introduced into thermodynamics in

- B.H. Lavenda and C. Scherer, "Statistical inference in equilibrium and nonequilibrium thermodynamics," *Riv. Nuovo Cimento* **11**, # 6 (1988).

§4.4

In

- E.T. Jaynes, "The minimum entropy production principle," in *Annual Review of Physical Chemistry*, S. Rabinovitch, ed. (Annual Reviews Inc., Palo Alto, CA, 1980); reprinted in *E.T. Jaynes: Papers on Probability, Statistics and Statistical Physics*, R.D. Rosenkrantz, ed. (D. Reidel, Dordrecht, 1983), pp. 402-424,

Jaynes claims "the content of Gibbs' variational principle is that, given the measure $[\Omega]$ as a function of certain macroscopic quantities (energy, volume, mole numbers, etc.) the equilibrium properties of a system are determined." Yet, Jaynes has failed to realize that the volume and number of particles are complementary variables and the measure Ω cannot depend upon both of them simultaneously. Lorentz in his *Lectures* (§15) has clearly demonstrated the equivalence between the statistical and thermodynamic definitions for entropy in the case where the system depends upon the energy and volume and *not* upon the number of molecules in it. So Jaynes' approach, which categorically separates microscopic and macroscopic properties, is too simplistic and so too is his recipe of "predicting that behavior that can happen in the greatest number of ways, consistent with our data." The apparent success of his "maximum entropy" estimates in

- E.T. Jaynes, "Information theory and statistical mechanics. I, II," *Phys. Rev.* **106**, 620-630 (1957); **108**, 171-190 (1957); reprinted in *E.T. Jaynes: Papers on Probability, Statistics and Statistical Physics*, R.D. Rosenkrantz, ed. (D. Reidel, Dordrecht, 1983), pp. 4-38,

is due to the maximization of a concave function, subject to constraints, to obtain an exponential family of distributions.

§4.5

It is well known that exponential densities maximize entropy or minimize the discrimination information statistic which is at the basis of Kullback's approach:

- S. Kullback and R.A. Leibler, "On information and sufficiency," *Ann. Math. Stat.* **22**, 79-86 (1951);

- S. Kullback, "Certain inequalities in information theory and the Cramér-Rao inequality," *Ann. Math. Stat.* **25**, 88-102 (1954);

- S. Kullback, *Information Theory and Statistics* (Wiley, New York, 1959), Ch. 3.

And it is the exponential family of distributions which gives sufficient statistics for any size of the sample. This is what is behind Jaynes' maximum entropy approach—nothing more!

§4.8

Ehrenfest's work on the adiabatic principle, which was one of the cornerstones of the old quantum theory, can be found in

- P. Ehrenfest, "A mechanical theorem of Boltzmann and its relation to the theory of energy quanta," *Proc. Acad. Amsterdam* **16**, 591 (1916); reprinted in *Collected Scientific Papers*, M.J. Klein, ed. (North-Holland, Amsterdam, 1959), p. 340, or

- M.J. Klein, *Paul Ehrenfest* (North-Holland, Amsterdam, 1970), Ch. 11.

§4.9

In

- B.H. Lavenda, "Thermodynamic uncertainty relations," *Int. J. Theoret. Phys.* **26**, 1069-1084 (1987)

we used the second law to establish the inequality sign in the thermodynamic uncertainty relations.

We have developed the Bayesian formulation of statistical thermodynamics in

- B.H. Lavenda, "The Bayesian approach to thermostatics," *Int. J. Theoret. Phys.* **27**, 451-472 (1988)

which naturally leads to the dual distributions.

Jeffreys' use of improper prior probability densities to represent "knowing little" is discussed in

- H. Jeffreys, *Theory of Probability*, 3rd ed. (Clarendon, Oxford, 1961), p. 119.

Stigler, in

- S. Stigler, *The History of Statistics: The Measurement of Uncertainty before 1900* (Harvard Univ. Press, Cambridge, MA, 1986), pp. 127-129,

has criticized Jeffreys, along with Fisher and others, for having misinterpreted Bayes. According to Stigler, Bayes' argument of knowing nothing was applied to the empirical observable and not the *a priori* probability, which is not an observable. Reasoning from the binomial distribution, Bayes assumed that the number of successes is uniformly distributed rather than the probability of success. In other words, instead of assuming the principle of insufficient reason, where equal intervals of values of the *a priori* probability, p, are equally likely, Bayes assumed that the marginal distribution of N is uniform:

$$\Pr(0 < p < 1; N = n) = \int_0^1 \binom{m}{n} p^n (1-p)^{m-n} \, dp = \frac{1}{m+1}$$

since it is independent of n. Other distributions can also lead to a marginal distribution giving the same independence. The only way the binomial distribution can be singled out is to require that the marginal distribution holds for all n *and* all $m \geq 1$. Then the form of the marginal distribution is both necessary and sufficient in order that p be uniformly distributed. But this confines the argument to the binomial distribution which, according to Stigler, "puts a rather severe restriction on 'knowing nothing.' "

The asymptotic expansion of the likelihood function is described in

- D.V. Lindley, *Introduction to Probability and Statistics. Part 2: Inference* (Cambridge Univ. Press, Cambridge, 1970), pp. 128-130.

§4.10

Sufficiency in thermodynamics has been emphasized by Mandelbrot. Mandelbrot attributes his approach to a pioneering work of Szilard:

- L. Szilard, "Über die Ausdehnung der phänomenologischen Thermodynamik auf die Schwankungserscheinungen" ("On the extension of phenomenological thermodynamics to fluctuation phenomena"), *Z. Phys.* **32**, 753 (1925); reprinted and translated in *The Collected Works of Leo Szilard*, Vol. 1, B.T. Feld and G. Weiss-Szilard eds. (MIT Press, Cambridge, MA, 1972), pp. 70-102.

In this paper, Szilard shows how "phenomenological" and "statistical" thermodynamics can be made compatible. Phenomenological thermodynamics defines heat as a form of energy which always flows from hot to cold. Entropy always increases in such a process. The abstractness of the entropy concept led to the search for alternative formulations. Boltzmann proposed a definition of entropy in terms of the number of microscopic complexions that are compatible with a given macroscopic state [cf. §1.2]. To many this seemed replacing one abstraction by another.

Phenomenological and statistical thermodynamics differ in one important respect: In the phenomenological formulation thermodynamic equilibrium is the final resting place for all systems that have been left to themselves for a sufficiently long period of time. In other words, thermodynamic equilibrium is associated with a single, static state that persists in the course of time. Alternatively, according to the statistical formulation which takes a closer look at it, equilibrium is a distribution of constantly fluctuating configurations. One must explain why the system tends to such a distribution in time and why such a distribution is stable. Szilard attempted to answer these questions by showing that the second law "not only gives information about the mean value of these fluctuating parameters but also about the laws governing the deviations from these mean values."

Since the paper does not make for easy reading today, we can imagine that it was even more difficult to follow at the time it was written. Buried in the paper are many concepts, such as sufficiency, that anticipated mathematical statistics. The major result of the paper is that fluctuations can be included in a phenomenological formulation without any reference to a particular microscopic model. This surpised Szilard's thesis advisor, von Laue, who showed it to Einstein. With Einstein's support it was accepted for his Ph.D. thesis in 1922 and it took another three years before it was published.

With the hindsight of mathematical statistics, we can distill the following from Szilard's paper:

- the equivalence between the average over an ensemble and the average over one system over many experiments,

- the derivation of the exponential family of distributions which characterize equilibrium which Szilard refers to as the "normal" distribution,

- the realization that there exists a sufficient statistic for estimating the conjugate intensive parameters, and

- the expression for the entropy as the negative of the Kullback–Leibler expression for the minimum discrimination information statistic.

Szilard concludes his paper by saying

the second law does not lose anything of its exact character even in view of fluctuations and does not deteriorate to an approximation but evolves in some higher harmony thus determining the laws of the fluctuations as well.

Beginning in the mid-1950s, Mandelbrot picked up where Szilard had left off and showed how a purely phenomenological approach could be brought within the framework of mathematical statistics. In

- B. Mandelbrot, "An outline of a purely phenomenological theory of statistical thermodynamics: I. Canonical ensembles," *IRE Trans. Inform. Theory* **IT-2**, 190-203 (1956)

Mandelbrot argued that in order to obtain macroscopic behavior from microscopic dynamics, the latter must always be supplemented by some hypothesis of randomness. He then, drops the time bomb: If the kinetic foundations are not sufficient to explain macroscopic behavior with the additional hypothesis of randomness, in what sense are they necessary? In other words, is it possible to "short-circuit" the microscopic motions and focus attention on the element of randomness?

Moreover, since what we want to estimate is entirely macroscopic in nature, wouldn't it be sufficient to introduce randomness at the macroscopic level through the process of measurement of the different values that an extensive quantity can have on account of fluctuations? The fluctuations themselves are caused by the contact between system and reservoir which occurs when we attempt to make a measurement. For instance, to ask for the temperature of an isolated system is meaningless. In order to estimate the temperature, the system must first be placed in contact with a heat reservoir. This allows the energy to fluctuate and makes our measurement meaningful. If the heat reservoir is infinite, we can assign a definite temperature to the system but we cannot be sure of the system's energy since it will fluctuate about some mean value.

In this paper, Mandelbrot plots his future course: the role of sufficiency, the Cramér–Rao lower bound on the variance leading to the thermodynamic uncertainty relations, the method of maximum likelihood, and—most important—the realization that phenomenological statistical thermodynamics falls within the framework of the theory of mathematical statistics.

Mandelbrot develops the concept of sufficiency in thermodynamics in

- B. Mandelbrot, "The role of sufficiency and of estimation in thermodynamics," *Ann. Math. Statist.* **33**, 1021-1038 (1962).

His intent was to show that "it is a great pity that mathematical and physical statistics should have developed largely independently of each other, while using the same concepts." Mandelbrot concluded that Szilard can be considered as the co-inventor of the notion of sufficiency along with R. A. Fisher inasmuch as he showed that the Gibbs probability distribution for the canonical ensemble is the only one that has a single sufficient statistic.

It also predates Neyman's factorization theorem for a sufficient state and the work by Koopman and Pitman on the exponential family of distributions. Perhaps the almost two-year delay between the receipt and acceptance of the paper provides a hint to the difficulty of getting across an unorthodox view. His final paper on the subject,

- B. Mandelbrot, "On the derivation of statistical thermodynamics from purely phenomenological principles," *J. Math. Phys.* **5**, 164-171 (1964),

rephrased the zero, first, and second laws in order to permit a derivation of the canonical distribution. We note that in the last section he confused the Boltzmann and Gibbs definitions of temperature as well as their definitions of the entropy.

Spurred by the Szilard–Mandelbrot formulation, Tisza gave a postulatory approach to phenomenological statistical thermodynamics in

• L. Tisza and P.M. Quay, "The statistical thermodynamics of equilibrium,"
  Ann. Phys. (N.Y.) **25**, 48-90 (1963).

Szilard's method of deriving the canonical probability distribution is given in slightly modified form (i.e., distributions instead of densities are employed). Consider a composite system with a total energy $E = E_1 + E_2$. The conditional probability density that the first system will have energy E_1, given the total energy E, is

$$f(E_1; E) = \frac{f_1(E_1; \beta) \, f_2(E - E_1; \beta)}{f(E; \beta)}, \qquad (4.116)$$

which is assumed to be independent of β. (This is the criterion for a sufficient statistic.) It says that the probability of finding the first system with energy E_1 is independent of the temperature of the composite system: it depends only on the total energy of the composite system and nothing more.

Taking logarithms of both sides and introducing

$$\eta(E; \beta, \beta') = \ln f(E; \beta) - \ln f(E; \beta')$$

lead to

$$\eta_1(E_1; \beta, \beta') + \eta_2(E - E_1; \beta, \beta') = \eta(E; \beta, \beta'). \qquad (4.117)$$

This is because $f(E_1; E)$ is independent of β, or for that matter β', so that

$$f(E_1; E) = \frac{f_1(E_1; \beta') \, f_2(E - E_1; \beta')}{f(E; \beta')}$$

is also true. Differentiating this relation first with respect to E_1 at constant E_2 and then with respect to E_2 at constant E_1 give, respectively,

$$\frac{\partial}{\partial E_1} \eta(E_1; \beta, \beta') = \frac{\partial}{\partial E_1} \eta(E_1 + E_2; \beta, \beta') = \frac{\partial}{\partial E} \eta(E; \beta, \beta')$$

and

$$\frac{\partial}{\partial E_2} \eta(E_2; \beta, \beta') = \frac{\partial}{\partial E_2} \eta(E_1 + E_2; \beta, \beta') = \frac{\partial}{\partial E} \eta(E; \beta, \beta').$$

The left-hand sides are independent of E but depend upon β and β'. Therefore, we obtain

$$\eta(E; \beta, \beta') = B(\beta, \beta')E + A(\beta, \beta')$$

upon integrating. There are similar expressions for η_1 and η_2.

In order to satisfy (4.117), we require

$$A_1(\beta, \beta') + A_2(\beta, \beta') = A(\beta, \beta').$$

In terms of the logarithm of the original probability density we have

$$\ln f(E; \beta) = \ln f(E; \beta') - B(\beta, \beta')E - A(\beta, \beta').$$

Since the left-hand side is independent of β', the terms that contain β' on the right-hand side must add up to zero. Denoting the β' independent components of $\ln f(E; \beta')$, $B(\beta, \beta')$, and $A(\beta, \beta')$ by $\Omega(E)$, $\varphi(\beta)$, and $\ln \mathcal{Z}(\beta)$, respectively, we find the exponential form

$$f(E; \beta) = \frac{\Omega(E)}{\mathcal{Z}(\beta)} e^{-\varphi(\beta)E}$$

as the most general form of the probability density.

Szilard then pointed out that the temperature independence of the conditional probability density, (4.116), can be maintained if and only if $\varphi(\beta)$ is the same function of β for both subsystems. Hence, it cannot depend upon the specific characteristics of the system and must be a universal function of β or the temperature. Szilard also gave

$$
\begin{aligned}
S(\beta) &= -\int f(E; \beta) \ln \left(\frac{f(E; \beta)}{\Omega(E)} \right) dE \\
&= B(\beta)\bar{E} + \ln \mathcal{Z}(\beta)
\end{aligned}
$$

which, apart from the sign, will later be known as the Kullback–Leibler discrimination information statistic. Taking the differential of the entropy and using the second law results in setting $B(\beta) = \beta$.

Dutta in

- M. Dutta, "An essentially statistical approach to the thermodynamic problem," *Proc. Nat. Inst. Sci. Ind.* **19**, 109-126 (1953)

espoused the view that "an essentially statistical method, free from mechanical hypotheses about the nature of the constituent parts, for the interpretations of thermodynamic problems may be considered desirable and overdue." In this paper, which was communicated by S. N. Bose, Dutta built on Szilard's base by expressing the entropy as the negative of the logarithm of the likelihood function with parameters having their values at their maximum likelihood estimates. Note that it predates both Mandelbrot's and Jaynes' contributions, showing that the time was ripe for developing a phenomenological theory of thermostatistics. Even without Szilard's contribution, someone was bound to bridge the gap between mathematical and physical statistics since the former had reached maturity.

In

- M. Dutta, "An essentially statistical approach to thermodynamics," *Proc. Nat. Inst. Sci. Ind.* **21**, 373-381 (1956),

which was also communicated by Bose, Dutta extended his phenomenological approach to black radiation and systems composed of different constituents. Finally, in

- M. Dutta, "On maximum (information-theoretic) entropy estimation," *Sankhya* **28A**, 319-328 (1966)

Dutta compared the statistical estimates obtained from Jaynes' maximum entropy formalism, the method of maximum likelihood, and Gauss' method of estimation using the arithmetic mean as the statistic. Although he observed that the essence of Jaynes' work lies in the use of a probabilistic argument attached to a statistical model, he rightly emphasized that he (as far back as 1951!) "also proposed an essentially statistical discussion of problems of statistical thermodynamics for a statistical model."

Furthermore, he correctly pointed out that estimates based on maximum entropy are limited to the exponential family of distributions, as shown by Kullback in his monograph. The method of maximum likelihood has no such limitation, but the distribution must be given in advance rather than the minimum discrimination statistic, or the maximum entropy formalism, which specifies the distribution as well as providing for statistical estimates. Gauss' principle also suffers from the same limitations since it always leads to a probability distribution belonging to the exponential family. Hence, when the method of maximum likelihood is applied to a probability distribution of the exponential type, all three methods are equivalent. It is a pity that Dutta has not received the credit he rightly deserves.

§4.11

Our discussion follows

- B.H. Lavenda, "The statistical basis of nonequilibrium thermodynamics", in *Nonequilibrium Thermodynamics*, Vol. 3, S. Sieniutycz and P. Salamon, eds. (Taylor and Francis, New York, 1990), pp. 175-211.

- B.H. Lavenda and J. Dunning-Davies, "Probability distributions of thermodynamic intensive variables," *Int. J. Theoret. Phys.* **30**, 907-922 (1991).

Chapter 5

Origins of the Canonical Distribution

5.1 Introduction

In Chapter XV of *Elementary Principles in Statistical Mechanics*, Gibbs asked the following question: "If two phases differ only in that certain entirely similar particles have changed places with one another, are they to be regarded as identical or different phases?" His answer was that "if the particles are regarded as indistinguishable, it seems in accordance with the spirit of the statistical method to regard the phases as identical." With these words, Gibbs not only introduced the concept of *indistinguishability* of identical, nonlocalizable particles but also foresaw the advent of quantum statistics.

Gibbs defined two types of phases: *generic* and *specific* phases. The former remains unaltered under the exchange of positions of similar particles whereas the latter does not. For N indistinguishable particles, Gibbs claimed that the probability coefficient of the specific phase must be cut down by a factor of $N!$ to give the probability coefficient of the generic phase. Undoubtedly, his choice of the generic phase was based on his analysis of the change in entropy when gases in two adjacent chambers are allowed to interdiffuse. If the gases are different, there should be an increase in the entropy, while if they are the same, there should be no increase. This is the famous Gibbs paradox that we discussed in §1.13 and which he resolved by concluding that "it is equilibrium with respect to the generic phases, and not with respect to specific, with which we have to do in the evaluation of the entropy." The remedy which Gibbs offered was to divide the product of the single-particle partition functions by $N!$ because there are $N!$ ways of permuting the particles among themselves which are indistinguishable complexions.

We recall from §4.3 that the probability density of the canonical ensemble,

$$f(E; \beta) = \frac{e^{-\beta E}}{\mathcal{Z}(\beta)} \Omega(E), \tag{5.1}$$

belongs to the exponential family of distributions. We found that these distributions maximize the entropy, subject to the imposed constraints, and have some very important properties regarding statistical inference. The structure function, $\Omega(E)$, can increase no faster than a fixed power of the energy, E, for, otherwise, it would not be compatible with the central limit theorem [cf. §4.2]. This is the property that will most concern us here, since it makes (5.1) a gamma density. It implies that the law of large numbers is responsible for the success of the canonical ensemble.

Moreover, it appears that the canonical density is a continuous form of a discrete distribution. In fact, in Chapter II of his monograph, Gibbs derived the canonical density from the cumulative distribution of the number of degrees-of-freedom which is far more rigorous than the heuristic justification given in Chapter IV. Although Gibbs observed that many properties of the function he derived belong to a *law of error*, he did not pursue the analogy further.

In this chapter we will follow through this analogy and show that if the particle number instead of the volume is allowed to fluctuate, it becomes the probability distribution of the grand-canonical ensemble, from which the partition function follows by normalization. There is no need to introduce the property of indistinguishability that requires the division of the "thermodynamic" probability by $N!$. In fact, we have no such leeway since we are dealing with *proper* probability distributions that cannot be tampered with without, of course, destroying the probabilistic interpretation.

In the derivation of the canonical distribution for the continuous random variables of energy and volume from their parent distributions of the number of degrees-of-freedom and particle number, we will transfer our attention from the single-particle canonical distribution, discussed in the last chapter, to the N-particle canonical distribution. Furthermore, if we want to analyze the discrete fluctuations in the particle number rather than the continuous fluctuations in the volume, we obtain, in a straightforward manner, the grand-canonical distribution. The major attribute of the derivation is that there is no *ad hoc* deletion of terms or division by $N!$ to take into account that the particles are indistinguishable. It is already incorporated into our probabilistic description. Moreover, the derivation will clearly point out that fluctuations in the volume and particle number cannot occur simultaneously. In §3.7 we have given a thermodynamic argument why the two quantities cannot fluctuate simultaneously; here we will give a probabilistic argument why the two descriptions are equivalent, leaving till the next chapter a full analysis of this statistical equivalence principle.

5.2   Derivation of the Canonical Distribution

The cut-and-dried manner of introducing the canonical density (5.1) in statistical mechanics masks its statistical significance. In this section, we rederive the *continuous* density (5.1) from its parent probability distribution for a *discrete* random variable. The single-particle canonical density (5.1) is independent of the volume of the system.

We preface our discussion by noting that (5.1) is actually a gamma density, which seems to imply that the law of large numbers is what is actually behind the amazing success of thermostatistics. Actually, the relation between the canonical distribution and the theory of errors can be found in Chapter II of Gibbs' treatise, although it appears there in a very cryptic way. And although he noted the analogy with the theory of errors and that the probability distribution depends only on the number of degrees-of-freedom "being in other respects independent of its dynamic nature," Gibbs stopped short of asking what is the random variable and its distribution which give rise to the gamma density? We will now provide an answer to this question.

5.2.1 The Temperature Ensemble

Consider a phase space with $2m$ degrees-of-freedom without bothering to give them any mechanical interpretation. As we remarked in §4.1, a system can be represented by a point with coordinates $x_1, x_2, \ldots, x_{2m}$ in this phase space. The volume element,

$$dV(E;m) = dx_1 \cdots dx_{2m} = \Omega(E; m-1)\, dE, \qquad (5.2)$$

is a special case where all other extensive parameters needed to specify the state of the system are held constant. The surface area, Ω, is explicitly given by

$$\Omega(E; m-1) = \frac{dV}{dE} = C^m \frac{E^{m-1}}{\Gamma(m)},$$

where C is a numerical constant that depends on the nature of the physical system under investigation.

This is an asymptotic expression for large m. It assumes that the energy can be expressed as the sum of squares of the coordinates in the phase space,

$$E = x_1^2 + x_2^2 + x_3^2 + \cdots + x_{2m}^2,$$

which is the square of the distance from the origin to any given point in a multidimensional space of $2m$ dimensions. The volume of the hypersphere swept out by the radius $\sqrt{E}$ is proportional to E^m. Consequently, the fraction of the volume which lies in a hypersphere of half the radius is $(E/2)^m / E^m = (1/2)^m$, and since m is a very large number, this is a very small fraction. We are thus led to the conclusion that in the case of a hypersphere of very high dimensionality, essentially all the volume lies near to the surface!

Suppose that we divide the phase space into two parts: one with volume

$$V_1(E_1; k) = \int \cdots \int dx_1 \cdots dx_{2k} = C^k \frac{E_1^k}{k!},$$

where the integration extends over all phases for which the energy does not exceed E_1, and another with volume

$$\mathcal{V}_2(E_2; m - k) = \int \cdots \int dx_{2k+1} \cdots dx_{2m} = C^{m-k} \frac{E_2^{m-k}}{(m - k)!}$$

for all phases whose energies are not greater than E_2. The total energy of the composite system is

$$E = E_1 + E_2.$$

The energies, E_1 and E_2, are not functions of the instantaneous values of the degrees-of-freedom in each subvolume; rather, they are related to the *expected* number of degrees-of-freedom in each of the subvolumes.

We can never be absolutely certain that there are exactly $2k$ degrees-of-freedom in the first subvolume for we have no way of controlling precisely the microscopic coordinates. Therefore, we must formulate it probabilistically and ask for the probability that out of the entire phase space of $2m$ degrees-of-freedom, one of the two subvolumes will contain k pairs of degrees-of-freedom.

The probability of finding a pair of degrees-of-freedom in the first sub-volume is $p = E_1/E$. Since the presence of one pair does not alter the presence of there being another pair, the probability that there are k pairs of degrees-of-freedom present in the first subvolume is p^k. Similarly, the probability that a pair of degrees-of-freedom is not present in the first subvolume is $q = (E - E_1)/E$, which is the probability that it will be present in the complementary subvolume. Hence, the probability of their being $m - k$ pairs of degrees-of-freedom in the second subvolume is q^{m-k}. The probability of finding k pairs of degrees-of-freedom in the first subvolume is the product of these two expressions multiplied by the number of ways that k pairs of degrees-of-freedom can be chosen from m pairs of degrees-of-freedom, or

$$f(k|m) = \binom{m}{k} p^k q^{m-k} = \frac{\mathcal{V}_1(E_1; k)\, \mathcal{V}_2(E - E_1; m - k)}{\mathcal{V}(E; m)}. \tag{5.3}$$

Newton's binomial theorem guarantees that the binomial distribution, (5.3), is normalized. We can state it in the form of a "composition" law:

$$\mathcal{V}(E; m) = \sum_{k=0}^{m} \mathcal{V}_1(E_1; k)\, \mathcal{V}_2(E - E_1; m - k) \tag{5.4}$$

for the phase space volume elements. This composition law is the discrete form whose continuous analog has been given in §4.6. The sum is over pairs of degrees-of-freedom and *not* over the energy since the former are the random variables. It is precisely the transfer from the former to the latter, as the relevant random variable, that enables us to recover the usual composition rule for the structure function.

Thus far, we have considered the total number of pairs of degrees-of-freedom to be a constant. But suppose that it isn't fixed so that m can actually fluctuate. If the degrees-of-freedom are "well mixed" in the phase space (like raisins in a dough or misprints in a book when their total number is large but their individual probability is small), the probability of there being m pairs of degrees-of-freedom, with an expected value βE, is

$$f(m) = \frac{(\beta E)^m}{m!} e^{-\beta E}. \tag{5.5}$$

The property of being well mixed has led us to the Poisson distribution, (5.5), with parameter β. This parameter is a measure of the density of pairs of degrees-of-freedom in the phase space. The replacement of a fixed number by a random variable having a well defined probability distribution is a common practice in statistics which goes under the name of "randomized sampling."

Since m is no longer constant, the probability distribution (5.3) must be considered as a *conditional* probability for k, given the value of m; the latter can assume any value from k to infinity. The probability of m is given by the Poisson distribution (5.5) which when multiplied by (5.3) and the product is summed over all possible values that m can take, given the total probability,

$$f(k) = \sum_{m=k}^{\infty} f(k|m)f(m) = \frac{(\beta E_1)^k}{k!} e^{-\beta E_1}, \tag{5.6}$$

of finding k pairs of degrees-of-freedom regardless of whatever the value of m may be.

Expression (5.6) shows that under randomized sampling from a varying population, the random variable k is no longer a binomial variate, as it would be for sampling with replacement from a constant population; rather, it is a Poisson variate whose expectation βE_1 depends on the scale factor, β. However, it is quite evident from the form of (5.6) that β is *not* the scale parameter of the Poisson distribution (5.6). *It is the random variable which must be scaled in order to make it a standardized variable* and the energy, E_1, thus far is no random variable. We now have to find the distribution for which β is the true scale factor.

We have kept the subvolumes constant and asked for the probability of there being k pairs of degrees-of-freedom in one of them. We can do something else that will turn out to be equivalent to this: we can vary the subvolume V_1 until it contains precisely k pairs of degrees-of-freedom. By increasing this subvolume, we are increasing the energy E_1 until the subvolume can accommodate exactly k pairs of degrees-of-freedom. This energy, $\mathbf{E}$, is the required random variable. The probability that the phase space volume $V_1(E_1)$ will contain less than k pairs of degrees-of-freedom will be given by

$$\Pr(\mathbf{E} > E_1) = \sum_{j=0}^{k-1} \frac{(\beta E_1)^j}{j!} e^{-\beta E_1}. \tag{5.7}$$

In other words, the probability that the energy will exceed E_1 is equal to the probability that the random variable $\mathbf{K}$, having a Poisson distribution of mean

βE_1, will be less than k,

$$\Pr(\mathbf{E} > E_1) = \Pr(\mathbf{K} < k).$$

Since

$$\Pr(E_1 \leq \mathbf{E}) = 1 - \Pr(\mathbf{E} > E_1)$$

coincides with the cumulative distribution function, $F_\beta(E_1)$, we obtain the probability density function

$$f(E_1; \beta) = \frac{\partial F_\beta}{\partial E_1} = \beta \frac{(\beta E_1)^{k-1}}{\Gamma(k)} e^{-\beta E_1}. \tag{5.8}$$

This follows from the well known identity

$$\sum_{j=0}^{k-1} \frac{(\beta E_1)^j}{j!} e^{-\beta E_1} = \int_{E_1}^{\infty} \beta \frac{(\beta x)^{k-1}}{(k-1)!} e^{-\beta x} \, dx,$$

which can easily be proved directly by integrating by parts. Thus β is the scale parameter in the distribution of E_1 which is a gamma density with $2k$ degrees-of-freedom.

This is precisely Gibbs' result, and as he noted, it depends only on the number of degrees-of-freedom of the system "being in other respects independent of its dynamical nature." Actually, Gibbs worked backward: beginning with the gamma density, (5.8), he integrated to get the cumulative distribution

$$F_\beta = 1 - \sum_{j=0}^{k-1} \frac{(\beta E_1)^j}{j!} e^{-\beta E_1}.$$

Then he assumed that there is a value of the energy for which $F_\beta = \frac{1}{2}$ and concluded that the probability of a phase falling between these limits is greater than the probability of it falling in any other limits enclosing an equal extension-in-phase. He observed that it is equal to the probability of the phase falling outside these limits.

The generality of the canonical density, (5.1), is due to its being a gamma density which, as we have seen in §4.2, is intimately connected to the law of large numbers. It also possesses the important "reproductive" property that the sum of two independent gamma variates with $2k$ and 2ℓ degrees-of-freedom is also a gamma variate with $2(k + \ell) = 2m$ degrees-of-freedom and whose density is again a gamma density,

$$f(E; \beta) = \beta \frac{(\beta E)^{m-1}}{\Gamma(m)} e^{-\beta E}, \tag{5.9}$$

where E is the sum of the energies of the two systems with a common value of the scale parameter, β. We will soon see that this reproductive property of the gamma density is responsible for the composition law of the structure function which we derived in §4.6.

Recalling our discussion of the method of maximum likelihood in §4.3, we remark that the mean of the gamma density,

$$\bar{E} = m/\beta, \tag{5.10}$$

coincides with the maximum likelihood estimate of the parameter β in terms of the mean energy $\bar{E}$. Furthermore we gave, in §4.4, a thermodynamic argument of why β is proportional to the inverse temperature. And thus we established (5.10) as the law of equipartition for a system with $2m$ degrees-of-freedom.

We can also appreciate the beautiful symmetry between the mean of the gamma density, (5.10), and

$$\bar{m} = \beta E, \tag{5.11}$$

which is the mean of the Poisson distribution, (5.5), under randomized sampling. The two descriptions are, in fact, statistically equivalent. We can either determine the mean number of pairs of degrees-of-freedom in terms of a fixed energy, (5.11), or, according to (5.10), the mean energy in terms of a fixed number of pairs degrees-of-freedom. This is a fundamental statistical equivalence principle that holds for nonconjugate variables like the particle number and volume which shall be discussed at length in the following chapter. If m fluctuates, E is a parameter, while if E fluctuates, m is fixed. In contrast to the *discrete* random variable, m, of the Poisson distribution, (5.5), the energy, E, is a *continuous* random variable having a gamma density, (5.9).

When the gamma density is cast in the form of the canonical density, (5.1), we obtain the expression

$$\mathcal{Z}(\beta) = (C/\beta)^m$$

for the single-particle partition function. Consequently, the ratio $\Omega/\mathcal{Z}$ together with the canonical density (5.1) is independent of the particular nature of the physical system which is contained in the constant C.

Let us recall that in order to obtain the probability that there are exactly $2k$ degrees-of-freedom in a certain subvolume of phase space, we divided the phase volume into two complementary subvolumes. We now divide phase space according to the number of degrees-of-freedom. A small subsystem, which we label by the index 1, has $2m_1$ degrees-of-freedom while the larger system which will act as the heat reservoir, labeled by the index 2, has $2m_2$ degrees-of-freedom. By virtue of the reproductive property of the gamma density, the composite system will have $2(m_1 + m_2) = 2m$ degrees-of-freedom.

In §4.6 we found that the probability density for the energy of the small subsystem is given by

$$\frac{\Omega_1(E_1; m_1 - 1)\, \Omega_2(E - E_1; m_2 - 1)}{\Omega(E; m - 1)} \tag{5.12}$$

subject to the condition that the energy of the composite system is E. The composition law for the structure function,

$$\Omega(E) = \int_0^E \Omega_1(E_1; m - 1)\, \Omega_2(E - E_1; m - m_1)\, dE_1, \tag{5.13}$$

ensures that (5.12) is a proper probability density. In contrast to summing over a discrete random variable to get the composition law of the phase volumes, (5.4), we integrate over a continuous random variable to get an analogous law for the surface areas. In the next section, we will give still another derivation of the composition law, (5.13).

The usual procedure to derive the canonical density from the composition law (5.13) is to exploit the fact that $E_1 \ll E$. The structure function of the heat reservoir, Ω_2, is developed in a Taylor series,

$$\frac{\Omega_1(E_1; m_1 - 1)\,\Omega_2(E - E_1; m_2 - 1)}{\Omega(E; m - 1)} = \frac{\Omega_1(E_1)}{\Omega(E)}\left\{\Omega_2(E) - \frac{\partial \Omega_2}{\partial E}E_1 + \cdots\right\},$$

in which higher powers of E_1 are neglected. Introducing the reservoir parameter,

$$\beta = \frac{\partial \ln \Omega_2}{\partial E},$$

and setting

$$\mathcal{Z}(\beta) = \Omega_2(E)/\Omega(E)$$

give

$$\frac{\Omega_1(E_1; m_1 - 1)\,\Omega_2(E - E_1; m_2 - 1)}{\Omega(E; m - 1)} \simeq \frac{\Omega_1(E_1)}{\mathcal{Z}(\beta)}e^{-\beta E_1},$$

where we have used the approximation $e^{-x} \simeq 1 - x$.

Although the derivation may seem impeccable, there is always a nonnegligible error that is introduced when two terms are used to approximate an exponential. This can be avoided simply by noticing that (5.12), for $m_1, m_2 \gg 1$, is approximately the beta density

$$f_1(x; \beta) = \frac{1}{B(m_1, m_2)}x^{m_1 - 1}(1 - x)^{m_2 - 1},$$

where $x = E_1/E$ and $B(m_1, m_2) = \Gamma(m_1)\Gamma(m_2)/\Gamma(m)$ is the beta function. The beta density can be derived from a product of gamma densities simply through a change of variables. Introducing the equipartition value for the total energy according to (5.10) and taking the limit as $m \to \infty$, with $m_2 \approx m$, the second term in the beta density becomes

$$\lim_{m \to \infty}\left(1 - \frac{\beta E_1}{m}\right)^m = e^{-\beta E_1}.$$

Consequently, in this limit, the beta density transforms into the gamma density characterizing energy fluctuations in a small system which is in thermal contact with a heat reservoir.

5.2.2   The Pressure Ensemble

The type of complementarity which we have already noticed in the number of degrees-of-freedom and energy also appears between the particle number and the volume. Indeed, this will lead us in the next chapter to raise it to a "statistical equivalence principle" between fluctuations in complementary variables, where one is always discrete while the other always continuous. Here, we will consider it insofar as these fluctuations contribute to the canonical ensemble.

We divide a given volume V into r subvolumes V_j and ask for the probability that a particle will be found in any one of these subvolumes. Since the gas of particles is uniformly distributed throughout the volume and the particles are pointlike so that they occupy essentially no volume, this probability of finding a particle in the jth subvolume will be

$$p_j = \frac{V_j}{V}.$$

Then the probability that out of a total of N particles there will be N_j particles present in this subvolume is $p_j^{N_j}$. Consequently, the probability of finding N_1 particles in the subvolume V_1, N_2 particles in the subvolume V_2, and so on, will be given by the multinomial distribution

$$f(N_1, N_2, \ldots, N_r | N) = \frac{N!}{N_1! N_2! \cdots N_r!} p_1^{N_1} p_2^{N_2} \cdots p_r^{N_r} \tag{5.14}$$

subject to the condition that the total number of particles,

$$\sum_{j=1}^{r} N_j = N,$$

be constant.

According to Boltzmann [cf. §1.2], who thought large numbers were more impressive than "mere" probabilities, each p_j represents the cell volume ω_j. The whole (molecular) μ-space is divided up into very small, but finite, parallelopipeds ω_j. Since we have no a priori reason for discriminating among them, they are all given a common volume ω. Then instead of the multinomial distribution, (5.14), Boltzmann was led to consider the volume of a so-called Z-star

$$Z = \frac{N!}{N_1! N_2! \cdots N_r!} \omega^N. \tag{5.15}$$

As the phase point is varied continuously, the system will pass from one Z-star to another with the set of integers N_j changing in a discontinuous manner. The value of Z is taken as the "thermodynamic" probability of a given state. The contact between Z and the real world is made by setting the entropy equal to the logarithm of the thermodynamic probability. But the entropy must be an extensive quantity, so that Z must be corrected by dividing it by $N!$. We recall from §1.13 that the rationale behind this "correction" is that the number of configurations have been overcounted exactly $N!$ times since permutations of the *indistinguishable* particles do not give rise to new configurations. But the multinomial distribution, (5.14), *already* assumes that the particles are indistinguishable; we are only interested in the number of particles in a given cell or volume elements and not which particles are in it. The apparent similarity between the multinomial distribution, (5.14), and Boltzmann's Z-star, (5.15), is deceiving.

Thus it is not at all clear why one should divide by $N!$ —if not only to obtain an entropy which is extensive. If we hold on to the fact that it is a proper probability distribution, then this degree of arbitrariness cannot be allowed. It is precisely the strength of the probabilistic approach that prevents us from taking the liberty of inserting extraneous considerations. All that is necessary is to appeal to the method of randomized sampling.

If we allow for randomized sampling, we must release the constraint that the total number of particles, N, be constant. Since the particles are "well mixed," N becomes a Poisson variate whose distribution is

$$f_V(N; \varrho) = \frac{(\varrho V)^N}{N!} e^{-\varrho V}. \tag{5.16}$$

The scale factor, ϱ, is the density. Since N is now a random variable, the multinomial distribution (5.14) is a *conditional* probability distribution. The total probability will then be obtained by multiplying (5.14) by (5.16) and summing over all possible values of N. We then obtain

$$
\begin{aligned}
f(N_1, N_2, \ldots, N_{r-1}; \varrho) &= \sum_N f(N_1, N_2, \ldots, N_r | N) f_V(N; \varrho) \\
&= \prod_{j=1}^{r-1} \frac{(\varrho V_j)^{N_j}}{N_j!} e^{-\varrho V_j}.
\end{aligned}
$$

Thus randomized sampling from a multinomial distribution, given that the total number is a Poisson variate, induces a transformation from $r - 1$ independent multinomial variates to $r - 1$ independent Poisson variates.

And just as β was not the scale parameter for the number of degrees-of-freedom, so too ϱ is not the scale parameter for the number of particles. This suggests that there is another random variable for which ϱ is the true scale parameter.

Instead of asking for the probability of finding N particles in a fixed volume, we allow the volume to vary until it contains precisely N particles. By the same line of reasoning used in the last subsection, we construct the cumulative distribution function and differentiate it to obtain

$$f_N(V; \varrho) = \varrho \frac{(\varrho V)^{N-1}}{\Gamma(N)} e^{-\varrho V}. \tag{5.17}$$

Analogous expressions are obtained for each of the subvolumes, V_j. Hence, ϱ is the scale parameter in the distribution for V, which turns out to be a gamma distribution.

The mean value of the Poisson distribution (5.16),

$$\bar{N} = \hat{\varrho}V, \tag{5.18}$$

coincides with the most likely value of the scale parameter, ϱ, in the presence of fluctuations in the particle number. Alternatively, the mean value of the gamma density (5.17),

$$\bar{V} = \frac{\hat{\varrho}}{N}, \tag{5.19}$$

coincides with the most likely value of the density, $\hat{\varrho}$, in the presence of volume fluctuations.

The symmetry between these two maximum likelihood estimates of the density is apparent: *either V fluctuates at constant N or N fluctuates at constant V*. The particle number is a discrete random variable while the volume is a continuous variate. Although this seems entirely reasonable from everyday experience, we will see in §6.5 that it corresponds to the classical limit of quantum statistics.

The gamma density for the volume, (5.17), can be written in the canonical form

$$f_N(V) = \frac{\Omega(V; N-1)}{\mathcal{Z}(\varrho)} e^{-\varrho V},$$

where

$$\Omega(V; N-1) = \frac{V^{N-1}}{\Gamma(N)} \tag{5.20}$$

is the configuration space structure function and

$$\mathcal{Z}(\varrho) = \varrho^{-N}$$

is the configuration space partition function. The logarithm of the partition function, $\mathcal{Z}(\varrho)$, is the generating function for the moments of V; in particular, the mean value is

$$-\frac{d}{d\varrho} \ln \mathcal{Z}(\hat{\varrho}) = \frac{N}{\hat{\varrho}} = \bar{V}. \tag{5.21}$$

Integrating the expression for the configurational space structure function, (5.20), gives

$$\mathcal{V}(V; N) = \frac{V^N}{N!}, \tag{5.22}$$

which is the volume of configuration space occupied by the system. Gibbs could only recover the numerator by integrating over all $3N$ positional coordinates and referred to it as the *extension-in-configuration*. The division by $N!$ had to be introduced in a completely *ad hoc* manner to make theory agree with fact. Here we have shown, just as in §1.13, that one need not have recourse to the quantum result that the particles are really indistinguishable in order to justify the division by $N!$. A valid theory shouldn't contain any extraneous elements.

5.3   Complementary Canonical Distributions

The canonical density, (5.9), refers to a single particle. We want to generalize it to a system containing N particles. We recall from §3.4.1 that the generating function of N random variables is the Nth power of the generating function for a single variable.

In view of the definition of the single-particle partition function,

$$\mathcal{Z}(\beta) = \int_0^\infty e^{-\beta E} \Omega(E)\, dE,$$

the N-particle partition function is given by

$$\tilde{\mathcal{Z}}(\beta) \equiv \mathcal{Z}^N(\beta) = \int_0^\infty e^{-\beta U} \tilde{\Omega}(U)\, dU = \left(\frac{C}{\beta}\right)^{mN},$$

where the total energy $U = NE$. With the expression for the structure function as

$$\tilde{\Omega}(U) = C^{mN} \frac{U^{mN-1}}{\Gamma(mN)},$$

the probability density can be written in the canonical form

$$\tilde{f}(U; \beta) = \frac{e^{-\beta U}}{\tilde{\mathcal{Z}}(\beta)}\, \tilde{\Omega}(U). \qquad (5.23)$$

The "total" canonical distribution of the combined phase space for N particles in a closed system will be the product of the gamma densities, (5.23) and (5.17):

$$
\begin{aligned}
f_N(U, V) &= \tilde{f}(U; \beta) f_N(V; \varrho) \\
&= \frac{\tilde{\Omega}(U)}{\tilde{\mathcal{Z}}_N(\beta)} e^{-\beta U} \varrho \frac{(\varrho V)^{N-1}}{\Gamma(N)} e^{-\varrho V} \\
&= \frac{\tilde{\Omega}(U)}{\tilde{\mathcal{Z}}_N(\beta, \varrho)} e^{-\beta U} \frac{V^{N-1}}{\Gamma(N)} e^{-\varrho V},
\end{aligned}
\qquad (5.24)
$$

where the moment generating function

$$
\begin{aligned}
\tilde{\mathcal{Z}}_N(\beta, \varrho) &= \int_0^\infty \int_0^\infty e^{-\beta U - \varrho V} \tilde{\Omega}(U)\, \Omega(V)\, dU\, dV \\
&= \mathcal{Z}^N(\beta) \mathcal{Z}(\varrho) = \left(\frac{1}{\lambda^{2m}}\right)^N \left(\frac{1}{\varrho}\right)^N
\end{aligned}
\qquad (5.25)
$$

is the product of the partition functions of the two phase spaces. The quantity $\lambda = \sqrt{\beta/C}$ will soon be given a physical meaning.

We have considered the properties of the single-particle partition function with respect to β in §4.4, and there is nothing to add in its generalization to N particles. As we have already mentioned, it is a well known property that the generating function of r events is the product of the generating functions of the individual events. Therefore we need only concentrate on the properties of the generating function with respect to ϱ.

Consider the function

$$\tilde{Z}_N(\beta, \varrho)\, e^{\beta U + \varrho V}. \tag{5.26}$$

It is quite clear that since $\tilde{Z}_N$ is a positive and monotonically decreasing function of ϱ tending to infinity as $\varrho \to 0$, (5.26) too will tend to infinity as $\varrho \to 0$. And because of the exponential factor, it will also tend to infinity as $\varrho \to \infty$. Since $\ln \tilde{Z}_N$ is a convex function of β and ϱ, so too will be the logarithm of (5.26). Hence, it must necessarily possess a single minimum at

$$V + \left(\frac{\partial \ln \tilde{Z}_N}{\partial \varrho} \right)_\beta = 0.$$

And since the second derivative does not vanish, the maximum likelihood value $\hat{\varrho}$ can be obtained by inverting this implicit equation.

In thermodynamics U, V, and N are usually taken to be the independent extensive variables in the entropy representation. Once fluctuations are taken into account, this is no longer true and we must choose between a continuous description in terms of the volume or a discrete characterization in terms of the number of particles.

If we choose U and N as the independent variables, we must use the phase volume (5.22), rather than its surface area (5.20), since we are dealing with a discrete distribution in N in contrast to a continuous density in V. The canonical distribution of an open system, with rigid walls which are in contact with a heat reservoir,

$$
\begin{aligned}
f_V(U, N) &= \tilde{f}(U) f_V(N) \\
&= \frac{\tilde{\Omega}(U)}{\tilde{Z}(\beta)} e^{-\beta U} \frac{(\varrho V)^N}{N!} e^{-\varrho V} \\
&= \frac{\tilde{\Omega}(U)}{\tilde{Z}_V(\beta, z)} \frac{(V)^N}{N!} e^{-\beta U + N \ln z},
\end{aligned} \tag{5.27}
$$

is the product of the continuous gamma density (5.23) and the discrete Poisson distribution (5.16). The partition function for such a grand-canonical ensemble is

$$
\begin{aligned}
\tilde{Z}_V(\beta, z) &= e^{\varrho V} \\
&= \sum_{N=0}^{\infty} \int_0^{\infty} e^{-\beta U} \tilde{\Omega}(U)\, dU\, \frac{(zV)^N}{N!} \\
&= \sum_{N=0}^{\infty} \frac{(zV/\lambda^{2m})^N}{N!} \\
&= \exp\left(\frac{zV}{\lambda^{2m}} \right),
\end{aligned} \tag{5.28}
$$

where the fugacity is defined as $z \equiv e^{\mu/T} = \varrho \lambda^{2m}$.

Again, we may limit our remarks about the generating function (5.28) to those concerning the new variable z. Consider

$$\tilde{\mathcal{Z}}_V(\beta, z)\, e^{\beta U - N \ln z} \qquad (5.29)$$

as a function of $\ln z$. As $\ln z \to -\infty$, expression (5.29) $\to \infty$ because of the exponential factor, and as $\ln z \to \infty$, it again tends to infinity because $\tilde{\mathcal{Z}}_V \to \infty$. The second derivative of the logarithm of (5.29), with respect to $\ln z$, coincides with that of $\ln \tilde{\mathcal{Z}}_V$, and since the latter is positive, (5.29) will be a convex function of $\ln z$. Hence, (5.29) must necessarily possess a single minimum at

$$\left(\frac{\partial \ln \tilde{\mathcal{Z}}_V}{\partial \ln z} \right)_\beta - N = 0.$$

This likelihood equation determines implicitly the fugacity, or chemical potential, as a function of the temperature, T, and density, ϱ. The expression obtained for the fugacity corresponds to the maximum likelihood value, $\hat{z}$, because

$$\frac{\partial^2 \ln \tilde{\mathcal{Z}}_V}{\partial \ln z^2} > 0.$$

We are now in a position to generalize our minimax principle of the entropy, discussed in §4.5, to other conjugate thermodynamic variables. The characterizing properties of the two canonical distributions, (5.24) and (5.27), are determined by the entropies

$$S(X, \gamma) = \beta U + \gamma X + \ln \tilde{\mathcal{Z}}(\beta, \gamma), \qquad (5.30)$$

where $\tilde{\mathcal{Z}} = \tilde{\mathcal{Z}}_N(\beta, \varrho)$ when $\gamma = \varrho$ and $X = V$ or $\tilde{\mathcal{Z}} = \tilde{\mathcal{Z}}_V(\beta, z)$ when $\gamma = \ln z$ and $X = -N$. Since $\ln \tilde{\mathcal{Z}}$ is a strictly convex function of γ, the saddle function, (5.30), will be strictly convex in γ and weakly concave in X. This follows from the definition of a saddle function: a function is said to be saddle shaped at a point if the Hessian

$$\frac{\partial^2 S}{\partial X^2} \frac{\partial^2 S}{\partial \gamma^2} - \left(\frac{\partial^2 S}{\partial X \, \partial \gamma} \right)^2 < 0$$

at that point. Since our interest is in deriving criteria for equilibria other than thermal, we hold the pair (β, U) constant.

Then, according to the implicit function theorem, we may solve

$$\frac{\partial S}{\partial \gamma} = \frac{\partial \ln \tilde{\mathcal{Z}}}{\partial \gamma} + X = 0$$

for the path $\hat{\gamma} = \gamma(X)$ on the surface $S(X, \gamma)$ since $\partial^2 S / \partial \gamma^2 \neq 0$. The slope of this curve satisfies

$$\frac{\partial^2 S}{\partial X \, \partial \gamma} + \frac{\partial^2 S}{\partial \gamma^2} \frac{\partial \hat{\gamma}}{\partial X} = 0. \qquad (5.31)$$

The path on the saddle function has height

$$S(\gamma(X), X) = S(X),$$

which is the thermodynamic entropy. By the chain rule we have

$$\frac{\partial S(X)}{\partial X} = \frac{\partial S(X,\gamma)}{\partial X} = \hat{\gamma}, \qquad \frac{\partial^2 S(X)}{\partial X^2} = \frac{\partial^2 S(X,\gamma)}{\partial X^2} + \frac{\partial^2 S(X,\gamma)}{\partial X\,\partial\gamma}\frac{\partial\hat{\gamma}}{\partial X},$$

and introducing (5.31) into the latter expression, we find

$$\frac{\partial^2 S(X)}{\partial X^2} = \frac{\partial^2 S(X,\gamma)}{\partial X^2} - \left(\frac{\partial^2 S(X,\gamma)}{\partial X\,\partial\gamma}\right)^2 \left(\frac{\partial^2 S(X,\gamma)}{\partial\gamma^2}\right)^{-1}$$

$$= -\left(\frac{\partial^2 \ln \tilde{Z}}{\partial\gamma^2}\right)^{-1} < 0$$

because the saddle entropy, $S(X,\gamma)$, is strictly convex in γ and weakly concave in X. From this inequality we conclude that $S(X)$ is concave in X in addition to being concave in the energy, U.

Since the entropy tends to a maximum as a function of the extensive variables, subject to the constraints imposed upon them, we have a generalization of the minimax principle for the entropy. This minimax principle asserts that the thermodynamic entropy is given by

$$\sup_{U,X} S(U,X) = \sup_{U,X}\inf_{\beta,\gamma} \left[\beta U + \gamma X + \ln \tilde{Z}(\beta,\gamma)\right]$$

subject to the constraints imposed upon U and X, where X stands for either V or $-N$ but not both simultaneously.

5.4   Thermodynamic Identifications

We have already made it clear that the system will not support both fluctuations in the volume and particle number *simultaneously*. The next chapter will be devoted to elevating this observation to a statistical equivalence principle. However, of more immediate interest is the thermodynamic identifications of the various symbols that have been introduced.

In order to proceed further we must estimate the scale parameters, β and ϱ, from observations made on their conjugate extensive variables and give them thermodynamic meanings. We shall find that this can be done only in a certain order so that there is a definite hierarchy in securing different types of equilibria, with the highest priority given to thermal equilibrium where all subsystems arrive at a common value of β. The analysis will add support to the conclusions we arrived at in §3.7.4 regarding the nature of this hierarchy.

From our past experience in §4.3, we know that the maximum likelihood values of these parameters coincide with thermodynamic quantities so that we will work only with those quantities. The appropriate partition function for the analysis of volume fluctuations, at a fixed particle number, is (5.25).

The total derivative of its logarithm, evaluated at $\hat{\beta}$ and $\hat{\varrho}$, is

$$
\begin{aligned}
d\ln\tilde{Z}_N(\hat{\beta},\hat{\varrho}) &= \left(\frac{\partial\ln\tilde{Z}_N}{\partial\beta}\right)_{\hat{\varrho}} d\hat{\beta} + \left(\frac{\partial\ln\tilde{Z}_N}{\partial\varrho}\right)_{\hat{\beta}} d\hat{\varrho}\\
&= -\bar{U}\,d\hat{\beta} - \bar{V}\,d\hat{\varrho}, \qquad\qquad (5.32)
\end{aligned}
$$

where [cf. Eq.(5.21)]

$$
\left(\frac{\partial\ln\tilde{Z}_N}{\partial\varrho}\right)_{\hat{\beta}} = -\bar{V}.
$$

Evaluating the left-hand side with the aid of expression (5.25) gives $\bar{V} = N/\hat{\varrho}$. This can be taken as the definition of the density in a system which undergoes volume fluctuations for a constant value of the particle number. Adding the total differential of $\hat{\beta}\bar{U} + \hat{\varrho}\bar{V}$ to both sides of (5.32) results in

$$
d(\ln\tilde{Z}_N + \hat{\beta}\bar{U} + \hat{\varrho}\bar{V}) = \hat{\beta}\left[d\bar{U} + (\hat{\varrho}/\hat{\beta})\,d\bar{V}\right]. \qquad\qquad (5.33)
$$

Thus, we see that the quantity

$$
\hat{\beta}\left[d\bar{U} + (\hat{\varrho}/\hat{\beta})\,d\bar{V}\right]
$$

is the total derivative of a certain thermodynamic function. This statement actually contains the second law of thermodynamics. Setting $\hat{\varrho}/\hat{\beta} = P$, we can identify the term $-P\,d\bar{V}$ as the work done by the compressional forces on the system. Positive work means that the increment in the volume is negative.

According to the first law of thermodynamics, the difference in the energy and external work is the "amount of heat," δQ, received by the system during the elementary transition from one equilibrium state to a neighboring one. The second law then tells us that the quantity $\delta Q/T$ is always a total differential, and comparing it with the above expression gives the classical thermodynamic definition of temperature, $\hat{\beta} = 1/T$. With the second law, $\delta Q/T = dS$, and (5.32), expression (5.33) can be written as the double Massieu transform

$$
\begin{aligned}
S &= \ln\tilde{Z}_N(\hat{\beta},\hat{\varrho}) - \hat{\beta}\left(\frac{\partial\ln\tilde{Z}_N}{\partial\beta}\right)_{\hat{\varrho}} - \hat{\varrho}\left(\frac{\partial\ln\tilde{Z}_N}{\partial\varrho}\right)_{\hat{\beta}}\\
&= \ln\tilde{Z}_N + \frac{\bar{U} + P\bar{V}}{T}.
\end{aligned}
$$

Comparing this with the Euler relation results in

$$
T\ln\tilde{Z}_N(T,P) = -\mu N.
$$

This identifies the left-hand side with the negative of the Gibbs potential.

Now consider fluctuations in the particle number at constant volume. The partition function adapted to such an analysis is the grand-canonical partition function, (5.28). The total derivative of its logarithm, at the maximum likelihood estimates of its variables, is

$$
\begin{aligned}
d\ln \tilde{Z}_V(\hat{\beta}, \ln \hat{z}) &= \left(\frac{\partial \ln \tilde{Z}_V}{\partial \beta}\right)_z d\hat{\beta} + \left(\frac{\partial \ln \tilde{Z}_V}{\partial \ln z}\right)_{\hat{\beta}} d\ln \hat{z} \\
&= -\bar{U} d\hat{\beta} + \bar{N} d\ln \hat{z},
\end{aligned}
\tag{5.34}
$$

where

$$
\left(\frac{\partial \ln \tilde{Z}_V}{\partial \ln z}\right)_{\hat{\beta}} = \bar{N}.
$$

Using (5.28) to evaluate the left side gives the expression

$$
\bar{N} = \hat{z}V/\hat{\lambda}^{2m}
\tag{5.35}
$$

for the maximum likelihood estimate of the fugacity, where $\hat{\lambda} = \sqrt{\hat{\beta}/C}$.

The inverse of the maximum likelihood estimate $\hat{z}$ appears as a ratio of the volume occupied by a molecule to a volume of a hypersphere of $2m$ dimensions which is proportional to $\hat{\lambda}^{2m}$. The optimal value, $\hat{\lambda}$, appears as a characteristic radius. Adding $d(\hat{\beta}\bar{U} - \bar{N}\ln \hat{z})$ to both sides of (5.34) results in

$$
d(\ln \tilde{Z}_V + \hat{\beta}\bar{U} - \bar{N}\ln \hat{z}) = \hat{\beta}\left[d\bar{U} - (\ln \hat{z}/\hat{\beta})\, d\bar{N}\right].
\tag{5.36}
$$

Again, we have a quantity,

$$
\hat{\beta}\left[d\bar{U} - (\ln \hat{z}/\hat{\beta})\, d\bar{N}\right],
$$

which is the total differential of a certain thermodynamic function. By setting $\ln \hat{z}/\hat{\beta} = \mu$, the second term becomes the quasi-static chemical work, $\mu\, d\bar{N}$. The term within the brackets then becomes the difference between the internal energy and the chemical work which, if there are no other forms of quasi-static work, is the amount of heat added to the system, δQ.[1]

Thus, the second law identifies $\beta\, \delta Q$, to within a multiplicative constant, as dS, where the entropy, S, is given by the double Massieu transform:

$$
\begin{aligned}
S &= \ln \tilde{Z}_V(\hat{\beta}, \ln \hat{z}) - \hat{\beta}\left(\frac{\partial \ln \tilde{Z}_V}{\partial \beta}\right)_{\ln \hat{z}} - \ln \hat{z}\left(\frac{\partial \ln \tilde{Z}_V}{\partial \ln z}\right)_{\hat{\beta}} \\
&= \ln \tilde{Z}_V(T, \mu) + \frac{\bar{U} - \mu\bar{N}}{T}.
\end{aligned}
$$

---

[1]In classical thermodynamics there is nothing to prevent us from adding on the work term, $P\, d\bar{V}$, since there is no connection with the other work term, $-\nu\, d\bar{N}$. However, when fluctuations are taken into account, they are no longer independent contributions in the first law and the appearance of one excludes that of the other.

Comparison to the Euler relation identifies

$$T \ln \tilde{Z}_V(T, \mu) = P\bar{V} \qquad (5.37)$$

as the negative of the grand-canonical potential, W, which we introduced in §3.7.4.

The foregoing analysis has shown that the scale parameters, ϱ and $\ln z$, are more primitive notions than their maximum likelihood estimates. The latter are proportional to the thermodynamic intensive variables of density and chemical potential, respectively. Thermodynamics emerges in the attempt to optimize the scale parameters β, ϱ, and $\ln z$ by estimating them in terms of the statistics $\bar{U}$, $\bar{V}$, and $\bar{N}$, respectively. These average values are assumed to be identical to the thermodynamic extensive variables of internal energy, volume, and number of particles, respectively.

Let us take a pause here to compare our treatment with the conventional derivation of the grand-canonical ensemble. The expression for the Gibbs potential of an ideal gas is [cf. §3.6.1]

$$G = NT \ln \left(\frac{N \hat{\lambda}^{2m}}{\bar{V}} \right), \qquad (5.38)$$

where $\hat{\lambda} = h\sqrt{\hat{\beta}/2\pi M}$ is referred to as the mean thermal wavelength of a particle of mass M. This is the characteristic length of the system or what we have previously seen to be the radius of a characteristic hypersphere in phase space. Expression (5.38) can be written as [cf. §3.8]

$$\mu = T \left\{ \ln N - \ln[\mathcal{Z}(\hat{\beta})V] \right\},$$

where $\mathcal{Z}(\hat{\beta})$ is the single-particle partition function at $\hat{\beta} = 1/T$.

This expression is seen to be identical to (5.35) once N has been identified as the average number of particles, $\bar{N}$. But conventional statistical mechanics does not make such subtle distinctions, nor does it recognize $\hat{\beta}$ as an optimal value of the scale factor β. For a vanishing chemical potential, the single-particle partition function is equal to the number of particles. But if μ vanishes, there is no longer a conservation of particles. The implications of this will be discussed in §6.3.

In order to derive the partition function of the grand-canonical ensemble, we must choose (5.19), in preference to (5.18), for the density ϱ. Since the Helmholtz potential $A = G - PV$ and $PV = NT$ for an ideal gas, we have the expression

$$A = -NT \ln \left(\frac{N \hat{\lambda}^{2m}}{Ve} \right)$$

for the Helmholtz potential of an ideal gas, where e is the base of the Napierian logarithm.

The Helmholtz potential of an ideal gas can be written as $A/T =$ $\ln\left[\left(V/\hat{\lambda}^{2m}\right)^{N}/N!\right]$, provided Stirling's approximation can be applied. The factorial, $N!$, in the denominator usually has to be written in by hand based on the argument that what was apparently distinguishable is actually indistinguishable! But $\tilde{\mathcal{Z}} = \mathcal{Z}^{N}$ has nothing to do with the permutation of the particles among themselves so the argument for dividing $\mathcal{Z}^{N}$ by $N!$ is contrived.

The grand-canonical partition function is *defined* conventionally as the moment generating function of $\exp(A/kT)$:

$$\tilde{\mathcal{Z}}_{V}(T,\hat{z}) = \sum_{N=0}^{\infty} \hat{z}^{N}\left(\frac{\bar{V}e}{\hat{\lambda}^{2m}N}\right)^{N}$$
$$= \sum_{N=0}^{\infty}\frac{(\hat{z}V/\hat{\lambda}^{2m})^{N}}{N!} = \exp\left(\frac{\hat{z}V}{\hat{\lambda}^{2m}}\right).$$

We leave it to the reader to decide which of the derivations of the grand-canonical partition function is preferable.

5.5   Statistical Equivalence

The maximum values of the canonical distributions, (5.24) and (5.27), are obtained by evaluating them at the maximum likelihood estimates $\hat{\beta}$ and $\hat{\varrho}$ for the scale parameters. We thus obtain

$$\left.\begin{array}{c}\hat{f}_{N}(U,V)\\\hat{f}_{V}(U,N)\end{array}\right\} = \tilde{\Omega}(U)\,e^{-\hat{\beta}[U-\mu N+PV]}\left\{\begin{array}{c}V^{N-1}/\Gamma(N)\\V^{N}/N!\end{array}\right. \tag{5.39}$$

In the first case the partition function is (5.25), while in the second case it is (5.37).

We will now show that (5.39) represents Gauss' law of error for which the average value of the extensive fluctuating quantity is its most probable value. The extensive variable which is held constant is set equal to its average value for, otherwise, we would not be able to make contact with thermodynamics. In the case of volume fluctuations, we add and subtract the quantity $\hat{\beta}[\bar{U} + P\bar{V} - \mu N]$ in the exponent of the first expression in (5.39). We then obtain

$$\hat{f}_{N}(U,V) = \exp\{-\hat{\beta}[(U-\bar{U}) - P(V-\bar{V})]$$
$$- S(\bar{U},\bar{V},N) + S(U,V,N)\}. \tag{5.40}$$

The stochastic entropy,

$$S(U,V,N) = N\ln\left(\frac{VU^{m}}{N^{m+1}}\right) + N\left[m\ln\left(\frac{C}{m}\right) + m + 1\right],$$

is manifestly extensive and is the same function of the fluctuating variables U and V as the thermodynamic entropy of an ideal gas is of $\bar{U}$ and $\bar{V}$. With the aid of the relations

$$\left(\frac{\partial S}{\partial U}\right)_{V,N} = \frac{mN}{U} = \beta \quad \text{and} \quad \left(\frac{\partial S}{\partial V}\right)_{U,N} = \frac{N}{V} = \varrho,$$

it may be tranformed into the canonical or Gibbsian form

$$S(U, V, N) = \beta U + \varrho V - N \ln z.$$

It is interesting to observe that had we used the relation

$$\left(\frac{\partial S}{\partial N}\right)_{U,V} = \ln\left(\frac{V}{\lambda^{2m} N}\right) = -\ln z, \qquad (5.41)$$

where $\lambda = \sqrt{\beta/C}$, we would have obtained

$$S(U, V, N) = N[(m + 1) - \ln z].$$

Therefore, if use is made of (5.41), the stochastic entropy coincides with the thermodynamic entropy of an ideal gas since $\hat{\beta}\bar{U} = mN$ and $\hat{\varrho}\bar{V} = N$. Consequently, N must be treated as a fixed parameter in the expression for the stochastic entropy; otherwise (5.41) will wipe out fluctuations in the volume as well as those in the energy.

In order to analyze fluctuations in the particle number, we add and subtract the quantity $\hat{\beta}[\bar{U} + PV - \mu\bar{N}]$ in the exponent of the second expression in (5.39). Keeping V as a fixed parameter, we obtain

$$\begin{aligned} \hat{f}_V(U, N) &= \exp\{-\hat{\beta}[(U - \bar{U}) + \mu(N - \bar{N})] \\ &\quad - S(\bar{U}, V, \bar{N}) + S(U, V, N)\} \end{aligned} \qquad (5.42)$$

as the error law for simultaneous fluctuations in the energy and particle number.

In §3.7.2 we came to the conclusion that there is really no purely *thermodynamic* argument against simultaneous fluctuations in the number and volume at constant energy. Indeed, since thermodynamics considers N and V to be independent variables, one would expect it. But the physical situation of holding the energy constant is indeed suspicious. Nevertheless, the functional dependencies of the thermodynamic and stochastic entropies in (5.40) and (5.42) do not provide the slightest hint that the two forms of fluctuations are simultaneously incompatible. Rather, it is the *probabilistic* interpretation of (5.39) which gives grounds for inferring that one variable, but not both, will fluctuate at a constant value of the other. It is also to be accredited for their statistical equivalency, as we now show.

In order to demonstrate the equivalency of fluctuations in the particle number and volume, we analyze their mean square fluctuations. Although there will necessarily be fluctuations in the energy, they will not directly impinge on our analysis. It will therefore suffice to consider the marginal probability densities

$$\hat{f}_N(V) = \int_0^\infty \hat{f}_N(U, V)\, dU$$

$$= \zeta \frac{(\zeta V)^{N-1}}{\Gamma(N)} e^{-\hat{\beta}PV} \qquad (5.43)$$

and

$$\hat{f}_V(N) = \int_0^\infty \hat{f}_V(U, N)\, dU$$

$$= \frac{(\zeta V)^N}{N!} e^{-\hat{\beta}PV} \qquad (5.44)$$

for the temperature–pressure and grand-canonical ensembles, respectively, where $\zeta = \hat{z}/\hat{\lambda}^{2m}$.

In order that (5.43) and (5.44) be proper probability distributions, it is necessary that

$$\zeta = \hat{z}/\hat{\lambda}^{2m} = \hat{\beta}P. \qquad (5.45)$$

This ensures that (5.43) is a gamma density with expectation

$$\bar{V} = N/\zeta = N/\hat{\beta}P. \qquad (5.46)$$

The first equality in (5.46) is $\hat{z} = N\hat{\lambda}^{2m}/\bar{V}$, corresponding to the definition of the fugacity, (5.41), when the volume is taken at its mean value. The second equality in (5.46) gives $P\bar{V} = N/\hat{\beta}$, which is the equation of state of an ideal gas. Alternatively, (5.45) ensures that (5.44) is a Poisson distribution with expectation

$$\bar{N} = \zeta V = \hat{\beta}PV. \qquad (5.47)$$

The first equality in (5.47), again, corresponds to the definition of the fugacity, (5.41), only this time the volume is fixed and the particle number is set at its mean value.

Consequently, we may eliminate ζ in (5.43) in favor of the density, which according to (5.46) is $\hat{\varrho} = N/\bar{V}$. We then obtain the gamma density

$$\hat{f}_N(V) = \hat{\varrho} \frac{(\hat{\varrho}V)^{N-1}}{\Gamma(N)} e^{-\hat{\varrho}V}. \qquad (5.48)$$

Alternatively, if we introduce the density, defined from (5.47) as $\hat{\varrho} = \bar{N}/V$, into (5.44), we obtain the Poisson distribution

$$\hat{f}_V(N) = \frac{(\hat{\varrho}V)^N}{N!} e^{-\hat{\varrho}V}. \qquad (5.49)$$

These two probability distributions correspond to the two types of experiments that we performed in §5.2.2.

With the aid of the gamma density (5.48), we find

$$\overline{(\Delta V)^2} = \frac{\bar{V}^2}{N} = \frac{N}{\hat{\varrho}^2} \tag{5.50}$$

for the mean square fluctuation in the volume. In analogy with our discussion of the wave nature of fluctuations in black radiation in §2.3.2, we can conclude that the proportionality between the dispersion in the volume and the square of the average volume is the hallmark of fluctuation of a continuous random variable.

In contrast, the Poisson distribution (5.49) gives

$$\overline{(\Delta N)^2} = \bar{N} = \hat{\varrho}V \tag{5.51}$$

for the mean square fluctuation in the particle number. Again, in analogy with particle fluctuations in black radiation, the fact that the dispersion, (5.51), is proportional to $\bar{N}$ itself is the hallmark of fluctuations in a discrete random variable. On the one hand we have a *continuum* behavior while, on the other hand, the behavior is entirely *corpuscular*.

The number of particles is fixed in the volume dispersion (5.50) while the volume is fixed in the dispersion in the number of particles (5.51). Setting the former equal to the average number of particles and the latter equal to the average volume, we come out with

$$\hat{\varrho} = \sqrt{\overline{(\Delta N)^2} \Big/ \overline{(\Delta V)^2}} \tag{5.52}$$

when (5.51) is divided by (5.50) and the positive square root is taken. The ratio of the standard deviation in the number of particles to that of the volume is the density, $\hat{\varrho}$. So the dispersion in one random variable can be converted into the dispersion of the other simply by multiplying by the optimum value of the scale parameter, $\hat{\varrho}$. Hence, albeit the physical situations where V fluctuates at constant N, or vice versa, appear to be seeming different, they are nevertheless statistically equivalent to one another. Given the mean square fluctuation in the particle number, (5.52) can be used to convert it into the mean square fluctuation in the volume, or vice versa. We shall return to (5.52) in our analysis of quantum fluctuations in §6.7.

5.6   Properties of the Stochastic Entropy

The properties of the stochastic entropy may be generalized beyond those discussed in §4.6, where we considered them a function only of the energy. We shall see that the second law makes statements regarding the evolution of the system, in addition to specifying the direction that heat will flow when two bodies are put into thermal contact at different temperatures. Our problem is to determine how far the probabilistic foundations of statistical physics can go in providing the basis for the assertions made by the second law.

Again consider the function (5.26) where $\tilde{Z}_N$ is given by (5.25). It possesses a single minimum with respect to β at $\hat{\beta}$ and a single minimum with respect to ϱ at $\hat{\varrho} = \hat{\beta}P$. It is therefore clear that the establishment of thermal equilibrium must precede all other forms of equilibria. For a system in thermal equilibrium but perhaps not in mechanical equilibrium, the partition function can be written as

$$\tilde{Z}_N(\hat{\beta}, \varrho) = \frac{1}{\lambda^{2mN}} \int_0^\infty e^{-\varrho V} \varrho \frac{(\varrho V)^{N-1}}{\Gamma(N)} \, dV.$$

A composite system is formed from two subsystems which are separated by a movable, diathermal wall.[2] The wall is also impermeable to particles. The volumes of the two chambers are denoted by V_1 and V_2. There are N_1 particles in the first chamber while in the second chamber there are $N_2 = N - N_1$ particles. Since V_1 and V_2 are independent random variables with gamma densities of the form (5.43), the probability that $V_1 + V_2 = V$ is

$$
\begin{aligned}
f_N(V; \varrho) &= \int_0^V \varrho \frac{(\varrho V_1)^{N_1-1}}{\Gamma(N_1)} e^{-\varrho V_1} \varrho \frac{[\varrho(V - V_1)]^{N-N_1-1}}{\Gamma(N - N_1)} e^{-\varrho(V - V_1)} \, dV_1 \\
&= \varrho \frac{(\varrho V)^{N-1}}{\Gamma(N_1)\Gamma(N - N_1)} e^{-\varrho V} \int_0^1 x^{N_1-1}(1 - x)^{N-N_1-1} \, dx \\
&= \varrho \frac{(\varrho V)^{N-1}}{\Gamma(N)} e^{-\varrho V},
\end{aligned}
$$

where we used the beta integral

$$\int_0^1 x^b(1 - x)^c \, dx = \frac{\Gamma(b + 1)\Gamma(c + 1)}{\Gamma(b + c + 2)}$$

to go from the second to the third equality.

From the reproductive property of the gamma density, the composition law for the structure function,

$$\Omega(V) = \int_0^V \Omega_1(V_1)\,\Omega_2(V - V_1) \, dV_1,$$

follows together with the product law for the partition function [cf. §4.6]. In other words, since V_1 and V_2 are independent random variables with partition functions $\tilde{Z}_{N_1}(\hat{\beta}, \varrho)$ and $\tilde{Z}_{N_2}(\hat{\beta}, \varrho)$, the partition function of the sum $V_1 + V_2$ is

$$\tilde{Z}_N(\hat{\beta}, \varrho) = \tilde{Z}_{N_1}(\hat{\beta}, \varrho)\,\tilde{Z}_{N_2}(\hat{\beta}, \varrho). \tag{5.53}$$

By virtue of (5.53), the entropy, (5.30), for the composite system is

$$
\begin{aligned}
S &= \hat{\beta}\bar{U} + \varrho V + \ln \tilde{Z}_N(\hat{\beta}, \varrho) \\
&= \hat{\beta}\bar{U}_1 + \varrho V_1 + \ln \tilde{Z}_{N_1}(\hat{\beta}, \varrho) + \hat{\beta}\bar{U}_2 + \varrho V_2 + \ln \tilde{Z}_{N_2}(\hat{\beta}, \varrho) \tag{5.54}
\end{aligned}
$$

---

[2] Recall that a diathermal wall permits the flow of heat, in contrast to an adiabatic wall which prohibits it.

since $\bar{U} = \bar{U}_1 + \bar{U}_2$ and $V = V_1 + V_2$. The composite system has already reached thermal equilibrium at a temperature $\hat{\beta}^{-1}$. The entropy (5.54) is the stochastic entropy since it is a function of the random variable V. The functions $\varrho V_1 + \ln \tilde{Z}_{N_1}(\hat{\beta}, \varrho)$ and $\varrho V_2 + \ln \tilde{Z}_{N_2}(\hat{\beta}, \varrho)$ reach a minimum when $\varrho = \hat{\varrho}_1$ and $\varrho = \hat{\varrho}_2$. The stationary conditions

$$\bar{V}_i = -\left(\frac{\partial \ln \tilde{Z}_{N_i}}{\partial \varrho}\right)_{\hat{\beta}} = \frac{N_i}{\hat{\varrho}_i}, \quad i = 1, 2, \tag{5.55}$$

provide the maximum likelihood estimates of $\hat{\varrho}_i$. In view of (5.33), we identify the term $(\hat{\gamma}/\hat{\beta})\, d\bar{X}$ as the negative of the compressional work. Thus $\hat{\varrho}_i/\hat{\beta} = P_i$ are the pressures of the subsystems.

Since S is a minimum at (5.55), it follows that

$$\begin{aligned} S &\geq \hat{\beta}(\bar{U}_1 + P_1\bar{V}_1) + \ln \tilde{Z}_{N_1}(\hat{\beta}, P_1) + \hat{\beta}(\bar{U}_2 + P_2\bar{V}_2) + \ln \tilde{Z}_{N_2}(\hat{\beta}, P_2) \\ &= S_1(\bar{U}_1, \bar{V}_1, N_1) + S_2(\bar{U}_2, \bar{V}_2, N_2), \end{aligned}$$

where S_i are the thermodynamic entropies of the subsystems since they are functions of average values that coincide with the thermodynamic extensive variables. The number of particles is fixed in both subsystems and, consequently, for the composite system as well.

From this inequality we conclude that the entropy of the system, obtained by bringing into mechanical interaction two previously mechanically isolated systems, which have already reached thermal equilibrium, can never be smaller than the sum of the entropies of the two subsystems. The equality sign applies to the case where the two subsystems were initially at the same pressure as well as having the same temperature. In the final state of mechanical equilibrium, both subsystems come to a common pressure which, according to (5.55), also implies that the composite system is homogeneous.

We now consider the nature of the equilibrium that ensues when two previously isolated subsystems are connected by a rigid, permeable diathermal wall. The first subsystem has a fixed volume V_1 while the second subsystem has a fixed volume $V_2 = V - V_1$. There are N_1 in the first subsystem and N_2 in the second. Since they are independent random variables with Poisson distributions, the probability that $N_1 + N_2 = N$ is

$$\begin{aligned} f_V(N_1 + N_2 = N; \varrho) &= \sum_{N_1+N_2=N} \frac{(\varrho V_1)^{N_1}}{N_1!} e^{-\varrho V_1} \frac{(\varrho V_2)^{N_2}}{N_2!} e^{-\varrho V_2} \\ &= \frac{\varrho^N e^{-\varrho V}}{N!} \sum_{N_1=0}^{N} \binom{N}{N_1} V_1^{N_1} V_2^{N_2} \\ &= \frac{(\varrho V)^N}{N!} e^{-\varrho V}, \end{aligned}$$

where we have used Newton's binomial theorem.

The reproductive property of the Poisson distribution is responsible for both the composition law for the configuration phase volume,

$$\frac{V^N}{N!} = \frac{(V_1 + V_2)^N}{N!} = \frac{1}{N!} \sum_{N_1=0}^{N} \binom{N}{N_1} V_1^{N_1} V_2^{N-N_1}$$

$$= \sum_{N_1+N_2=N} \frac{V_1^{N_1} V_2^{N_2}}{N_1! \; N_2!},$$

and the multiplicative property of the partition function,

$$\tilde{Z}_V(\hat{\beta}, z) = \tilde{Z}_{V_1}(\hat{\beta}, z) \, \tilde{Z}_{V_2}(\hat{\beta}, z).$$

On the strength of this product law, the stochastic entropy, (5.30), of the composite system is

$$\begin{aligned} S &= \hat{\beta}\bar{U} - N \ln z + \ln \tilde{Z}_V(\hat{\beta}, z) \\ &= \hat{\beta}\bar{U}_1 - N_1 \ln z + \ln \tilde{Z}_{V_1}(\hat{\beta}, z) + \hat{\beta}\bar{U}_2 - N_2 \ln z + \ln \tilde{Z}_{V_2}(\hat{\beta}, z). \end{aligned}$$

The stochastic entropy is now a function of the random variable N.

The functions

$$\ln \tilde{Z}_{V_1}(\hat{\beta}, z) - N_1 \ln z$$

and

$$\ln \tilde{Z}_{V_2}(\hat{\beta}, z) - N_2 \ln z$$

reach their minima for $\ln z = \ln \hat{z}_1$ and $\ln z = \ln \hat{z}_2$, respectively. The maximum likelihood estimates, $\hat{z}_i$, are derived from the implicit formulas

$$\bar{N}_i = \left(\frac{\partial \ln \tilde{Z}_{V_i}}{\partial \ln z} \right)_{\hat{\beta}} = \frac{\hat{z}_i V_i}{\hat{\lambda}^{2m}}, \quad i = 1, 2. \tag{5.56}$$

These are the stationary conditions for the likelihood function.

Consulting (5.36), we identify the term $(\hat{\gamma}/\hat{\beta}) \, d\bar{X}$ as the negative of the chemical work so that $\ln \hat{z}_i / \hat{\beta} = \mu_i$, where the μ_i are the chemical potentials of the two subsystems.

Since S is a minimum at these values, it follows that

$$\begin{aligned} S &\geq \hat{\beta}(\bar{U}_1 - \mu_1 \bar{N}_1) + \ln \tilde{Z}_{V_1}(\hat{\beta}, \mu_1) + \hat{\beta}(\bar{U}_2 - \mu_2 \bar{N}_2) + \ln \tilde{Z}_{V_2}(\hat{\beta}, \mu_2) \\ &= S_1(\bar{U}_1, \bar{N}_1, V_1) + S_2(\bar{U}_2, \bar{N}_2, V_2). \end{aligned}$$

The thermodynamic entropies are now functions of the average energies, $\bar{U}_i$, and the average number of particles, $\bar{N}_i$, in each of the subsystems for fixed values of the volumes. From this inequality we conclude that the entropy of a composite system, obtained by converting two previously closed systems that are in thermal equilibrium into an open one by introducing a rigid, permeable wall, cannot be inferior to the sum of the entropies of the two subsystems; the two become equal in the event that the two subsystems have initially the same chemical potential as well as the same temperature. In the final state of equilibrium, the chemical potentials become equal which, according to (5.56), again implies that the composite system is homogeneous.

Having already attained thermal equilibrium, the criterion for mechanical equilibrium is the equality of the pressures. The condition for equilibrium with respect to matter flow requires the equality of the chemical potentials of the two subsystems. Mechanical equilibrium and equilibrium with respect to matter flow cannot be established simultaneously—but both require thermal equilibrium.

The constraint which enables a comparison to be made between a less and a more constrained state of equilibrium is the diathermal wall: it is either movable and impenetrable or rigid and permeable, but it cannot be both at the same time! This allows either the volume to vary for a fixed number of particles in each of the subsystems or the number of particles to vary for fixed volumes. When the state of mechanical equilibrium has been established, the composite system will have a common pressure. Alternatively, when equilibrium with respect to matter flow has been achieved, the composite system will have a common chemical potential. Although these appear to be distinct thermodynamic criteria, both imply that the density should be uniform, which provides an ulterior connection between the two forms of equilibria.

5.7   Classical Limit of Quantum Statistics

In this section we show that the marginal distribution, (5.49), of the grand-canonical ensemble has the same form as the classical limit of the probability distribution for both Bose–Einstein and Fermi–Dirac statistics in each energy interval. In other words, in place of the average number of particles, (5.35), there will appear the average number of particles, $\bar{N}_\varepsilon$, in the energy interval from ε to $\varepsilon + d\varepsilon$. The integral of $\bar{N}_\varepsilon$ over all particle energies, ε, is simply (5.35).

In §3.3 we saw how Bose–Einstein statistics can be derived from the negative binomial distribution,

$$f(N;p) = \binom{mV + N - 1}{N} p^{mV} q^N, \tag{5.57}$$

where p and q are the *a priori* probabilities and m is the density of states in momentum space divided by h^3.[3] As we have mentioned in §2.3.3, the usual derivation of Bose–Einstein statistics involves the thermodynamic probability which utilizes only the binomial coefficient. Maximizing it with respect to the total number of particles, or simply setting N equal to its average value, and equating its logarithm with the entropy give Bose–Einstein statistics when the temperature is introduced through the second law.

---

[3]m should not to be confused with the same symbol used earlier in this chapter to denote half degrees-of-freedom.

However, the thermodynamic probability *per se* cannot predict the magnitude of the deviations or fluctuations from the average or most probable number of particles. The probability distribution (5.57) could not be used in the derivation because one did not know what the *a priori* probabilities were. Claiming complete ignorance, Boltzmann and his followers were led to postulate equal *a priori* probabilities which essentially reduce the distribution to a binomial coefficient [cf. §1.2].

However, we know from our discussion in §3.3.2 that it is possible to ask for the most likely values of the *a priori* probabilities which maximize the negative binomial distribution, (5.57). In §4.3, we employed the method of maximum likelihood, which essentially inverts the function $f(N; p)$ by taking p as the variable and N as the parameter. But what value of the parameter are we supposed to take?

Recall that we are to make observations, and since they are independent of one another and identically distributed, the probability will be a product of negative binomial distributions. The statistic, then, turns out to be the arithmetic mean,

$$\bar{N} = \frac{1}{r} \sum_i^r N_i,$$

which we assume to be the most probable value of the number of particles that is measured. The extremum condition of the log-likelihood function, which is proportional to the product of negative binomial distributions, yields

$$\hat{p} = \frac{mV}{\bar{N} + mV}$$

as the most likely value of the *a priori* probability, p.

Introducing $\hat{p}$ and $\hat{q} = 1 - \hat{p} = \bar{N}/(\bar{N} + mV)$ into (5.57) converts it into the Gaussian error law,

$$\hat{f}(N; \bar{N}) = A \exp \left\{ S(N) - S(\bar{N}) - \left(\frac{\partial S}{\partial \bar{N}} \right)_V (N - \bar{N}) \right\}, \qquad (5.58)$$

leading to the average value, $\bar{N}$, as the most probable value of the number of particles observed. The statistical entropy of the negative binomial distribution is

$$S(\bar{N}, V) = \bar{N} \ln \left(\frac{mV + \bar{N}}{\bar{N}} \right) + mV \ln \left(\frac{mV + \bar{N}}{mV} \right),$$

and the stochastic entropy, $S(N)$, has the same functional form in compliance with the Greene–Callen principle [cf. §3.6].

Employing the second law in the form

$$\left(\frac{\partial S}{\partial \bar{N}} \right)_V = \left(\frac{\partial S}{\partial \bar{N}} \right)_{V, \bar{U}} + \left(\frac{\partial S}{\partial \bar{U}} \right)_{V, \bar{N}} \frac{d\bar{U}}{d\bar{N}}$$

$$= \ln \left(\frac{mV + \bar{N}}{\bar{N}} \right) = -\frac{\mu - \varepsilon}{T},$$

where ε is the energy per particle, leads to the Bose–Einstein "distribution"

$$\bar{N}_\varepsilon = \frac{mV}{e^{\hat{\beta}(\varepsilon-\mu)} - 1},$$

which is really the expectation of the distribution. Introducing this distribution into the maximum likelihood value of the negative binomial distribution leads to

$$\hat{f}(N;\bar{N}_\varepsilon) = \binom{mV + N - 1}{N} e^{-N\hat{\beta}(\varepsilon-\mu)} \left(1 - e^{-\hat{\beta}(\varepsilon-\mu)}\right)^{mV}, \qquad (5.59)$$

which is the result we found in §2.5.1.

In the classical limit, $mV \gg N$, and in this limit,[4]

$$\binom{mV + N - 1}{N} \simeq \frac{(mV)^N}{N!}.$$

For a nonrelativistic particle of mass M, m is proportional to the density of states in the momentum range from p to $p + dp$, namely,

$$\begin{aligned}
m\, dp &= \frac{4\pi p^2}{h^3}\, dp = 4\pi M \frac{\sqrt{2M\varepsilon}}{h^3}\, d\varepsilon \\
&= \frac{C^{3/2}}{\Gamma(\frac{3}{2})} \varepsilon^{1/2}\, d\varepsilon = \Omega(\varepsilon)\, d\varepsilon,
\end{aligned}$$

where $C = 2\pi M/h^2$. Observing that in the classical limit the logarithm of the last term in (5.59) is approximated by

$$-mV \ln\left(1 - e^{-\hat{\beta}(\varepsilon-\mu)}\right) \simeq mV e^{-\hat{\beta}(\varepsilon-\mu)} = \bar{N}_\varepsilon \qquad (5.60)$$

enables us to write the negative binomial distribution in the classical limit as the Poisson distribution (5.44) of the grand-canonical ensemble. The only difference is that the mean number of particles,

$$\bar{N} = zV \int_0^\infty \Omega(\varepsilon)e^{-\hat{\beta}\varepsilon}\, d\varepsilon = \hat{z}V/\hat{\lambda}^3,$$

is replaced by the mean number of particles, (5.60), in the energy range $d\varepsilon$. This is a particular case of (5.35) with $m = \frac{3}{2}$ and thermal wavelength given by $\hat{\lambda} = h\sqrt{\hat{\beta}/2\pi M}$.

Likewise, we have shown in §3.3.3 how Fermi–Dirac statistics is derived from the binomial distribution,

$$f(N;p) = \binom{mV}{N} p^N q^{mV-N},$$

---

[4]The explicit calculation is performed in §6.5.

by evaluating the *a priori* probabilities at their maximum likelihood values, $\hat{p}$ and $\hat{q}$, where

$$\hat{p} = \frac{\bar{N}}{mV}$$

and $\hat{q} = 1 - \hat{p}$.

The maximum likelihood value of the binomial distribution can now be cast as the error law (5.58) leading to the average number of particles as the most probable value. The statistical entropy is now given by the expression

$$S(\bar{N}) = \bar{N} \ln \left(\frac{mV - \bar{N}}{\bar{N}} \right) - mV \ln \left(\frac{mV - \bar{N}}{mV} \right),$$

which has the form of an entropy-of-mixing. Introducing the temperature by means of the second law, we obtain the Fermi–Dirac distribution

$$\bar{N}_\varepsilon = \frac{mV}{e^{\hat{\beta}(\varepsilon - \mu)} + 1}.$$

Evaluating the *a priori* probabilities at their maximum likelihood values converts the binomial distribution into the error law

$$\hat{f}(N; \bar{N}_\varepsilon) = \binom{mV}{N} e^{-\hat{\beta}N(\varepsilon - \mu)} \left(1 + e^{\hat{\beta}(\varepsilon - \mu)} \right)^{-mV}. \tag{5.61}$$

In the classical limit $mV \gg N$, we again find the marginal distribution of the grand-canonical distribution (5.44) under the approximations

$$\binom{mV}{N} \simeq \frac{(mV)^N}{N!}$$

and

$$mV \ln \left(1 + e^{-\hat{\beta}(\varepsilon - \mu)} \right) \simeq mV e^{-\hat{\beta}(\varepsilon - \mu)} = \bar{N}_\varepsilon. \tag{5.62}$$

The average number of particles, (5.62), in the energy range $d\varepsilon$ replaces the total mean value, (5.35), in the marginal distribution (5.44).

Both the maximum likelihood values of the negative binomial and binomial distributions, (5.59) and (5.61), can be cast in the canonical form

$$\hat{f}_V(N) = \frac{(mV)^N}{N!} e^{-\hat{\beta}[N(\varepsilon - \mu) + P_\varepsilon V]}, \tag{5.63}$$

analogous to the canonical distribution (5.39). The only—but crucially important—difference is that the energy, $\bar{U} = \bar{N}\varepsilon$, has now become a *dependent* rather than an independent variable, as it is in thermodynamics. The average energy is a linear function of the number of particles. This is the Planck quantization rule, and it is for this reason that we considered the marginal distribution (5.44) rather than the grand-canonical distribution (5.27).

By comparing (5.63) with the maximum likelihood value of the negative binomial distribution, (5.59), we determine the pressure P_ε in the energy range $d\varepsilon$ as

$$P_\varepsilon = -\frac{m}{\hat{\beta}} \ln \left(1 - e^{-\hat{\beta}(\varepsilon - \mu)}\right).$$

Similarly, a comparison between (5.63) and the maximum likelihood value of the binomial distribution (5.61) leads to the expression

$$P_\varepsilon = \frac{m}{\hat{\beta}} \ln \left(1 + e^{-\hat{\beta}(\varepsilon - \mu)}\right)$$

for the pressure in the energy interval $d\varepsilon$. In the classical limit, both go over into the equation of state of an ideal gas which is valid in each energy interval.

Bibliographic Notes

This chapter is an extension of our article

- B.H. Lavenda and J. Dunning-Davies, "Underlying probability distributions of the canonical ensemble," *Int. J. Theoret. Phys.* **29**, 85-99 (1990).

The basic "trick" involved in converting a discrete Poisson distribution into a gamma density is described in

- D. Blackwell and M.A. Girshick, *Theory of Games and Statistical Decisions* (Wiley, New York, 1954), §11.4.

The relation between the sum of term in the Poisson distribution and the incomplete gamma integral is widely used in renewal theory,

- D.R. Cox, *Renewal Theory* (Methuen, London 1962),

where the sum (5.7) is known as the survivor function and the gamma densities are referred to as the Erlangen distributions. It is also used in the theory of random walks,

- D.R. Cox and H.D. Miller, *The Theory of Stochastic Processes* (Chapman and Hall, London, 1978),

where the realizations of the stochastic process are jumps which have a high probability of being zero and a small probability of being $+1$.

To the best of our knowledge, the first time the saddle function properties of the stochastic entropy, together with its the minimax theorem, have been formulated explicitly is in

- B.H. Lavenda, "Black-hole versus black-body thermodynamics," *Z. Naturforsch.* **45A**, 879-882 (1990),

although it is implicit in Khinchin's approach,

• A.I. Khinchin, *Mathematical Foundations of Statistical Mechanics*, transl. by G. Gamow (Dover, New York, 1949), §33.

In our article we showed that the putative expression for the entropy of a black hole, which has caused so much commotion in cosmological circles, is really not an entropy at all. The minimax property of the entropy must always give rise to a concave function in the extensive variable when the stochastic entropy is evaluated at its stationary minimum with respect to the conjugate intensive variable. Since the Bekenstein–Hawking expression for the entropy of a black hole is convex in the extensive variable, it meets none of these requirements and, consequently, violates the second law.

Chapter 6

Statistical Equivalence Principle

6.1 Particle or Field?

Einstein loved to couch general principles in terms of equivalences between two seemingly unconnected quantities. His equivalence principle of mass and energy, asserting that *any* change in the energy of a body implies a corresponding change in its inertial mass, and the equivalence between a uniform, stationary gravitational field and a system moving with constant acceleration without any gravitation are beautiful examples of principles which predict that two ways of looking at a phenomenon represent physically indistinguishable situations. In this chapter, we will discover an equivalence principle of a statistical nature between fluctuations in the particle number and volume. In analogy with the way that the square of the speed of light links changes in the energy and the mass, so the average density will be found to relate the standard deviations in the number of particles and volume.

Volume fluctuations or those in the *field* are equivalent to *particle* fluctuations. We have seen in §2.3.2 that such a type of equivalence had already been implied by Einstein in his 1904 paper on the determination of the characteristic wavelength of black radiation. That paper represents a bold step in applying fluctuation theory, which up to that time belonged exclusively to the realm of molecular mechanics, to a cavity containing black radiation which could hardly be considered as housing a system of molecules. Einstein's program, at that time, was to show how the molecular theory of heat does, in fact, apply to thermal radiation. The essential dualism in a theory such as Lorentz's electron theory, where particle and field concepts maintain positions of equal prominence, always posed a worry to Einstein, who thought such a dualism is unsatisfactory. Only in later years was Einstein to reverse his position and argue vigorously for the field concept in which the field equations contain particlelike solutions.

As we have seen in Chapter 2, the particle nature of black radiation has always dominated over the field aspect. This is undoubtedly due to Planck's reliance on Boltzmann's method for calculating the entropy. The field enters through Planck's determination of the number of oscillators there are in any given frequency interval. It makes use of Maxwell's theory, which Einstein argues was incongruent with the fact that the elementary oscillators can absorb and emit discontinuously. According to Einstein's explanation, Maxwell's theory could only account for the average, as opposed to the instantaneous, energy possessed by an oscillator. And when the average energy is introduced into the theory from the law of equipartition, it leads to catastrophic consequences.

Einstein's thoughts on fluctuations are in marked contrast to the opinions of both Boltzmann and Gibbs. Boltzmann, writing in his *Theory of Gases*, argued that

> even in the smallest neighborhood of the tiniest particles suspended in a gas, the number of molecules is already so large that it seems futile to hope for any observable deviation, even in a very small time, from the limits that the phenomena would approach in the case of an infinite number of molecules.

Gibbs was equally as pessimistic about the observations of deviations from the average.

In §2.3.2 we have shown that Einstein's derivation of the empirical constant entering into Wien's displacement law is marred by the fact that he equated an extensive quantity, the mean square fluctuation in the energy, with a nonextensive one, the square of the average energy. Rather than reducing the volume of the cavity to the scale of the wavelength of the radiation, the volume should be scaled to the number of particles per frequency interval that are expected to be found in it. We saw in §5.5 that density of particles is the proportionality factor relating the standard deviation of the number of particles to that of the volume. For black radiation we have to qualify this by replacing the standard deviation of the number of particles by the corresponding expression *per frequency interval*. In the classical limit it tends to zero, and the equivalence between particle and volume fluctuations is no longer perceptible.

After a brief discussion of a Stefan gas in §6.3, we will contrast it with an ideal gas. Under very general conditions, we will show that these two types of gases form two general classes: one in which the energy is a function of the volume independent of the number of particles and another in which the energy is a function of the number of particles independent of the volume.

In §6.4 we will show that the entropy of an ideal gas gives rise to a law of error for either the number of particles or volume, depending upon which fluctuates. In other words, two probability distributions are generated by the same law of error, and it only matters which variable is considered to be fluctuating at a fixed value of the other.

In §5.5 we had our first inkling of the type of complementarity that exists between fluctuations in the volume and particle number. The former is governed by the *continuous* gamma density while the latter is ruled by a *discrete* Poisson distribution. By considering a simple experiment of watching particles

in solution under a microscope, we appreciate that it is equivalent to counting the number of particles in a given area or enlarging the aperture of the
microscope until the area contains a predetermined number of particles.

Unlike the entropy of an ideal gas, the entropy of a Stefan gas does not
lead to an error law for fluctuations in the volume. In the limit where the
number of accessible states far outweighs the number of particles, the system
will tend either to a classical ideal gas or a Stefan gas depending upon whether
the "particles" are conserved or not. In §6.5 we will take our lead from the
fact that the gamma and Poisson distributions are limiting forms which are
achieved only in the classical limit. Of the two, it is the gamma density which
indicates the form the entropy should have since the Poisson distribution is
a limiting form of two distributions, the binomial distribution in which the
number of particles is conserved and the negative binomial distribution in
which they are not conserved.

We then go on to investigate the limiting forms of the error laws in the
particularly illuminating case of black radiation. The classical, or Wien, limit,
corresponding to low intensity or high frequency radiation at low temperatures,
is achieved by letting the density of oscillators per frequency interval increase
without limit. The vast number of oscillators far outweighs the small number
of energetic photons. In the opposite limit, as the characteristic parameter
tends to zero, corresponding to the Rayleigh–Jeans region of the spectrum,
the reverse situation occurs: the vast number of low energetic photons vastly
overwhelms the number of oscillators. The oscillators represent the field, and
their fluctuations are taken into account by fluctuations in the volume, whereas
fluctuations in photons are described by fluctuations in the number of particles. Just as the equivalence between energy and mass stops at the point of
quantization, so too, the energy, associated with the number of particles, can
be quantized while the pressure, which is the intensive conjugate variable of
the volume, is and must remain a continuous variable.

It is well known that the isotherms of a photon gas are flat and, consequently, the gas is metastable. However, it is not generally acknowledged that
all systems obeying quantum statistics are metastable: the discriminant of
the quadratic form of the entropy vanishes. In §6.6 we shall attribute this to
the fact that the energy is relegated to the position of a dependent variable,
being defined solely in terms of the number of particles. This leaves the average number of particles and the volume as the independent variables, and
the vanishing of the discriminant of the quadratic form simply says that they
cannot fluctuate simultaneously. The local form of Gauss' principle, valid for
small fluctuations, vanishes when both the number of particles and volume are
allowed to fluctuate simultaneously. In other words, one can say that the correlations in the volume and particle fluctuations exceed any finite magnitude.

In §6.8.1 we shall use the expression for the mean square fluctuation in
the particle number, which will be derived in §6.7, to estimate the characteristic wavelength of black radiation, where the energy density has its maximum
value. This corrects Einstein's estimate based on setting the mean square fluc-

tuation in the energy equal to the square of the average energy. In addition, the expression for the mean square fluctuation in the volume will enable us in §6.8.2 to obtain an estimate of the Stefan–Boltzmann constant. This will be achieved by relating fluctuations in the total energy of black radiation to fluctuations in the volume, analogous to the way in which fluctuations in the energy per frequency interval were related to fluctuations in the number of photons in the same frequency interval to estimate the characteristic wavelength.

This shows that fluctuations in the volume have a *global* character whereas fluctuations in the particle number in any given frequency interval have a *local* character. Furthermore, the analysis of fluctuations in the volume give the same order of magnitude for the characteristic wavelength as that for fluctuations in the particle number.

6.2   The Generalized Stefan Law

In a paper published in the *Philosophical Magazine* in 1902 on the pressure of vibrations, Lord Rayleigh queried whether the relation between the pressure and energy density of radiation, $P = u/3$, could not be generalized to

$$P = u/\eta, \tag{6.1}$$

where η is some numerical quantity. Introducing this generalized relation into the relation [cf. §2.1]

$$\frac{dP}{dT} = \frac{h}{T}, \tag{6.2}$$

where the enthalpy density $h = u + P$, Lord Rayleigh derived the generalized Stefan law

$$u = \sigma\, T^{\eta+1} \tag{6.3}$$

upon integration.

The physical significance of the parameter η can be deduced by considering a quasi-static change, just as we did in §2.1. During such a change, the heat added to the system in an infinitesimal transition from one equilibrium state to a neighboring one is given by the first law,

$$dQ = d(uV) + P\, dV.$$

Introducing the generalized equation of state (6.1) and considering an adiabatic change $(dQ = 0)$, we find

$$V\, du + \left(1 + \frac{1}{\eta}\right) u\, dV = 0.$$

Separating variables and integrating give

$$uV^{1+1/\eta} = \text{const.}$$

Upon introducing the generalized Stefan law, (6.3), we get[1]

$$TV^{1/\eta} = \text{const.,} \qquad (6.4)$$

which is comparable to the classical relation

$$TV^{\gamma-1} = \text{const.,} \qquad (6.5)$$

along an adiabatic curve of an ideal gas.

We denote by $\gamma = C_P/C_V$ the ratio of the specific heats at constant pressure and volume. For an ideal gas, the difference between the specific heats per molecule is $C_P - C_V = 1$ when T is measured in energy units. And since $C_V = m$, where m is half the number of degrees-of-freedom of a molecule, we find

$$\gamma = \frac{m+1}{m} = \frac{\eta+1}{\eta}$$

upon equating the exponents in (6.4) and (6.5). It is therefore apparent that

$$\eta = m, \qquad (6.6)$$

or one-half of the number of degrees-of-freedom of the elementary constituent of the system.

6.3   Ideal versus Stefan Gases

Lord Rayleigh based himself on Boltzmann's derivation of Stefan's law. Boltzmann's derivation employed a Carnot cycle in which the radiation constituted the working substance that is enclosed within movable reflecting walls. The rather unusual part of the demonstration was Lord Rayleigh's use of Eq. (6.2), which has the form of a Clapeyron equation rather than a Gibbs–Duhem equation. In other words, there appears a total rather than a partial derivative of the pressure.

Lord Rayleigh interpreted the quantity $h\,dV$ as the heat which must be added to a system when it undergoes a volume change, dV, at constant temperature. Although Lord Rayleigh assumed that the pressure was a function of both V and T, it is precisely because *the pressure is independent of the volume* that Eq. (6.2) can be integrated without the appearance of an arbitrary function of the volume in Stefan's law (6.3). The Clapeyron equation, (6.2), asserts that the pressure is a function only of the temperature, independent of the volume. Therefore, when the volume is increased, at constant temperature,

---

[1]This adiabatic condition has an interesting counterpart in cosmology. If we set the volume $V \propto R^3$, where R is the "cosmic scale factor," the condition of adiabaticity is $T = \text{const.}/R$. If the universe is expanding, then R is increasing and consequently the temperature is decreasing. This decrease is precisely what is needed to compensate the cosmological red-shift in the frequency so that, miraculously enough, the Planck radiation profile is preserved as the universe continuously expands.

both the density of the energy and pressure will not vary. This is analogous to a phase equilibrium between saturated vapor and liquid. When the volume is increased while the temperature is kept constant, the liquid takes care that the pressure should not vary. Since black bodies transform all incident rays into heat, the tiny speck of coal dust ensures that the density and pressure shall not vary by regulating the heat supply.

It is possible to speak in terms of an equilibrium between rays of different frequencies in black radiation that all possess a common temperature. It is the tiny speck of coal dust that mediates the exchanges of energy between the modes of the electromagnetic field. Thus when photons pass from one range of frequencies to another, we have something that is analogous to the passage of a substance from one phase to another. The equality of the fugacities of the two phases is required for the existence of a phase equilibrium. If the fugacity is a function of the pressure and temperature, the condition that the fugacity of the two phases be equal provides a relation between P and T such that one cannot be varied independently of the other. We say that such a relation decreases the number of "thermodynamic degrees-of-freedom" of the system by 1. But a fugacity or chemical potential is completely foreign to the description of black radiation since the number of photons is not conserved.

In contrast to Eq. (6.2), the Gibbs–Duhem equation for a single phase is

$$n \, d \ln z = u \, d(1/T) + d(P/T), \qquad (6.7)$$

where z is the fugacity, $\ln z \equiv \mu/T$, and μ is the chemical potential. If

$$z = \text{const.}, \qquad (6.8)$$

the Gibbs–Duhem equation, (6.7), reduces to the Clapeyron equation, (6.2). Condition (6.8) is analogous to the establishment of a phase equilibrium which we interpret as a lack of conservation of "particles" (e.g., photons or phonons), since particles can pass from one "phase" to another.

Introducing (6.1) into the Gibbs–Duhem relation (6.7) results in

$$\left(\frac{\eta n T}{u} \right) d \ln z = d \ln \left(\frac{u}{T^{\eta+1}} \right). \qquad (6.9)$$

Since the right-hand side of (6.9) is a total differential so, too, must be the left-hand side. This implies that

$$u = \eta n T. \qquad (6.10)$$

We are therefore able to integrate (6.9) to obtain

$$z = \vartheta^{-1} u / T^{\eta+1}, \qquad (6.11)$$

where ϑ is a constant of integration. Comparing (6.10) with (6.11) leads to

$$n = \frac{\vartheta z}{\eta} T^{\eta}. \qquad (6.12)$$

Hence, there are *two* solutions to the Gibbs–Duhem relation (6.9): solution (6.8) corresponds to what we shall refer to as a "Stefan" gas, while the other solution, (6.11), describes a classical ideal gas. If the fugacity is constant, then the density is proportional to T^η, while if the density is constant, the fugacity is proportional to $T^{-\eta}$. These solutions constitute two general classes of *perfect* gases. From (6.10) we see that in the former, the energy is proportional to the volume independent of the number of particles, while in the latter, it is proportional to the number of particles independent of the volume.

Whereas the thermodynamic relations of an ideal gas appear more naturally in *molar* form, those of a Stefan gas show themselves up more naturally as *density* relations. For an ideal gas, we consider the fugacity as a function of the temperature and molar volume, $v_m = n^{-1}$. The logarithm of the fugacity, (6.11), may be expressed equivalently in terms of the molar volume as

$$\ln z = -\ln(T^\eta v_m) + \text{const.} \tag{6.13}$$

if we remember that $n \propto T^\eta$ for a Stefan gas.

Converting the Gibbs–Duhem relation,

$$N \, d\ln z = U \, d(1/T) + V \, d(P/T), \tag{6.14}$$

into its molar form and then introducing (6.13) in place of the logarithm of the fugacity lead to

$$-\eta \frac{dT}{T} - \frac{dv_m}{v_m} = -\frac{u_m}{T^2} \, dT + v_m \, d\left(\frac{P}{T}\right),$$

where $u_m = U/N$ is the molar energy.

Noting that the energy density of an ideal gas is given by (6.10), the Gibbs–Duhem relation reduces to

$$d\left(\frac{1}{v_m}\right) = d\left(\frac{P}{T}\right).$$

This can be integrated immediately to give the equation of state,

$$P = T/v_m, \tag{6.15}$$

of an ideal gas, where we have set the arbitrary constant of integration equal to zero.

At constant density, equations (6.15) and (6.10) are the equations of state of an ideal gas and provide all the thermodynamic information known about it. They are equivalent to a knowledge of the fundamental relation that expresses the entropy as a function of all the independent extensive variables of the system.

In order to derive the fundamental relation, we convert the differential form of Euler's relation,

$$dS = \frac{1}{T} \, dU + \frac{P}{T} \, dV - \ln z \, dN, \tag{6.16}$$

for the entropy into its molar form,

$$ds_m = \frac{1}{T} du_m + \frac{P}{T} dv_m, \tag{6.17}$$

where $s_m = S/N$ is the molar entropy. It is interesting to note that the negative increment in the chemical work per unit temperature, $-\ln z\, dN$, does not appear in the molar relation (6.17). The fugacity will appear as a derived quantity from the fundamental relation.

Introducing the equations of state of an ideal gas, (6.10) and (6.15), renders the right-hand side of (6.17) a total differential:

$$ds_m = \eta\, d\ln u_m + d\ln v_m.$$

This can be immediately integrated to give

$$s_m = s_0 + \ln\left(u_m^\eta v_m\right),$$

where s_0 is an undetermined constant of integration. Reverting to extensive quantities, the fundamental relation becomes

$$S = N s_0 + N \ln\left(\frac{U^\eta V}{N^{\eta+1}}\right). \tag{6.18}$$

The intensive quantities are derived from the fundamental relation through differentiation with respect to their conjugate extensive quantities. For instance, the negative of the logarithm of the fugacity is

$$\left(\frac{\partial S}{\partial N}\right)_{U,V} = -\ln z = s_0 - (\eta + 1) + \ln\left(\frac{U^\eta V}{N^{\eta+1}}\right),$$

which can be seen to be identical to (6.13) when the equation of state (6.10) is substituted for the energy.

All that we have said about an ideal gas can now be contrasted with what we will say about a Stefan gas. In place of *molar* quantities, *densities* are now to be preferred. The fugacity relation (6.13) becomes

$$\ln z = \ln(n/T^\eta) + \text{const.} \tag{6.19}$$

Introducing (6.19) into the Gibbs–Duhem relation (6.7) gives

$$dn - \eta \frac{n}{T} dT = -\frac{u}{T^2} dT + d\left(\frac{P}{T}\right).$$

If the left-hand side is to vanish, $n \propto T^\eta$, and this together with (6.10) give the generalized Stefan law, (6.3).

Therefore, condition (6.8) implies a lack of particle conservation and reduces the Gibbs–Duhem equation (6.7) to the Clapeyron equation (6.2) which applies to any system whose equation of state is of the form $P = P(T)$. Photons in thermal equilibrium satisfy condition (6.8) trivially: the photon chemical potential vanishes. But photons do not interact with one another, so in order to come to thermal equilibrium, other particles which they interact with must be taken into account.

It may turn out that the chemical potential does not vanish. We have referred to such radiation as nonthermal in §2.4, in contrast with thermal black radiation. Notwithstanding such a difference, condition (6.8) must always hold when there is no conservation of "particles." The absolute value of the photon chemical potential is strictly proportional to the temperature.

In order to derive the fundamental relation of a Stefan gas, we convert the differential of the Euler relation (6.16) into the density relation

$$ds = \frac{1}{T} du - \ln z \, dn. \tag{6.20}$$

Let us take note of the disappearance of the increment in the compressional work term per unit temperature, $-(P/T)\,dV$. The pressure will be a derived quantity from the fundamental relation in exactly the same way that the fugacity was derived from the fundamental relation (6.18). Solving (6.11) for $1/T$, we find

$$\frac{1}{T} = \left(\frac{\vartheta z}{u}\right)^{1/(\eta+1)}.$$

Introducing this into (6.20) and integrating the resulting expression give

$$s = \left(\frac{\eta+1}{\eta}\right)(\vartheta z)^{1/(\eta+1)} u^{\eta/(\eta+1)} - n \ln z.$$

But from (6.12), we see that the density is a function of the temperature and consequently of the energy, namely,

$$n = (\vartheta z)^{1/(\eta+1)} u^{\eta/(\eta+1)},$$

so that the expression for the entropy density becomes

$$s = \frac{1}{\eta}(\eta + 1 - \ln z)(\vartheta z)^{1/(\eta+1)} u^{\eta/(\eta+1)}.$$

As a check to see whether this is, in fact, the correct expression for the entropy density, we differentiate with respect to the energy density. We then obtain

$$\frac{\partial s}{\partial u} = \frac{\eta + 1 - \ln z}{\eta + 1}\left(\frac{\vartheta z}{u}\right)^{1/(\eta+1)} = \frac{1}{T},$$

showing that the last term in molar Euler should not be there. Therefore, the correct expression for the entropy of a Stefan gas is

$$S = \left(\frac{\eta+1}{\eta}\right)(\vartheta z)^{1/(\eta+1)} U^{\eta/(\eta+1)} V^{1/(\eta+1)}. \tag{6.21}$$

Finally, expressing the entropy in terms of the number of particles gives the simple expression

$$S = (\eta + 1)N. \tag{6.22}$$

Differentiating it with respect to N leads to

$$\frac{\partial S}{\partial N} = \eta + 1 = -\ln z.$$

We thus obtain the expression

$$z = e^{-(\eta+1)}$$

for the fugacity of a Stefan gas, which is simply our condition (6.8).

Thus far, there is nothing to make us think that there is anything less "fundamental" about the fundamental relation of a Stefan gas, (6.21), in contrast to that of an ideal gas, (6.18). Fundamental differences will, nevertheless, show up when we proceed to investigate deviations from most probable behavior on the basis of Gauss' principle.

6.4   Complementarity

As usual, when we study fluctuations in the extensive thermodynamic variables, we identify average values of the extensive quantities with their thermodynamic values. But if deviations from these values occur, then they must also be the most probable values and, because of the additivity of the extensive variables, the most probable values with the arithmetic means or the means of the distribution. As we have seen on numerous occasions, it is the entropy, or more precisely its property of concavity, which determines the law of error. The concavity property is couched in the logarithmic form of the entropy, having either the form $-X \ln X$ or $\ln X$. It is for this reason that the entropy of an ideal gas, (6.18), will give rise to a *bona fide* error law for fluctuations in the energy and the number of particles or volume.

Prior to deriving the error laws, we must convert the entropy into its parametric form:

$$S(\bar{V}, \bar{N}) = \bar{N}s_1 + \bar{N}\ln\left(\frac{T^\eta \bar{V}}{\bar{N}}\right), \tag{6.23}$$

with the temperature playing the role of a parametric variable. We have collected all the constants into a single constant s_1 and identified the thermodynamic extensive variables with their average values. We hold off a discussion concerning *why* the entropy must be converted into its parametric form before it can be used in Gauss' principle till later. For the moment, our aim is to provide a contrast between it and the parametric form of the entropy of a Stefan gas, (6.22).

It is quite clear that the entropy of a Stefan gas will not lead to any law of error precisely because it is linear in the average number of particles. This observation might well have been the motivation behind Planck's search for a new "fundamental" relation in which the entropy has the required logarithmic form in order for it to be the potential for a law of error [cf. §2.5.1].

The parametric form of the entropy, (6.23), together with the equation of state for the energy, (6.10), contain the same information as the fundamental relation, (6.18). However, the introduction of the equation of state into the fundamental relation, in order to get the parametric entropy, reduces the number of thermodynamic degrees-of-freedom of the system or the number of intensive variables that can be varied independently. The reason why the fundamental relation has to be written in the parametric form is that *the energy cannot be kept constant when the volume or number of particles fluctuate while the temperature can.* Even in this case, we will find that there is something prohibitive in the simultaneous fluctuations in the particle number and volume. We will discover that they are, in fact, complementary descriptions of the same phenomenon.

Introducing the parametric form of the entropy, (6.23), and a functionally identical expression for the stochastic entropy, $S(T, V, \bar{N})$, into Gauss' principle,

$$f(V) = A \exp \left\{ S(V) - S(\bar{V}) - \left(\frac{\partial S}{\partial \bar{V}} \right)_{\bar{N}} (V - \bar{V}) \right\},$$

results in the error law,

$$f(V) = A \left(\frac{V}{\bar{V}} \right)^{\bar{N}} e^{-\bar{N}(V - \bar{V})/\bar{V}},$$

for fluctuations in the volume.

At first sight this does not appear to be a fundamental law of error. However, when we set $A = \bar{N}/V,$[2] multiply numerator and denominator by $\bar{N}^{\bar{N}}$, and use Stirling's approximation, we come out with the gamma density,

$$f(V) = \varrho \frac{(\varrho V)^{\bar{N}-1}}{\Gamma(\bar{N})} e^{-\varrho V}, \qquad (6.24)$$

as the law of error for fluctuations in the volume, where the average density $\varrho = \bar{N}/\bar{V}$. It is well known that the gamma density represents the continuous analog of the negative binomial distribution. This will serve as a clue in deriving Planck's expression for the entropy.

If we now allow the particle number to fluctuate, at constant volume and temperature, and introduce the stochastic entropy, $S(T, \bar{V}, N)$, together with the thermodynamic entropy, (6.23), into Gauss' principle,

$$f(N) = A \exp \left\{ S(N) - S(\bar{N}) - \left(\frac{\partial S}{\partial \bar{N}} \right)_{\bar{V}} (N - \bar{N}) \right\},$$

---

[2]We recall that the norming function can be a function of the random variable but must be independent of its average value.

we obtain the Poisson distribution,

$$f(N) = \frac{\bar{N}^N}{N!} e^{\bar{N}}, \tag{6.25}$$

as the error law for fluctuations in the number of particles. Thus, depending upon whether the volume or number of particles fluctuate, we get a continuous probability density, (6.24), or a discrete probability distribution, (6.25), respectively.

We may well ask what happens when both V and N fluctuate simultaneously. Gauss' principle for two variables is [cf. §1.8]

$$\ln f(V, N) = S(V, N) - S(\bar{V}, \bar{N}) - (V - \bar{V}) \left(\frac{\partial S}{\partial \bar{V}} \right)_{\bar{N}}$$
$$- (N - \bar{N}) \left(\frac{\partial S}{\partial \bar{N}} \right)_{\bar{V}} + \text{const.} \tag{6.26}$$

But when the entropies are introduced in their parametric forms, we get

$$f(V, N) = A \frac{(\varrho V)^N}{N!} e^{-\varrho V}. \tag{6.27}$$

It is thus evident that *both N and V cannot fluctuate simultaneously*. On the one hand, if V fluctuates, (6.27) is the gamma density with $A = \varrho$, while on the other hand, if N fluctuates, (6.27) is the Poisson distribution with $A = 1$.

The two distributions exhibit a certain complementarity and provide equivalent descriptions of the same physical process. Suppose we want to estimate the average concentration of particles per unit volume in a liquid. Let us assume that the particles are well mixed so that the distribution of the number of particles in any portion of the liquid is Poisson. Our problem is to estimate the number of particles per unit volume, ϱ.

In any given subvolume, the distribution of the number of particles N is given by the Poisson distribution, (6.25), which we can write as

$$f(N) = \frac{(\varrho V)^N}{N!} e^{-\varrho V}. \tag{6.28}$$

The density, $\varrho = \bar{N}/V$, appears in the form of a scale parameter, but it is *not* the scale parameter for the distribution (6.28) because it must always accompany the random variable and not some fixed quantity like V. This persuades us into believing that there exists another distribution for which ϱ is the true scale parameter.

Now, in any given portion of liquid we can count the number of particles that enters the volume; the probability of N entrants will be seen to be given by the Poisson distribution (6.28). Alternatively, we can perform another type of experiment. We can increase the volume continuously until there are just N particles in the volume. This volume, $\mathbf{V}$, is the pertinent random variable.

The event $\mathbf{V} > V$ is equivalent to the statement that "the volume V has fewer than N particles." The volume V can have a maximum of $N-1$, but no more, so that

$$\Pr(\mathbf{V} > V) = \sum_{j=1}^{N-1} \frac{(\varrho V)^j}{j!} e^{-\varrho V}.$$

The cumulative distribution function of $\mathbf{V}$, $F_\varrho(V)$, coincides with

$$\Pr(\mathbf{V} \le V) = 1 - \Pr(\mathbf{V} > V)$$
$$= 1 - \sum_{j=1}^{N-1} \frac{(\varrho V)^j}{j!} e^{-\varrho V}. \qquad (6.29)$$

Using the identity

$$\sum_{j=1}^{N-1} \frac{(\varrho V)^j}{j!} e^{-\varrho V} = \int_V^\infty \varrho \frac{(\varrho x)^{N-1}}{(N-1)!} e^{-\varrho x} \, dx$$

that we used in §5.2.1 in the analogous case of energy and degrees-of-freedom, the cumulative distribution is easily seen to be the incomplete gamma function

$$F_\varrho(V) = \int_0^V \varrho \frac{(\varrho x)^{N-1}}{(N-1)!} e^{-\varrho x} \, dx.$$

The density of this distribution, $f(V) = dF_\varrho/dV$, is precisely the gamma density (6.24).

Thus, the cumulative distribution of the Poisson distribution may be expressed in terms of the cumulative distribution of the gamma density and vice versa. *Although the Poisson distribution governs a discrete random variable while the gamma density governs a continuous one, the two are statistically equivalent descriptions of the same process.* This is the basis of the principle of statistical equivalence.

6.5 Descent into Quantum Statistics: Planck's Resonators Resurrected

In a certain sense, we are in a real quandary. The fundamental relation of an ideal gas, (6.23), has led to the equivalent laws of error for fluctuations in the volume and particle number while that of a Stefan gas, (6.22), appears impotent to give an error law for fluctuations in the particle number. However, we know that excitations, like photons and phonons, undergo fluctuations, but at least from thermodynamic arguments, we are unable to predict their distributions and spectra.

Historically, the Poisson distribution was considered a limit law, and one of these limiting schemes is via the binomial distribution. It is thus reasonable to assume that both the entropies of an ideal and a Stefan gas are limiting forms of more fundamental expressions which give rise to probability distributions for N and V. Whether we arrive at the limiting form of an ideal or a Stefan gas depends entirely upon whether the fugacity is given by (6.11) or (6.8), respectively.

We cannot take a cue from the Poisson distribution because it is the limiting law for more than one distribution. However, this is not true of the gamma density for *gamma densities are the continuous analogs of the negative binomial distribution*. The entropy of the negative binomial distribution which leads to the law of error for fluctuations in particle number is

$$S(\bar{V}, \bar{N}_\varepsilon) = \bar{N}_\varepsilon \ln\left(\frac{\bar{N}_\varepsilon + m\bar{V}}{\bar{N}_\varepsilon}\right) + mV \ln\left(\frac{\bar{N}_\varepsilon + mV}{m\bar{V}}\right), \qquad (6.30)$$

where the index on $\bar{N}$ and the significance of the parameter m will become apparent in a moment. It suffices here to note that (6.30) is the *entropy of mixing* of two "substances." The question is what are these two substances?

The parameter m has dimensions of a density. In the limit as $m \to \infty$, the entropy of mixing, (6.30), transforms into a well known result. For in that limit we have

$$S(\bar{V}, \bar{N}_\varepsilon) = \bar{N}_\varepsilon \left[1 - \ln\left(\frac{\bar{N}_\varepsilon}{m\bar{V}}\right)\right], \qquad (6.31)$$

which has the same form as the entropy of an ideal gas. This was first emphasized by Einstein in support of his light-quantum hypothesis. In this limit, the average density in the interval between ε and $\varepsilon + d\varepsilon$ is given by the Maxwell–Boltzmann expression

$$\bar{N}_\varepsilon/\bar{V} = mze^{-\varepsilon/T}, \qquad (6.32)$$

and introducing it into (6.31) gives

$$S = \bar{E}_\varepsilon/T + \bar{N}_\varepsilon[1 - \ln z],$$

where $\bar{E}_\varepsilon = \varepsilon \bar{N}_\varepsilon$. Consequently, $\bar{N}_\varepsilon \, d\varepsilon$ is to be interpreted as the number of particles in the energy range $d\varepsilon$. Comparing this expression for the entropy to the Euler relation, (6.16), which we assume to be valid for every energy range $d\varepsilon$, leads to the equation of state of an ideal gas, $P_\varepsilon V = \bar{N}_\varepsilon T$.

In the opposite limit as $m \to 0$, the entropy of mixing (6.30) tends to

$$S(\bar{V}, \bar{N}_\varepsilon) = \bar{V}\left[1 - \ln\left(\frac{m\bar{V}}{\bar{N}_\varepsilon}\right)\right]. \qquad (6.33)$$

In comparison with (6.31), the terms $\bar{N}_\varepsilon$ and $m\bar{V}$ have been swapped in (6.33). In the former expression, the entropy is taken to be a concave function of $\bar{N}_\varepsilon$, having the form $-\bar{N}_\varepsilon \ln \bar{N}_\varepsilon$, while in the latter, it is a concave function of $\bar{V}$, having the form $-\bar{V} \ln \bar{V}$. For black radiation, the former corresponds to the Wien limit while the latter corresponds to the Rayleigh–Jeans region. It is rather ironical that the discovery of quantum theory occurred only after "classical" deviations from the quantum regime had been observed. These "deviations" are inherently nonclassical, as the expression for the entropy, (6.33), can testify.

If the volume fluctuates at a constant number of particles, (6.30) is the entropy of the maximum likelihood value of the negative binomial distribution:

$$f_{\bar{N}_\varepsilon}(V) = \binom{\bar{N}_\varepsilon + mV - 1}{mV} \left(\frac{m}{m + \varrho_\varepsilon}\right)^{mV} \left(\frac{\varrho_\varepsilon}{m + \varrho_\varepsilon}\right)^{\bar{N}_\varepsilon}, \tag{6.34}$$

where $\varrho_\varepsilon \equiv \bar{N}_\varepsilon/\bar{V}$ is the *local* density in the interval from ε to $\varepsilon + d\varepsilon$. The continuous gamma density,

$$f_{\bar{N}_\varepsilon}(V) = \varrho_\varepsilon \frac{(\varrho_\varepsilon V)^{\bar{N}_\varepsilon - 1}}{\Gamma(\bar{N}_\varepsilon)} e^{-\varrho_\varepsilon V}, \tag{6.35}$$

is recovered by allowing $m \to \infty$. The gamma density, (6.35), is independent of m in this limit.

In contrast, the error law for fluctuations in the particle number at constant volume is

$$f_{\bar{V}}(N_\varepsilon) = \binom{m\bar{V} + N_\varepsilon - 1}{N_\varepsilon} \left(\frac{m}{m + \varrho_\varepsilon}\right)^{m\bar{V}} \left(\frac{\varrho_\varepsilon}{m + \varrho_\varepsilon}\right)^{N_\varepsilon}. \tag{6.36}$$

In the limit as $m \to \infty$, the maximum likelihood value of the negative binomial distribution, (6.36), transforms into the Poisson distribution, (6.28), for N_ε since the binomial coefficient can be approximated by

$$\binom{m\bar{V} + N_\varepsilon - 1}{N_\varepsilon} \simeq \frac{(m\bar{V} + N_\varepsilon)^{m\bar{V}+N_\varepsilon}}{N_\varepsilon^{N_\varepsilon}(m\bar{V})^{m\bar{V}}}$$

$$\simeq \frac{(m\bar{V})^{N_\varepsilon}}{N_\varepsilon^{N_\varepsilon}} \left(1 + \frac{N_\varepsilon}{m\bar{V}}\right)^{m\bar{V}}$$

$$\simeq \frac{(m\bar{V})^{N_\varepsilon}}{N_\varepsilon^{N_\varepsilon}} e^{N_\varepsilon} \simeq \frac{(m\bar{V})^{N_\varepsilon}}{N_\varepsilon!},$$

where we have used

$$\lim_{m \to \infty} \left(1 + \frac{N_\varepsilon}{m\bar{V}}\right)^{m\bar{V}} = e^{N_\varepsilon}$$

and Stirling's approximation for N_ε. The last term in (6.36) is approximately $(\varrho_\varepsilon/m)^{N_\varepsilon}$, so that the terms in m cancel and we get (6.28).

In summary, we have shown that the gamma density and Poisson distribution emerge in the same limit as $m \to \infty$ and thus constitute different, yet statistically equivalent descriptions of the same phenomenon. This is the classical limit where, as we would expect, fluctuations in the particles are discontinuous while those in the volume are continuous. However, we cannot conclude that the Poisson distribution came from the negative binomial distribution since it is also the limiting form of the binomial distribution. Hence, volume, rather than particle number, fluctuations, which are classically governed by the gamma probability density, lead uniquely to the discrete negative binomial distribution.

In the opposite limit as $m \to 0$, the error law (6.34) merges with the Poisson distribution

$$f(V) = \frac{(m\bar{V})^{mV}}{(mV)!} e^{-m\bar{V}}. \tag{6.37}$$

The discrete nature of the volume fluctuations can hardly be considered as classical. For black radiation this means that we are now in the Rayleigh–Jeans region, where there are a lot of low energy photons and a few oscillators. In view of the statistical equivalence principle, this means that the particle number fluctuations must now be continuous, being governed by the gamma probability density

$$f(N) = \frac{m}{\varrho_\varepsilon} \frac{(mN/\varrho_\varepsilon)^{mV-1}}{\Gamma(mV)} e^{-mN/\varrho_\varepsilon}. \tag{6.38}$$

In terms of black radiation, the physical picture which emerges is that of a mixture of two perfect gases of the Stefan type: particles and oscillators. We shall now show that m is, indeed, the density of oscillators per frequency interval. Taking the derivative of the entropy of mixing, (6.30), with respect to $\bar{V}$ and equating it to the thermodynamic definition of the conjugate intensive variable give

$$\left(\frac{\partial S}{\partial \bar{V}} \right)_{\bar{N}_\varepsilon} = m \ln \left(\frac{\bar{N}_\varepsilon + m\bar{V}}{m\bar{V}} \right) = \frac{P_\varepsilon}{T}, \tag{6.39}$$

or equivalently,

$$\bar{N}_\varepsilon = m\bar{V} \left\{ e^{P_\varepsilon/mT} - 1 \right\}. \tag{6.40}$$

In the limit $P_\varepsilon/mT \ll 1$, (6.40) becomes the equation of state of a perfect gas, (6.15).

The break with classical thermodynamics comes when we assume that the average energy is a *dependent*, rather than an independent, extensive variable. Following Planck, we consider it to be a (linear) function of the average number of particles in the same frequency range. The derivative of (6.30) with respect to $\bar{N}_\varepsilon$ is

$$\begin{aligned}
\left(\frac{\partial S}{\partial \bar{N}_\varepsilon} \right)_{\bar{V}} &= \ln \left(\frac{\bar{N}_\varepsilon + m\bar{V}}{\bar{N}_\varepsilon} \right) \\
&= \left(\frac{\partial S}{\partial \bar{N}_\varepsilon} \right)_{\bar{V}, \bar{E}_\varepsilon} + \left(\frac{\partial S}{\partial \bar{E}_\varepsilon} \right)_{\bar{V}, \bar{N}_\varepsilon} \frac{d\bar{E}_\varepsilon}{d\bar{N}_\varepsilon} \\
&= -\ln z + \varepsilon/T, \tag{6.41}
\end{aligned}$$

where ε is the energy per particle.

Solving (6.41) for $\bar{N}_\varepsilon$ gives

$$\bar{N}_\varepsilon = \frac{m\bar{V}}{z^{-1} e^{\varepsilon/T} - 1}. \tag{6.42}$$

Eliminating the average density, $\bar{N}_\varepsilon/\bar{V}$, between (6.40) and (6.42) leads to

$$P_\varepsilon = -mT\ln\left(1 - ze^{-\varepsilon/T}\right). \tag{6.43}$$

When we set $\varepsilon = h\nu$, expression (6.43) is precisely the pressure of

$$m = \frac{8\pi\nu^2}{c^3}, \tag{6.44}$$

Planck "resonators" or "oscillators" per unit volume per the frequency interval. Hence, (6.30) is the entropy of mixing of two perfect gases: particles and oscillators.

6.6   Thermodynamic Structure

Stability criteria can be discussed in terms of the nature of the entropy or S-surface in the neighborhood of a given point D on its surface. Stability requires that the S-surface shall lie entirely below the tangential plane at that point (i.e., the property of concavity). If the homogeneous phase is stable, then this condition must be satisfied for all states on the S-surface. In such a case, the surface is said not to have a "plait."

Suppose that X, Y, S are the coordinates of the point D and $X + \Delta X, Y + \Delta Y, S + \Delta S$ are the coordinates of a neighboring point E on the surface. In addition, let $S + \Delta'S$ be that point E' where the tangential plane at D intersects the vertical line extending from E. Neglecting higher than second-order terms, we have

$$\begin{aligned} \Delta S &= \frac{\partial S}{\partial X}\Delta X + \frac{\partial S}{\partial Y}\Delta Y \\ &+ \frac{1}{2}\left\{\frac{\partial^2 S}{\partial X^2}(\Delta X)^2 + 2\frac{\partial^2 S}{\partial X\,\partial Y}(\Delta X)(\Delta Y) + \frac{\partial^2 S}{\partial Y^2}(\Delta Y)^2\right\} \end{aligned}$$

and

$$\Delta'S = \frac{\partial S}{\partial X}\Delta X + \frac{\partial S}{\partial Y}\Delta Y.$$

Now the conditions that E' shall lie above E is that $\Delta'S - \Delta S$ shall be positive for all values of ΔX and ΔY are

$$\frac{\partial^2 S}{\partial X^2} < 0, \qquad \frac{\partial^2 S}{\partial Y^2} < 0,$$

$$\frac{\partial^2 S}{\partial X^2}\frac{\partial^2 S}{\partial Y^2} - \left(\frac{\partial^2 S}{\partial X\,\partial Y}\right)^2 > 0. \tag{6.45}$$

The first or the second inequality is a consequence of the other two. Such a system has two degrees-of-freedom since two intensive variables are capable of independent variation; the third is fixed by the Gibbs–Duhem relation.

Inequalities (6.45) have important physical implications, like the positive definiteness of both the heat capacity and the isothermal compressibility. When the thermodynamic degrees-of-freedom are decreased, the second inequality in (6.45) becomes an equality implying a state of metastability. The curve determined by the equation

$$\frac{\partial^2 S}{\partial X^2}\frac{\partial^2 S}{\partial Y^2} - \left(\frac{\partial^2 S}{\partial X\,\partial Y}\right)^2 = 0 \quad .$$

is called a *spinodal* curve; it divides a region of stable phases from unstable phases. We now see that this pertains to the case of black radiation and, more generally, in all systems that obey quantum statistics.

6.6.1 Equations of State

There are only two equations of state for a Stefan gas,

$$T = T(U,V), \qquad\qquad P = P(U,V),$$

which are functionally dependent,

$$\frac{\partial(1/T, P/T)}{\partial(U,V)} = 0,$$

owing to condition (6.8) on the Gibbs–Duhem relation (6.7). Such a system is metastable, for on changing to the variables $1/T$ and V, we have

$$0 = \frac{\partial(1/T, P/T)}{\partial(U,V)} = \frac{\partial(1/T, P/T)/\partial(1/T, V)}{\partial(U,V)/\partial(1/T, V)} = -\frac{1}{T^3 C_V}\left(\frac{\partial P}{\partial V}\right)_T.$$

The equality can be satisfied by $C_V = \infty$, but this means that either of the first two stability criteria in (6.45) must be compromised. Hence

$$\left(\frac{\partial P}{\partial V}\right)_T = 0, \qquad\qquad\qquad (6.46)$$

implying that the isotherms of a Stefan gas are flat. The lack of particle conservation is responsible for this behavior. The isothermal compressibility is infinite and, consequently, so too is the heat capacity at constant pressure.

In the theory of first-order phase transitions, (6.46) is the condition for metastability which characterizes a locus of points where the isotherms of a liquid and gas terminate. Within this region, which has the form of an inverted parabola, a homogeneous system cannot exist. The nonconservation of photons can therefore be considered to be *formally* analogous to a two-phase equilibrium in which matter can pass from one phase to another [cf. §7.3].

In our analysis of the law of error for an ideal gas, we found it necessary to write the entropy in parametric form by introducing the equation of state for the energy, (6.10), into the fundamental relation (6.18). If we had attempted to introduce the entropy directly into Gauss' principle without first writing it in parametric form, we would not have come out with a probability distribution at all. But in its parametric form, the entropy satisfies

$$||S|| \equiv \left(\frac{\partial^2 S}{\partial \bar{N}^2}\right)\left(\frac{\partial^2 S}{\partial \bar{V}^2}\right) - \left(\frac{\partial^2 S}{\partial \bar{N}\,\partial \bar{V}}\right)^2 = 0. \tag{6.47}$$

The system is, again, metastable. Physically, this means that the particle number and volume cannot fluctuate simultaneously.

Likewise, the local entropy, (6.30), is a function of two independent variables, $\bar{N}_\varepsilon$ and $\bar{V}$. The equations of state, (6.41) and (6.39), have the form

$$T = T(\bar{N}_\varepsilon, \bar{V}), \qquad P_\varepsilon = P_\varepsilon[\bar{N}_\varepsilon, \bar{V}, T(\bar{N}_\varepsilon, \bar{V})],$$

with a possible additional equation of state

$$z = z[\bar{N}_\varepsilon, \bar{V}, T(\bar{N}_\varepsilon, \bar{V})], \tag{6.48}$$

linking the microscopic and macroscopic worlds.

Because of the implicit functional dependence, the partial derivative, say, of the local pressure P_ε with respect to the volume will be given by

$$\left(\frac{\partial P_\varepsilon}{\partial \bar{V}}\right)_{\bar{N}_\varepsilon} = \left(\frac{\partial P_\varepsilon}{\partial \bar{V}}\right)_{\bar{N}_\varepsilon, T} + \left(\frac{\partial P_\varepsilon}{\partial T}\right)_{\bar{V}, \bar{N}_\varepsilon}\left(\frac{\partial T}{\partial \bar{V}}\right)_{\bar{N}_\varepsilon}$$

$$= \left(\frac{\partial P_\varepsilon}{\partial \bar{V}}\right)_{\bar{N}_\varepsilon, T} + \frac{P_\varepsilon}{T}\left(\frac{\partial T}{\partial \bar{V}}\right)_{\bar{N}_\varepsilon}.$$

Let us also observe that there is nothing prohibiting the energy per particle, ε, from being a function of the temperature.

A more rigorous definition of the local entropy would be

$$S(\bar{N}_\varepsilon, \bar{V}) = \mathcal{S}[\bar{N}_\varepsilon, \bar{V}, \bar{E}_\varepsilon(\bar{N}_\varepsilon)], \tag{6.49}$$

so that the entropy S is the result of replacing $\bar{E}_\varepsilon$, the average energy in the range $d\varepsilon$, by its equivalent in terms of $\bar{N}_\varepsilon$ in the expression for $\mathcal{S}[\bar{N}_\varepsilon, \bar{V}, \bar{E}_\varepsilon]$. The equations of state (6.39) and (6.41) should be more correctly written as

$$\left(\frac{\partial S}{\partial \bar{V}}\right)_{\bar{N}_\varepsilon} = \left(\frac{\partial \mathcal{S}}{\partial \bar{V}}\right)_{\bar{N}_\varepsilon} \tag{6.50}$$

and

$$\left(\frac{\partial S}{\partial \bar{N}_\varepsilon}\right)_{\bar{V}} = \left(\frac{\partial \mathcal{S}}{\partial \bar{N}_\varepsilon}\right)_{\bar{V}, \bar{E}_\varepsilon} + \left(\frac{\partial \mathcal{S}}{\partial \bar{E}_\varepsilon}\right)_{\bar{V}, \bar{N}_\varepsilon}\frac{d\bar{E}_\varepsilon}{d\bar{N}_\varepsilon}, \tag{6.51}$$

respectively. The asymmetry of the two equations is due to the fact that monochromatic energy is defined in terms of the number of particles per frequency interval, and this is responsible for the fact that the integrated energy is a finite quantity. Rather, had we defined the monochromatic energy in terms of the number of oscillators per frequency interval, the integrated energy would have turned out to be infinite. In this sense, the two representations are not equivalent.

Can the average energy in any given frequency interval be a function of the volume? It does not appear so for then the entropy and average energy would stand on par, and it would be possible to invert (6.49) to obtain the energy as a function of the entropy. This may possibly be related to the fact that the *field*, represented by the volume of phase space occupied by the system, is a purely classical object. This is resonant with Planck's inconsistent use of classical electromagnetism in which the resonators absorb and emit energy continuously in order to determine the density of states, (6.44), and, in the sequel, to negate it by assuming that they do so discontinuously. This constitutes the basic paradox of the old quantum theory which, to the present day, still has no satisfactory explanation.

In the derivation of (6.41), we introduced the statistical expression for the partial derivative on the left-hand side of (6.51), used the second law in the form

$$\left(\frac{\partial S}{\partial \bar{E}_\varepsilon}\right)_{\bar{N}_\varepsilon, \bar{V}} = \frac{1}{T}, \tag{6.52}$$

and defined the energy per particle as

$$\frac{d\bar{E}_\varepsilon}{d\bar{N}_\varepsilon} = \varepsilon \tag{6.53}$$

on the right-hand side. In the derivation of (6.39), we assumed that the average energy is a function only of the average number of particles. Just as there is an equivalence between energy and mass, there is a statistical equivalence between the particle number and volume. However, it is the energy, rather than the mass, which is quantized just as it is the energy in terms of the number of particles, rather than the volume, that is quantizable.

6.6.2   Transformation of the Equations of State

Although the entropy is invariant under a change of variables, the equations of state (6.50) and (6.51) are not. As a first example, let us impose the constraint that average number of particles is never inferior to the average number of oscillators. In other words, each mode of the electromagnetic field must contain at least one photon.[3] The new independent variable,

$$\bar{N}'_\varepsilon = \bar{N}_\varepsilon + m\bar{V}, \tag{6.54}$$

will increase or decrease according to whether $\bar{N}_\varepsilon$ increases or decreases. The entropy, (6.30), is transformed into

$$S'(\bar{N}'_\varepsilon, \bar{V}) = m\bar{V} \ln\left(\frac{\bar{N}'_\varepsilon - m\bar{V}}{mV}\right) - \bar{N}'_\varepsilon \ln\left(\frac{\bar{N}'_\varepsilon - m\bar{V}}{\bar{N}'_\varepsilon}\right). \tag{6.55}$$

---

[3]See §7.2 for more details on this alternative radiation mechanism.

Although numerically equal to the entropy of the negative binomial distribution, (6.55) is the entropy of the Pascal distribution which we shall encounter in §7.2.1.

The equation of state (6.51) is explicitly given by

$$\frac{\varepsilon}{T} - \ln z = -\ln\left(\frac{\bar{N}'_\varepsilon - m\bar{V}}{\bar{N}'_\varepsilon}\right).$$

Solving for the average number of particles, we have

$$\bar{N}'_\varepsilon = m\bar{V}\left[\frac{1}{z^{-1}e^{\varepsilon/T} - 1} + 1\right] = \frac{m\bar{V}}{1 - ze^{-\varepsilon/T}}. \tag{6.56}$$

The explicit expression for the equation of state for the pressure, (6.50), is

$$\begin{aligned} P'_\varepsilon &= mT\ln\left(\frac{\bar{N}'_\varepsilon - m\bar{V}}{m\bar{V}}\right) \\ &= -mT\ln\left(z^{-1}e^{\varepsilon/T} - 1\right) \\ &= P_\varepsilon - m(\varepsilon - T\ln z). \end{aligned} \tag{6.57}$$

It differs from the pressure of (6.39) by a "zero-point" pressure term that will be associated with the capillary pressure in the analogy with the phase equilibrium between vapor and liquid [cf. §7.3].

As a second example, consider the constraint where the average numbers of particles and oscillators vary in such a way that their sum is constant. Eliminating $m\bar{V}$ in favor of

$$m\bar{V}' = \bar{N}_\varepsilon + m\bar{V} \tag{6.58}$$

in the entropy of mixing, (6.30), results in

$$S'(m\bar{V}', \bar{N}_\varepsilon) = \bar{N}_\varepsilon \ln\left(\frac{m\bar{V}' - \bar{N}_\varepsilon}{\bar{N}_\varepsilon}\right) - m\bar{V}'\ln\left(\frac{m\bar{V}' - \bar{N}_\varepsilon}{m\bar{V}'}\right).$$

We easily recognize this as the entropy of the maximum likelihood value of the binomial distribution. The equation of state (6.41) is now

$$\bar{N}_\varepsilon = \frac{m\bar{V}'}{z^{-1}e^{\varepsilon/T} + 1}, \tag{6.59}$$

which is the Fermi–Dirac "distribution" which we derived in §3.3.3. However, in terms of our original variables, it would be the Maxwell–Boltzmann law, (6.32). The equation of state for the pressure, (6.50), is

$$P_\varepsilon = mT\ln\left(1 + ze^{-\varepsilon/T}\right). \tag{6.60}$$

This is a good example of how the equations of state can be transformed into one another by imposing particular constraints.

6.6.3   Metastability of Systems Obeying Quantum Statistics

The local Gibbs–Duhem relation,

$$\bar{N}_\varepsilon \, d(\varepsilon/T - \ln z) + \bar{V} \, d(P_\varepsilon/T) = 0,$$

makes the two equations of state (6.39) and (6.41) functionally dependent so that

$$\frac{\partial[(\varepsilon/T - \ln z), P_\varepsilon/T]}{\partial(\bar{N}_\varepsilon, \bar{V})} = 0. \tag{6.61}$$

Therefore, systems obeying quantum statistics are only metastable. Moreover, if the fugacity is constant, as it is for a Stefan gas, the discriminant reduces to

$$0 = \frac{\partial(1/T, P_\varepsilon/T)/\partial(1/T, \bar{V})}{\partial(\bar{N}_\varepsilon, \bar{V})/\partial(1/T, \bar{V})} = -\frac{1}{T^3}\left(\frac{\partial \bar{N}_\varepsilon}{\partial T}\right)_{\bar{V}}^{-1}\left(\frac{\partial P_\varepsilon}{\partial \bar{V}}\right)_T$$

upon changing to the variables $1/T$ and $\bar{V}$. Since $(\partial^2 S/\partial \bar{N}_\varepsilon^2)_{\bar{V}} \neq 0$, we have

$$\left(\frac{\partial P_\varepsilon}{\partial \bar{V}}\right)_T = 0, \tag{6.62}$$

which is a local form of (6.46). This is valid for photons and phonons as well as for all systems in which the particle number is not conserved.

Furthermore, from the equation of state for the pressure, (6.43), we have

$$\frac{H_\varepsilon}{T}\left(\frac{\partial T}{\partial \bar{V}}\right)_{P_\varepsilon} = -T\bar{N}_\varepsilon\left(\frac{\partial \ln z}{\partial \bar{V}}\right)_{P_\varepsilon},$$

where $H_\varepsilon = \bar{N}_\varepsilon \varepsilon + P_\varepsilon \bar{V}$ is the enthalpy in the energy range $d\varepsilon$. In the case of a constant or vanishing fugacity, $(\partial T/\partial \bar{V})_{P_\varepsilon} = 0$, which sends the heat capacity at constant pressure, in the same energy interval, to infinity. However, in the case where we have an equation of state for the fugacity of the form (6.48), $\bar{N}_\varepsilon$ and $\bar{V}$ are independent variables so that the heat capacity at constant pressure is

$$\begin{aligned}
C_{P_\varepsilon} &= C_{\bar{V}} + \frac{T}{\bar{N}_\varepsilon}\left(\frac{\partial P_\varepsilon}{\partial T}\right)_{\bar{V}}\left(\frac{\partial \bar{V}}{\partial T}\right)_{P_\varepsilon} \\
&= C_{\bar{V}} + \frac{T}{\bar{N}_\varepsilon \bar{V}}\left[S + \bar{N}_\varepsilon\left(\frac{\partial \mu}{\partial T}\right)_{\bar{N}_\varepsilon, \bar{V}}\right]\left(\frac{\partial \bar{V}}{\partial T}\right)_{P_\varepsilon} \\
&= C_{\bar{V}} + \frac{T}{\bar{N}_\varepsilon \bar{V}}\left[S - \bar{N}_\varepsilon\left(\frac{\partial S}{\partial \bar{N}_\varepsilon}\right)_{\bar{V}}\right]\left(\frac{\partial \bar{V}}{\partial T}\right)_{P_\varepsilon} \\
&= C_{\bar{V}} + \frac{P_\varepsilon}{\bar{N}_\varepsilon}\left(\frac{\partial \bar{V}}{\partial T}\right)_{P_\varepsilon},
\end{aligned}$$

where we have used a Maxwell relation in passing from the second to the third line.

In the case where the fugacity is given by (6.48), the discriminant

$$\frac{\partial[(\varepsilon/T - \ln z), P_\varepsilon/T]}{\partial(\bar{N}_\varepsilon, \bar{V})} = 0. \tag{6.63}$$

Regardless of whether the equation of state for the pressure is given by (6.43) or (6.60), it is seen that the conditions for (6.63) to hold are

$$\left(\frac{\partial(P_\varepsilon/T)}{\partial \bar{X}}\right)_{\bar{Y}} = -\varrho_\varepsilon \left[\frac{\partial}{\partial \bar{X}}\left(\frac{\varepsilon}{T} - \ln z\right)\right]_{\bar{Y}}, \tag{6.64}$$

where $\bar{X}$ and $\bar{Y}$ stand for $\bar{N}_\varepsilon$ and $\bar{V}$ or vice versa. Changing to the variables $(1/T, \bar{V})$ and $(1/T, \bar{N}_\varepsilon)$, the conditions for the vanishing of the discriminants are

$$\left(\frac{\partial(P_\varepsilon/T)}{\partial T}\right)_{\bar{X}} = -\varrho_\varepsilon \left[\frac{\partial}{\partial T}\left(\frac{\varepsilon}{T} - \ln z\right)\right]_{\bar{X}} \tag{6.65}$$

and

$$\left(\frac{\partial P_\varepsilon}{\partial \bar{X}}\right)_T = \varrho_\varepsilon T \left(\frac{\partial \ln z}{\partial \bar{X}}\right)_T, \tag{6.66}$$

where, again, $\bar{X}$ denotes either $\bar{N}_\varepsilon$ or $\bar{V}$. In the case that the fugacity is a constant, condition (6.66) reduces to the condition of metastability, (6.62), and (6.65) becomes the Clapeyron equation, (6.2), indicating a phase equilibrium or, more generally, nonconservation of the particle number.

The fact that the discriminant (6.61) vanishes lends additional support to the conclusion that N and V cannot fluctuate simultaneously; the two processes provide complementary descriptions of the same phenomena. The discriminant enters into our probabilistic description in the following way.

Any function which is twice differentiable can be expressed as a truncated Taylor series expansion

$$S(V, N) = S(\bar{V}, \bar{N}_\varepsilon) + \left(\frac{\partial S}{\partial \bar{V}}\right)(\Delta V) + \left(\frac{\partial S}{\partial \bar{N}_\varepsilon}\right)(\Delta N)$$

$$+ \frac{1}{2}\left\{\frac{\partial^2 S}{\partial \hat{N}^2}(\Delta N)^2 + 2\frac{\partial^2 S}{\partial \hat{N}\,\partial \hat{V}}(\Delta N)(\Delta V) + \frac{\partial^2 S}{\partial \hat{V}^2}(\Delta V)^2\right\},$$

where $\hat{N} \in [\bar{N}_\varepsilon, N]$ and $\hat{V} \in [\bar{V}, V]$. For small fluctuations, the second derivatives of the entropy can be evaluated at $\bar{N}_\varepsilon, \bar{V}$. Then introducing this truncated Taylor series expansion into Gauss' principle (6.26) leads to

$$f(\Delta V, \Delta N) = \frac{\sqrt{\|S\|}}{2\pi} \exp\left\{\frac{1}{2}\frac{\partial^2 S}{\partial \bar{N}_\varepsilon^2}(\Delta N)^2\right.$$

$$\left. + \frac{\partial^2 S}{\partial \bar{N}_\varepsilon \,\partial \bar{V}}(\Delta N)(\Delta V) + \frac{1}{2}\frac{\partial^2 S}{\partial \bar{V}^2}(\Delta V)^2\right\},$$

where $\|S\|$ is the discriminant defined in (6.47). Since it vanishes, simultaneous fluctuations in the volume and particle number cannot be realized.

One single system is sufficient to show that the discriminant (6.47) vanishes. Take an ideal gas whose entropy is given in the parametric form (6.23). We recall that the parametric form is necessary since temperature, unlike the energy, can be kept constant in the presence of fluctuations in the particle number and volume. Since $\partial^2 S/\partial \bar{V}^2 = -\bar{N}/\bar{V}^2$, $\partial^2 S/\partial \bar{N}^2 = -1/\bar{N}$, and $\partial^2 S/\partial \bar{N}\,\partial \bar{V} = 1/\bar{V}$, the discriminant vanishes identically.

In order to convince ourselves that the vanishing of the discriminant is not due to some quirk behavior of the parametric form of the entropy, consider the entropy of the maximum likelihood value of the negative binomial distribution (6.30). The energy no longer enters into the fundamental (local) relation because it is now a dependent variable. Since the partial derivatives are $\partial^2 S/\partial \bar{N}_\varepsilon^2 = -m\bar{V}/\bar{N}_\varepsilon(\bar{N}_\varepsilon + m\bar{V})$, $\partial^2 S/\partial \bar{V}^2 = -\bar{N}_\varepsilon/\bar{V}(\bar{N}_\varepsilon + m\bar{V})$, and $\partial^2 S/\partial \bar{N}_\varepsilon \partial \bar{V} = m/(\bar{N}_\varepsilon + m\bar{V})$, the discriminant of the quadratic form (6.61) is again seen to vanish indentically.

6.7   Fluctuations

We have seen that the two characteristic limits, as $m \to \infty$ and $m \to 0$, give rise to complementary descriptions in terms of the volume and number of particles where one is a continuous while the other is a discrete random variable. To further elucidate the physical significance of this form of complementarity, we turn to an analysis of the fluctuations in the two limits.

The mean square fluctuations in the number of particles and volume can be determined either from the probability distributions (6.28) and (6.35), respectively, or from the relations

$$\overline{(\Delta N)_\varepsilon^2} = -\left(\frac{\partial^2 S}{\partial \bar{N}_\varepsilon^2}\right)_{\bar{V}}^{-1} = -\left[\frac{\partial(\varepsilon/T - \ln z)}{\partial \bar{N}_\varepsilon}\right]_{\bar{V}}^{-1}, \tag{6.67}$$

$$\overline{(\Delta V)_\varepsilon^2} = -\left(\frac{\partial^2 S}{\partial \bar{V}^2}\right)_{\bar{N}_\varepsilon}^{-1} = -\left(\frac{\partial(P_\varepsilon/T)}{\partial \bar{V}}\right)_{\bar{N}_\varepsilon}^{-1}, \tag{6.68}$$

respectively. Noting from (6.41) that

$$\left[\frac{\partial}{\partial \bar{N}_\varepsilon}\left(\frac{\varepsilon}{T} - \ln z\right)\right]_{\bar{V}} = -\frac{m\bar{V}}{\bar{N}_\varepsilon + m\bar{V}}$$

$$= -\frac{\left(z^{-1}e^{\varepsilon/T} - 1\right)^2}{m\bar{V}z^{-1}e^{\varepsilon/T}} \tag{6.69}$$

and from (6.39) that

$$\left(\frac{\partial(P_\varepsilon/T)}{\partial \bar{V}}\right)_{\bar{N}_\varepsilon} = -\frac{\varrho_\varepsilon m}{\bar{N}_\varepsilon + m\bar{V}}$$

$$= -\frac{m}{\bar{V}}ze^{-\varepsilon/T}, \tag{6.70}$$

the expression for the mean square fluctuations in the particle number and volume are

$$\overline{(\Delta N)^2_\varepsilon} = \bar{N}_\varepsilon \left(1 + \frac{\bar{N}_\varepsilon}{m\bar{V}}\right) = \frac{m\bar{V}z^{-1}e^{\varepsilon/T}}{\left(z^{-1}e^{\varepsilon/T} - 1\right)^2} \tag{6.71}$$

and

$$\overline{(\Delta V)^2_\varepsilon} = \bar{V}\left(\frac{\bar{V}}{\bar{N}_\varepsilon} + \frac{1}{m}\right) = \frac{\bar{V}}{m}z^{-1}e^{\varepsilon/T}, \tag{6.72}$$

respectively.

Recalling our discussion in §2.3.2, it was Einstein who first interpreted the individual terms in the expression for the mean square fluctuation in the number of particles based on energy fluctuations.[4] He would have been hard pressed to adapt the same interpretation to the expression for the mean square fluctuation in the volume (6.72); in particular, a relation between average energy and volume, like that between average energy and number of particles, is lacking.

Moreover, from (6.46), one would be led to believe that the dispersion in the volume is infinite. But the derivative in (6.70) is taken at constant $\bar{N}_\varepsilon$. There is nothing in the fundamental relation (6.30) that would lead us to believe that the particle number has a more priviledged role. The equation of state (6.41) determines $\bar{N}_\varepsilon = \bar{N}_\varepsilon(V, T)$ while (6.39) gives $\bar{V} = \bar{V}(\bar{N}_\varepsilon, T)$. This is why $\bar{V}$ must be kept constant in deriving (6.71), just as $\bar{N}_\varepsilon$ must not be varied in deriving (6.68). Otherwise both would be infinite. It is purely a matter of choice whether we choose to vary V or N in the fundamental relation, but varying both leads to disaster.

The two expressions for the mean square fluctuations, (6.71) and (6.72), are, in a certain sense, mirror images of one another, and their physical interpretations rely on the special connection between the Poisson distribution and the gamma density which give rise to the individual terms. As $m \to \infty$, the discreteness in the particle fluctuations corresponds to continuous volume fluctuations. The first terms in (6.71) and (6.72) correspond to the second central moments of the Poisson, (6.28), and gamma, (6.35), distributions, respectively. This is the classical, or Wien, region.

Alternatively, as $m \to 0$, the second terms dominate, and those terms are the second central moments of the gamma density for fluctuations in the number of particles and the Poisson distribution for fluctuations in the volume, respectively. The characters of the fluctuations have been interchanged: fluctuations in the particles are continuous while those in the volume are discontinuous. We are now in the nonclassical, or Rayleigh–Jeans, region.

The relation

$$\sqrt{\overline{(\Delta N)^2_\varepsilon}} = \varrho_\varepsilon \sqrt{\overline{(\Delta V)^2_\varepsilon}}, \tag{6.73}$$

obtained by dividing (6.71) by (6.72), is universally valid and embodies the statistical equivalence principle, as we have indicated in §5.5. The standard

---

[4]Recall that in order to convert (6.67) into the dispersion in energy, we multiply both sides by ε^2.

deviations in the particle number volume are proportional to one another. The local density, ϱ_ε, sets the scale and determines the relative magnitude of the two standard deviations. Although physically distinct, fluctuations in the number of particles at constant volume and fluctuations in the volume at a fixed number of particles are statistically equivalent.

For an ideal gas, the "integrated" mean square fluctuation in the number of particles, in which particles of all energies contribute, is

$$
\begin{aligned}
\overline{(\Delta N)^2} &= \frac{2\pi(2M)^{3/2}}{h^3}V\int_0^\infty \frac{\varepsilon^{1/2}ze^{-\varepsilon/T}}{(1-ze^{-\varepsilon/T})^2}\,d\varepsilon \\
&= \frac{V}{\lambda^3}g_{1/2}(z),
\end{aligned}
\tag{6.74}
$$

where

$$
g_n(z) \equiv \sum_{\ell=1}^\infty \frac{z^\ell}{\ell^n}.
\tag{6.75}
$$

The thermal wavelength is again denoted by λ. In the case that $z = 1$, $g_n(z)$ reduces to the Riemann zeta function,

$$
\zeta(n) \equiv \sum_{\ell=1}^\infty \frac{1}{\ell^n}.
$$

For small values of z, the sum (6.75) can be approximated by z itself, and since we are in the classical limit, we may use the expression [cf. Eq. (6.13)]

$$
z = \lambda^3 n
\tag{6.76}
$$

to evaluate the integrated mean square fluctuation in the number of particles. It reduces simply to

$$
\overline{(\Delta N)^2} = \bar{N},
$$

which is precisely what we would expect since the fluctuations are governed by the Poisson distribution (6.28).

In the so-called strongly degenerate case, in which $z \to 1$, the integrated mean square fluctuation in the number of particles diverges because of the divergence of $g_{1/2}(1)$. Fluctuations in the particle number, in any given energy interval, are now described by the gamma density (6.38), whose integrated second moment about the mean diverges. The strongly degenerate behavior of the particle statistics is thus to be associated with the continuum limit description. Therefore, expression (6.74) encompasses both the classical and nonclassical aspects of fluctuations in the particle number.

In order to derive the global mean square fluctuation in the volume, we must invert (6.72) prior to integrating over all values of the energy. We then obtain

$$
\begin{aligned}
1/\overline{(\Delta V)^2} &= \frac{2\pi z}{h^3}\frac{(2M)^{3/2}}{\bar{V}}\int_0^\infty \varepsilon^{1/2}e^{-\varepsilon/T}\,d\varepsilon \\
&= z/\lambda^3\bar{V},
\end{aligned}
$$

which upon inversion gives

$$\overline{(\Delta V)^2} = \lambda^3 \bar{V}/z \qquad\qquad (6.77)$$

for the integrated mean square fluctuation in the volume.

In the classical limit, we may introduce (6.76) into (6.77) to obtain

$$\overline{(\Delta V)^2} = \frac{\bar{V}^2}{\bar{N}} \qquad \text{(classical)},$$

which corresponds to the continuum limit where the fluctuations are governed by the gamma density (6.35). For strongly degenerate systems, (6.77) becomes

$$\overline{(\Delta V)^2} = \lambda^3 \bar{V} \qquad \text{(strongly degenerate)}.$$

There is no divergence in the global dispersion in the volume as there is for the particle number. The fluctuations become discrete in the strongly degenerate limit and are described by the Poisson distribution (6.37).

The first and second terms in (6.71) and (6.72) relate, respectively, to the high energy, low intensity and low energy, high intensity regions of the spectrum. The linear term in $\bar{N}_\varepsilon$ corresponds to the quadratic term in $\bar{V}$ which says that the particle nature of the radiation corresponds to the wave properties of the oscillators. We have seen that in the low intensity region fluctuations in the particle number are governed by the discrete Poisson distribution while fluctuations in the volume are characterized by a continuous gamma density. In the high temperature or high intensity limit, the situation is reversed. Therefore, the nature of fluctuations in the particles and oscillators appear diametrical opposite at the extreme ends of the spectrum.

6.8   Relevance of Fluctuations

We recall from §2.3.2 that Einstein wanted to determine the characteristic wavelength which he subsequently found comparable to Wien's displacement law. However, he set an extensive quantity—the mean square fluctuation in the energy—equal to a nonextensive quantity—the square of the mean energy. In this section, we will employ the expressions for the mean square fluctuation in the number of particles and volume to obtain estimates of the wavelength in Wien's displacement law and to get a numerical approximation of the Stefan–Boltzmann constant.

6.8.1   Characteristic Wavelength

Since Boltzmann's constant now becomes important, we will measure temperature in degrees rather than in energy units. Multiplying both sides of (6.71) by ε^2 and defining fluctuations in the energy in terms of fluctuations in the number of particles according to

$$\Delta E_\varepsilon = \varepsilon \, \Delta N_\varepsilon, \qquad\qquad (6.78)$$

we obtain

$$\overline{(\Delta E)_\varepsilon^2} = \varepsilon \bar{E}_\varepsilon + \bar{E}_\varepsilon^2/mV.$$

Integrating over all energies results in

$$\overline{(\Delta E)^2} = 8\pi V \frac{(kT)^5}{(hc)^3} \Gamma(5) \{\zeta(5) + [\zeta(4) - \zeta(5)]\}, \tag{6.79}$$

where we have kept the particle contribution coming from the first term distinct from the wave contribution, which is the term within the square brackets. It shows that the dominant contribution to the heat capacity will come from the corpuscular aspect of radiation.

Since the energy and average number of particles of black radiation are

$$E = 8\pi V \left(\frac{kT}{hc}\right)^3 kT\, \Gamma(4)\, \zeta(4) \equiv \sigma V T^4 \tag{6.80}$$

and

$$\bar{N} = 8\pi V \left(\frac{kT}{hc}\right)^3 \Gamma(3)\, \zeta(3), \tag{6.81}$$

respectively, we can express the right side of (6.79) in terms of the ratio of $E^2/\bar{N}$ as

$$\overline{(\Delta E)^2} = 1.48 E^2/\bar{N}. \tag{6.82}$$

With the aid of the relation between the mean square fluctuation in the energy and the heat capacity at constant volume,

$$\overline{(\Delta E)^2} = kT^2 C_V, \tag{6.83}$$

and Stefan's law (6.80), expression (6.82) reduces to

$$T^3/n = 4k/1.48\,\sigma,$$

where n is the number density.

Under the assumption that fluctuations will be most conspicuous where the mean distance between photons is of the order of the characteristic wavelength of black radiation, we set $\lambda_m = n^{-1/3}$, corresponding to the maximum in the energy density of the spectrum. Since the numerical value of

$$\sigma = 4\sigma_s/c = 7.56 \times 10^{-15} \text{ g/sec}^2 \text{ cm deg}^4,$$

we obtain

$$\lambda_m = 0.366/T \text{ [cm]}$$

for the characteristic wavelength.

Not only is this the same order of magnitude as the experimental value $\lambda_m = 0.29/T$, it moreover avoids having to set the mean square fluctuation in the energy equal to the square of the energy in order to bring out the importance of fluctuations in relating the scale of microscopic magnitudes to observable phenomena. The only assumption involved is that the characteristic wavelength is of the order of the average distance between photons, and this gives a correct order of magnitude of the wavelength for which the energy distribution, as a function of wavelength, has its maximum.

6.8.2   The Stefan–Boltzmann Constant

Just as the mean square fluctuation in the number of particles, (6.71), is physically relevant insofar as it gives a correct order of magnitude of the characteristic wavelength, so, too, does the mean square fluctuation in the volume, (6.72). We now show that it provides a correct order of magnitude for the Stefan–Boltzmann constant.

Suppose that fluctuations in the energy are caused by fluctuations in the volume according to

$$\Delta E = \sigma T^4 \, \Delta V.$$

This is permissible since we are dealing with Stefan's law, which is a global relation, as opposed to the local relation (6.78) between energy and particle number. With the number of oscillators in the frequency interval $d\nu$ given by (6.44), the inverse total mean square fluctuation in the volume is

$$\frac{1}{\overline{(\Delta V)^2}} = \frac{8\pi}{c^3 V} \int_0^\infty \nu^2 e^{-h\nu/kT} \, d\nu = \frac{8\pi}{V} \left(\frac{kT}{hc} \right)^3 \Gamma(3)$$

or inverting it,

$$\overline{(\Delta V)^2} = \left(\frac{hc}{kT} \right)^3 \frac{\overline{V}}{16\pi}.$$

Multiplying both sides by $\sigma^2 T^8$ to obtain

$$\overline{(\Delta E)^2} = \frac{1}{16\pi} \left(\frac{hc}{kT} \right)^3 \frac{E^2}{\overline{V}} \tag{6.84}$$

and using (6.83), we determine the Stefan–Boltzmann constant as

$$\sigma_s = \frac{16\pi k^4}{h^3 c^2} = 6.96 \times 10^{-5} \ \text{g/sec}^2 \, \text{cm} \, \text{deg}^4.$$

This is in remarkably good agreement with its actual value,

$$\sigma_s = \frac{2\pi^5 k^4}{15 h^3 c^2} = 5.67 \times 10^{-5} \text{g/sec}^3 \, \text{deg}^4,$$

considering the approximations that went into its estimation.

Alternatively, we may use the expression for the average number of particles, (6.81), to eliminate the average volume in (6.84). We then obtain

$$\overline{(\Delta E)^2} = 1.2 E^2 / \bar{N}.$$

Following the same procedure as before, we now come out with a characteristic wavelength given by

$$\lambda_m = 0.393/T \quad [\text{cm}],$$

which is the same order of magnitude as the experimental value and it, again, confirms Wien's displacement law. This supports the view that fluctuations in the particle number and volume provide a statistically equivalent description of the same phenomena.

Bibliographic Notes

§6.1

Einstein's 1904 paper is

- A. Einstein, "Zur allegemeinen molekularen Theorie der Wärme," *Ann. d. Phys.* **14**, 354-362 (1904).

See also the comments in

- R. McCormmach, "Einstein, Lorentz and the electron theory", *Historic. Stud. Phys. Sci.* **2**, 41-87 (1970)

concerning the fact that although Einstein found the same temperature dependence for the wavelength as Wien's displacement law, he did not claim that he had found a new derivation of that law but rather a partial confirmation of the existence of energy fluctuations. McCormmach describes in detail the influence of Lorentz's theory of the electron upon the thinking of the young Einstein. Although Lorentz's theory was unparalleled in terms of its clarity, it did try to wed two discordant branches of physics: the continuous field of Maxwell's equations with the particle mechanics of the electron. Since it led to the wrong black radiation formulas and created difficulties in explaining emission and absorption phenomena, one of the two had to go. MacCormmach concludes:

> Einstein was convinced by 1905 that the practice, exemplified by Lorentz' theory, of admitting side by side as equally primitive concepts the discrete point-masses of mechanics and the continuous field represented an unsatisfactory stage in the development of the theory, one which eventually had to go. He was never to lose this conviction, but he would soon reverse the prescription for the ills of a bifurcated physics that he had hinted in his 1905 remarks; he would soon argue for partial differential equations and continuous functions as the unifying language of physics.

A critical analysis of Einstein's work on fluctuation theory can be found in

- M.J. Klein, "Fluctuations and statistical physics in Einstein's early work," in the *Proceedings of the Einstein Centennial Symposium* (Addison-Wesley, Reading, MA, 1979), pp. 39-58.

§6.2

The reference to Rayleigh's paper is

- Lord Rayleigh, "On the pressure of vibrations," *Phil. Mag.* (6) **3**, 338-346 (1902).

§6.3

Our analysis of a Stefan gas can be found in

- B.H. Lavenda and J. Dunning-Davies, "Stefan–Boltzmann law for blackbodies and black-holes," *Int. J. Theoret. Phys.* **29**, 509-522 (1990).

§6.4 and §6.5

These sections follow our analysis in

- B.H. Lavenda, "Statistical equivalence and particle indistinguishability," *Phys. Essays* **5**, # 2 (1992).

§6.6

Thermodynamic stability criteria are discussed in

- H.A. Lorentz, *Lectures on Theoretical Physics*, Vol. II, transl. by L. Silberstein and A.P.H. Trivelli (Macmillan Press, London, 1927), §60 of "Thermodynamics."

Chapter 7

Radiative and Material Phase Transitions

7.1   Introduction

Planck's formula for black radiation had been around for nearly a quarter of a century before it was realized that it applies to an ideal quantum gas. Today, it seems like it is so obvious that it should have been realized immediately. In fact, shouldn't Einstein's analogy, drawn in his 1905 paper, between Wien's formula and an ideal gas which implied the existence of light quanta have been sufficient to conclude that radiation is a photon gas? But Einstein only used that part of Planck's formula which is equivalent to Boltzmann's statistics in the region $h\nu \gg T$. An additional contribution was needed which would divest Planck's formula of all the superfluous assumptions so that the particle picture would shine through.

This was accomplished by Satyendra Nath Bose, who kindled the spark that set Einstein in motion to derive all the properties of the quantum gas that bears both their names. For Einstein, already in his mid-40s, this was a mere interlude, although it was to be his last major achievement. For Bose, who compared himself to "a comet which came once and never returned again," it was his only major contribution to physics. Even Einstein's opinion of what Bose had done vacillated from one of high praise, as in an addendum to his translation of Bose's paper which appeared in *Zeitschrift für Physik*,

> In my opinion Bose's derivation of the Planck formula constitutes an important advance. The method used here also yields the quantum theory of an ideal gas, as I shall discuss elsewhere in more detail.

to one of marked reservation, as in a letter to Ehrenfest, where he admitted that Bose's "derivation is elegant but the essence remains obscure."

What actually did Bose do? As Bose admitted many years later, "I had no idea of what I had done was really novel," and he was not aware he was doing something that Boltzmann would have done on the same occasion. The surprising new feature was Bose's derivation of the constant of proportionality between the average energy of the radiation per unit frequency interval per unit volume and the average energy of one of Planck's oscillators. Instead of counting the number of standing waves in a given frequency interval, Bose counted the number m of "cells" in phase space in the same frequency interval that could be populated with n particles. Certainly the time was ripe for scientific discovery for his derivation utilized the expression for the momentum $p = h\nu/c$ which had only appeared a year and a half before in the Compton effect and slightly before in de Broglie's thesis, both of which Bose was undoubtedly unaware of at the time.

Bose's derivation of Planck's formula, using Boltzmann's method in which cells rather than "particles" are counted, opened up the way for Einstein to deduce the physical properties of an ideal quantum gas. In particular, Einstein predicted that a "condensation" would take place as the temperature is lowered. An application of such a condensation was proposed in 1938 by Fritz London to the λ transition in ^{4}He which was discovered by Willem Hendrick Keesom in 1928. Einstein, however, did not refer to the condensation phenomenon as a *phase transition*. The thorny point of this discussion was that the phase transition required what Einstein called the degeneracy parameter to tend to unity. This has the consequence of making the number of particles in the condensated state infinite. How is it possible to have started with a finite number of particles and wound up with an infinite number? Conventional wisdom tells us that a sharp phase transition can only occur in the so-called thermodynamic limit where both N and $V \to \infty$ such that their ratio, the density, remains constant. This remedy was first proposed by Hendrick Anton Kramers in a morning debate that took place during the van der Waals Centenary Conference in November 1938.

Although we will have our own views on whether or not one must go to the thermodynamic limit in order to realize a sharp phase transition, the application of the Bose–Einstein statistics to a phase transition phenomenon brings us to the theme of this chapter. Our purpose will be to illustrate the point that *probability distributions that belong to the same family, like different phases of the same matter, can enter into equilibrium with one another.* When we say probability distributions which "belong to the same family," we mean that they must give rise to the same statistics and differ only insofar as the number of states accessible the particles or a constraint imposed on the population of the states, or cells, in phase space. Our approach will be to derive the statistics from physical processes (à l'Einstein) rather than through formal, and hardly unique, counting procedures. These processes must give rise to stationary probability distributions from which the statistics may be derived. And it is precisely the limited number of probability distributions that give rise to either Bose–Einstein or Fermi–Dirac statistics which allows us to introduce a classification of the various phenomena.

We recall from §2.4 that Einstein derived Planck's formula as a stationary condition between competing processes of absorption and emission of photons.

But we may ask: is this the only mechanism that leads to Planck's formula and consequently to Bose–Einstein statistics? We know that Bose–Einstein statistics can be derived from the negative binomial distribution. Now, the negative binomial distribution is one member in a family of distributions to which the geometric and Pascal distributions also belong. And all yield Bose–Einstein statistics. This leads us to believe that there may be other physical mechanisms that give Planck's formula. In §7.2 we will propose an alternative mechanism to Einstein's that leads a modification of Planck's formula which was actually envisioned in a cryptic comment made by Einstein and Stern as far back as 1913. The modification involves an *integral* zero-point energy together with the cryptic remark that "it seems astonishing that the mere assumption of the zero point energy $h\nu$ (but not $h\nu/2$) permits us to derive Planck's radiation law, without the usual discontinuity assumptions." Exactly what they were referring to remains a mystery, but we will show what type of radiation mechanism is responsible for such an integral zero-point energy term.

A comparison of the two radiation mechanisms will show that only Einstein's radiation mechanism can span the entire spectrum. It is as if the alternative mechanism becomes unstable at a certain frequency like a material phase that becomes unstable beyond a certain temperature and begins to break up spontaneously. In §7.3 we investigate whether the probability distributions themselves, like material phases, can enter into a phase equilibrium which by varying the frequency can give rise to a phase transition. This will be accomplished by showing that the two probability distributions of black radiation satisfy a Clapeyron equation and hence play the role of two material phases. The pressure of one of these two distributions becomes negative at sufficiently high frequencies which is analogous to the negative pressure acting to break up a liquid. A similar type of critical phenomenon should be observed in the transition from thermal to nonthermal radiation.

What type of phenomena can we expect beyond the critical frequency where only nonthermal radiation can exist? To investigate such phenomena, we must generalize Einstein's radiative mechanism to include nonlinear effects which would allow for the possibility of population inversion. In §7.4 we provide such a generalization in which we take into account the possibility of two-photon absorption. This will be seen to have the effect of placing a nonlinear bound on the growth of a linear, unstable perturbation so that when the threshold has been surpassed, a transition occurs from chaotic light, described by the negative binomial distribution, to coherent light that is statistically characterized by a Poisson distribution. If Einstein had appreciated the stability condition that there is on his coefficients of absorption and stimulated emission, he would have been able to predict the effect of lasing, some 40 years prior to its discovery, solely by adding a nonlinear term into his mechanism to take into account two-photon absorption.

The concept of a phase equilibrium between two probability distributions is an enticing one, and possibly it may explain the transition that occurs in liquid ^{4}He. According to Fritz London's original interpretation, the cusp in the specific heat curve of an ideal Bose–Einstein gas had a vague similarity to the λ type discontinuity that is observed in the specific heat at saturation of liquid helium. Applying an ideal Bose–Einstein gas model to the transition between normal and superfluid helium, where the critical point is defined by the vanishing of the chemical potential, one comes out with a critical temperature which is a degree off from the experimentally determined one. The discrepancy was attributed to weak physical interactions between the particles which would hopefully lower the critical temperature to the observed value. But it is now known that in liquid ^{4}He physical interactions are much stronger than would be appropriate to treat the condensation as an ideal Bose–Einstein gas. This coupled to the facts that an asymptotic limit must be taken and the cusp in the specific heat curve of an ideal Bose–Einstein gas has absolutely nothing to do with the discontinuity in the heat capacity curve which is actually measured would lead us to believe that such a model must be abandoned. But for what must it be abandoned?

One highly successful model was proposed by Lev Davidovich Landau in 1941, who developed a phenomenological "quantized hydrodynamic" model of superfluidity which did not take into account the statistics at all! In fact when ^{3}He was condensed in 1948, it came as a blow to Landau's theory that it had no superfluid properties between 1.05 and 3.02 K since the original form of Landau's theory made no use of the assumption that the helium atoms obey Bose–Einstein statistics. What Landau disliked about London's theory (which was further developed by Laszlo Tisza) was how atoms in the condensed, or ground, state could move through the liquid without colliding with the excited atoms of the normal fluid and thereby exchange energy with them. In the proposed unified scheme of Nikolai N. Bogoliubov, both theories had to give up ground: the energy spectrum could be discussed in terms of elementary excitations, like Landau advocated, but no division into different types "can even be spoken of." Moreover, he found that at absolute zero not all the atoms have zero momentum, which is what London and Tisza predicted. In fact, because of this "depletion" effect, it is no longer possible to identify the superfluid phase with the fraction of particles in the zero momentum state. Hence we are no better off today at a genuine understanding of what the phase transition in liquid ^{4}He is really caused by than we were almost a half a century ago.

Undoubtely, the simplest thing is to seek a modification of the original ideal Bose–Einstein gas model that would eliminate the contradictory aspects of the phase transition. This will be done in §7.5 where we will propose a phase equilibrium between the two phases that, while they obey the same statistics, differ in respect to the volume of phase space which they can occupy. Borrowing on our development of the radiative phase equilibria in §7.3 by applying it to material phases, we note one exception and that is the nature of the critical point. It is well known that at the critical point of a first-order phase transition, like the one we are proposing, the discontinuity in the molar potentials vanishes since the two phases become identical at that point. Above the critical temperature, one phase becomes unstable and breaks up while below the critical phase there is a coexistence of the two phases. Unlike our analysis in §7.3, we have no indication that the superfluid phase becomes unstable above the critical temperature. Rather there is such a small fraction of the phase present that it cannot impose its properties upon the remaining fraction of normal fluid. We thus propose a type of hybrid transition between a first- and second-order one and discuss some of the predictions that the model makes as well as its limitations. This sets the stage for our next application.

Phase transitions may be classified into two broad classes: first-order transitions which are associated with phase equilibria between members of the same class of probability distributions to which the negative binomial belongs and second-order phase transitions which correspond to the coexistence of two distributions of the same type, like two binomial distributions. As we have previously shown, the former type of phase transition possesses a discontinuity in the molar entropy since there are two different probability distributions involved in the phase equilibrium while, in the latter, as we will show in §7.6, there results a discontinuity in the derivative of the entropy at the critical point. These two types of transitions exhaust all the possibilities. There do not exist probability distributions which belong to the exponential family whose entropies possess higher order discontinuities in their derivatives. Consequently, we conclude that there are no phase transitions of higher order than second.

Finally, in §7.7 we shall show how statistics can generate physical interactions. Recalling our discussion in §3.3.2, the exponential family of distributions alone satisfies Gauss' principle and comprises idealizations of actual distributions in which the physical dimensions of the objects being distributed are negligible with respect to the dimensions of the container. In this approximation, the *a priori* probability for finding a number of objects in a given container is independent of how many objects there already may be in the container. Taking the physical dimensions of the particles into account will lead to deviations from ideal behavior so that the *a priori* probabilities will change from one event to the next. By waiving the invariancy of the *a priori* probabilities, we will be able to generate physical interactions through the statistics. We will use the nonexponential distributions described §3.3.2 to derive van der Waals' equation which has, to the present day, defied all mechanical derivations.[1]

---

[1] For instance, the low density approximation is employed in the derivation of van der Waals' equation. Its application to determining the critical point condition between the number of particles and volume is in flagrant contradiction to this assumption.

7.2   An Alternative Radiation Mechanism

In §2.4, we saw how Einstein proposed a physical mechanism for the absorption and emission of radiation and out of a dynamical equilibrium condition obtained Planck's law. Absorption and spontaneous emission were already known at the time; the latter Einstein credits to Heinrich Hertz, who assumed that an "oscillating Planck resonator radiates energy in a well known way, regardless of whether or not it is excited by external radiation." The new term that Einstein introduced was the stimulated emission process which he undoubtedly found working backward because he knew the result that he had to obtain. Like the absorption process, stimulated emission is proportional to the intensity of the radiation. But unlike the absorption process, stimulated emission is not a classical phenomenon. In fact, in the next section, we will find that stimulated emission is necessary for population inversion leading to the process known as lasing.

The question is whether Einstein's radiation mechanism is the only one which is compatible with Planck's law. Planck's radiation law concerns only the thermal part of the spectrum; there is, in general, a contribution from an external source of radiation. The boundary separating thermal from non-thermal regions occurs roughly where the stimulated and spontaneous emission processes are of equal importance. At room temperature this occurs at a wavelength of approximately 50 μm which corresponds to a frequency of about 6×10^{12} Hz, lying in the far-infrared region of the spectrum. As the temperature is raised, this frequency is shifted toward the visible region. At smaller frequencies, the stimulated emission process dominates over spontaneous emission so if we want to consider radiation that is dominated by thermal excitation, is it possible to do away with spontaneous emission by replacing it with another process? In effect, what we are asking is whether Einstein's radiation mechanism is really responsible for thermal radiation.

Yet according to Einstein's mechanism, spontaneous emission is necessary in order to achieve thermal equilibrium in a cavity filled with radiation; otherwise, we would not be led to Planck's law. Moreover, is the picture of charged "resonators" lining the cavity walls true to the mechanism of thermal radiation? Atomic excited states in the optical frequency region have negligible thermal populations so that if they are involved in the establishment of thermal equilibrium, we would have to increase the temperature or use an external source of radiation whose frequency is in resonance with the transition between two energy levels. A mechanism which employs excited atoms or resonators hardly seems adapted to describe the physical processes that are involved in establishing thermal equilibrium in a cavity filled with radiation.

Einstein had Planck's law to work from; we have, in addition, the probability distribution from which it is derived. We can therefore work from the physical processes that give rise to the stationary probability distribution. Einstein's radiation mechanism gives the negative binomial distribution. But this distribution is not unique and belongs to a class that also contains the geometric and Pascal distributions. All three are error laws for Bose–Einstein statistics but differ in the zero-point energy term in Planck's formula. We will see that the zero-point term has to do with the constraints imposed in the occupancy of cells in phase space. In fact, if we consider the Pascal distribution, we will appreciate that the spontaneous emission process can be replaced by a constraint that each mode of the electromagnetic field must contain at least one photon without destroying the dynamical equilibrium between emission and absorption processes. In other words, empty energy cells are inadmissible. In this way we will obtain Planck's law in a state of dynamical equilibrium together with an *integral* zero-point energy term since photons cannot be chopped in half. The zero-point energy term has always raised havoc with electrodynamics by introducing infinities. Although conventional wisdom subtracts these off, claiming that it gives rise to no observable effects, perhaps there is more to the zero-point energy than what meets the eye.

7.2.1   Essence of the Zero-Point Energy

A simple probabilistic model that gives Bose–Einstein statistics assumes that the probability of occupying a cell is independent of the number of particles already in this cell; that is, $f(n) \propto q^n$, where q is the *a priori* probability for occupancy. Summing over all n gives the geometric series $\sum_{n=0}^{\infty} q^n = 1/(1 - q) = 1/p$ so that the normalized distribution

$$f(n) = pq^n \tag{7.1}$$

is known as the geometric distribution.

The *a priori* probabilities of success and failure may be estimated in the usual fashion by the method of maximum likelihood. To do so, we make a series of observations, say $n_1, n_2, \ldots, n_r$. The likelihood function is $(pq^{\bar{n}})^r$ whose stationary condition gives $\hat{q} = \bar{n}/(1 + \bar{n})$. The maximum likelihood value of the geometric distribution (7.1) is

$$f(n; \bar{n}) = \bar{n}^n/(1 + \bar{n})^{n+1}. \tag{7.2}$$

This distribution still depends upon the unknown parameter $\bar{n}$, the average number of particles. In order to evaluate it, we cast (7.2) in the form of an error law for which $\bar{n}$ is the most probable value of the quantity measured. The statistical entropy of this error law is

$$S(\bar{n}) = (1 + \bar{n}) \ln(1 + \bar{n}) - \bar{n} \ln \bar{n} \tag{7.3}$$

while the stochastic entropy is seen to vanish. According to the Greene–Callen principle of §3.6, such a system is not amenable to statistical considerations since the entropy of the microcanonical ensemble vanishes and we cannot claim that there is only a single thermodynamics valid for all ensembles. This can be seen by taking the derivative of (7.3) and equating it with the derivative of the thermodynamic entropy. We then obtain

$$\ln\left(\frac{1+\bar{n}}{\bar{n}}\right) = \frac{\varepsilon - \mu}{T},$$

which can be arranged to read

$$\bar{n} = \frac{1}{e^{(\varepsilon-\mu)/T} - 1}. \tag{7.4}$$

Although we have derived Bose–Einstein statistics from the geometric distribution, it is not thermodynamically admissible since neither the entropy, (7.3), nor the average number of particles, (7.4), is extensive. Information concerning the volume of the system is missing, and consequently we are not able to employ thermodynamic reasoning.

In comparison with the entropy of the negative binomial distribution [cf. §2.2.3], the number of cells has shrunk to 1 and an entropy of a single cell has no meaning! Increasing the number of cells to m would give us back the negative binomial distribution, but had we taken the geometric distribution to be $q^{n-1}p$ instead of (7.1), which is the probability of failure at the nth inspection, we would have arrived at the Pascal distribution

$$f(n) = \binom{n-1}{m-1} p^m q^{n-m}, \qquad n = m, m+1, \ldots, \tag{7.5}$$

where m must be an integer.[2] Normalization of the Pascal distribution (7.5) is ensured by the fact that (7.5) is p^m times the $(n-m+1)$th term in the series expansion of $(1-q)^m$ in powers of q:

$$\begin{aligned}
(1-q)^{-m} &= \sum_{n=0}^{\infty} \binom{-m}{n}(-q)^n \\
&= \sum_{n=0}^{\infty} \binom{m+n-1}{n} q^n \\
&= \sum_{n=m}^{\infty} \binom{n-1}{m-1} q^{n-m}.
\end{aligned}$$

The Pascal distribution (7.5) may be thought of as the probability distribution for a sum of random variables for the number of repetitions between any two given failures. In terms of an occupancy problem, $\binom{n-1}{m-1}$ is the number of ways of distributing n particles among m cells with no cell being vacant. We will now see that this constraint—that there should be at least one particle in each cell—is responsible for an *integral* zero-point energy.

---

[2]This is in contradistinction to the negative binomial distribution where m can take on any positive value. The difference is, however, really elusive since m is always very large.

Casting the maximum likelihood value of the Pascal distribution,

$$f(n; \bar{n}) = \binom{n-1}{m-1} \left(\frac{m}{\bar{n}}\right)^m \left(\frac{\bar{n}-m}{\bar{n}}\right)^{\bar{n}-m}, \tag{7.6}$$

in the form of a Gaussian law of error leads to the expression

$$S(\bar{n}) = \bar{n} \ln \left(\frac{\bar{n}}{\bar{n}-m}\right) - m \ln \left(\frac{m}{\bar{n}-m}\right) \tag{7.7}$$

for the statistical entropy. Since this expression is extensive and the stochastic entropy is the same function of n that (7.7) is of its average value, the Pascal distribution (7.6) is thermodynamically admissible whereas the geometric distribution (7.2) is not. Taking the derivative of (7.7) with respect to $\bar{n}$ and equating it to the same derivative of the thermodynamic entropy give

$$\ln \left(\frac{\bar{n}}{\bar{n}+m}\right) = \frac{\varepsilon - \mu}{T},$$

or equivalently,

$$\bar{n} = m \left[\frac{1}{e^{(\varepsilon-\mu)/T} - 1} + 1\right] = \frac{m}{1 - e^{-(\varepsilon-\mu)/T}}. \tag{7.8}$$

In comparison with the analogous expression for the average particle number derived from the negative binomial distribution, (7.8) contains a zero-point term, m, due to the constraint that there should be at least one photon in each of the m modes of the electromagnetic field. Further comparison between the two may be made by considering the dispersion in the particle number. We recall from §2.3.2 that the dispersion in the particle number for the negative binomial distribution is

$$\sigma^2 \left(= -\left(\frac{\partial^2 S}{\partial \bar{n}^2}\right)_V^{-1}\right) = \bar{n} + \frac{\bar{n}^2}{m}, \tag{7.9}$$

which can be compared with the corresponding expression,

$$\sigma^2 = \frac{\bar{n}^2}{m} - \bar{n}, \tag{7.10}$$

of the Pascal distribution, (7.6).

Although the two expressions for the dispersion in the particle number are numerically equal, expression (7.10) does not lend itself to the same interpretation that Einstein gave to expression (7.9) in terms of the independent and additive contributions resulting from the particle and wave properties of light. From (7.10) we would conclude that the two aspects of light interfere with one another and there is a limit to the particle aspect which is proportional to the average number of photons present. Moreover, the particle contribution in (7.9) can be traced back to spontaneous emission while the wave contribution is the result of stimulated emission. The former is dominant in the high frequency region where $\bar{n} \ll m$. This region cannot be attained by particle statistics which are described by the Pascal distribution (7.6). We now inquire into the physical processes that lead to such a stationary probability distribution.

7.2.2   The Photon Constraint

Radiation is introduced into a cavity with perfectly reflecting walls. In order that there be an exchange of energy between standing waves at different frequencies, we must introduce small specks of charcoal which can absorb and re-emit all possible frequencies so as to permit energy exchange among all possible vibrations. Without such a black body it would be impossible to achieve thermal equilibrium.

Since there must be at least m photons in each standing wave, the rate of absorption is $\alpha(n - m)$ while the rate of emission is βn. Stimulated emission is still present, but in place of spontaneous emission, we have introduced the constraint that there be a minimum number of photons in each mode. The master equation for such a process is

$$\dot{f}(n,t) = \left\{\alpha(\mathbf{E} - 1)(n - m) + \beta(\mathbf{E}^{-1} - 1)n\right\} f(n,t), \qquad (7.11)$$

since the coefficients of absorption and stimulated emission, α and β, are independent of n. The "step" operators, $\mathbf{E}$ and $\mathbf{E}^{-1}$, have been defined in §3.2.

The stationary distribution that satisfies the dynamical equilibrium condition, $\alpha(n - m)f(n) = \beta\mathbf{E}^{-1}nf(n)$, is

$$f(n) = \binom{n - 1}{m - 1} \left(\frac{\alpha - \beta}{\alpha}\right)^{m} \left(\frac{\beta}{\alpha}\right)^{n-m} \qquad (7.12)$$

This becomes identical to the maximum likelihood value of the Pascal distribution (7.6) when we set

$$\bar{n} = \frac{\alpha m}{\alpha - \beta}. \qquad (7.13)$$

The latter can be appreciated as the stationary solution to the average equation of motion

$$\dot{\bar{n}} = \beta\bar{n} - \alpha(n - m).$$

Recall that to obtain the average equation of motion, we multiply the master equation by n and sum over all $n \geq m$.

In order that such a stationary solution exist, we must impose

$$\alpha > \beta, \qquad (7.14)$$

which is the same condition required in the Einstein radiation mechanism. This condition is ensured by the second law of thermodynamics. Upon equating the derivative of the statistical entropy, $\ln(\alpha/\beta)$, with that of the thermodynamic entropy, $h\nu/T$ (assuming we are dealing with thermal radiation where $\mu \equiv 0$), at a frequency ν, we have

$$\frac{\beta}{\alpha} = \exp\left(-\frac{h\nu}{T}\right), \qquad (7.15)$$

which is always less than unity. Introducing (7.15) into the expression for the stationary solution (7.13) gives

$$\bar{n} = \frac{m}{1 - e^{-h\nu/T}}. \tag{7.16}$$

In the high frequency limit, $\bar{n} \to m$ and the modes become depopulated, while in the low frequency limit, $\bar{n} \to mT/h\nu$. We now turn to a comparison between (7.16) and the Bose–Einstein distribution

$$\bar{n} = \frac{m}{e^{h\nu/T} - 1}, \tag{7.17}$$

which differs from the former by the absence of the zero-point term, m.

7.2.3   Maxwell Distribution without Equipartition

According to his own account, Einstein's derivation of Planck's radiation law was motivated, in large part, by Wien's original justification for his "chromatic" distribution function that was based on an apparent analogy with the Maxwell velocity distribution,

$$f(v)\, dv = \frac{1}{\sqrt{2\pi\Theta}} e^{-v^2/2\Theta}\, dv, \tag{7.18}$$

in a single direction for the number of molecules in the velocity range dv. Referring to our discussion of §2.1, we recall that Wien considered the wavelength of radiation, λ, by any molecule to be a function of its velocity. Wien, basing himself on the law of equipartition of energy, set the modulus Θ proportional to the absolute temperature. Thus Wien applied equipartition to the *translatory* motion of the molecules while it was later appreciated that equipartition of energy could be applied only to the *vibrational* modes of electromagnetic radiation in the long wavelength limit.

In the same 1917 paper in which he derived Planck's radiation law, Einstein argued that the velocity distribution acquired through the interaction of molecules with a thermal radiation field, in which they are in equilibrium at a temperature T, must be the same as that distribution of velocities which arises from their mutual collisions. Hence, since the latter distribution is Maxwellian, the average kinetic energy of a molecule, per degree-of-freedom, will be $\frac{1}{2}T$. The law of equipartition of energy, he claimed, would be independent of the nature of the molecules and valid no matter what frequency of light is absorbed or emitted by the molecules.

Seven years prior to this, Einstein expressed his perplexity on the theorem of equipartition for translational motion when he calculated, on the basis of the usual electromagnetic theory, momentum fluctuations. He concluded that they are "*by far too small* compared with the *real* momentum fluctuations, if very high frequencies are involved." Although equipartition of energy will certainly not apply to the radiation field, why shouldn't it hold for translatory motion where it was "always brillantly corrobated?" In the seven years that intervened, Einstein convinced himself that it did.

In order to treat the interaction of molecules with a thermal radiation field, Einstein returned to one of his first loves—Brownian motion. Brownian motion is the continuous diffusion limit for small fluctuations [cf. §8.4]. Let us first consider fluctuations in the particle number.

If the rates of recombination and generation, $r(n)$ and $g(n)$, vary little between n and $n \pm 1$, the step operators $\mathbf{E}$ and $\mathbf{E}^{-1}$ in the general master equation

$$\dot{f}(n,t) = \left\{ (\mathbf{E} - 1)r(n) + (\mathbf{E}^{-1} - 1)g(n) \right\} f(n,t)$$

can be replaced by the Taylor series expansions

$$\mathbf{E} = 1 + \frac{\partial}{\partial n} + \frac{1}{2} \frac{\partial^2}{\partial n^2} + \cdots$$

and

$$\mathbf{E}^{-1} = 1 - \frac{\partial}{\partial n} + \frac{1}{2} \frac{\partial^2}{\partial n^2} - \cdots .$$

Then omitting all derivatives higher than second results in the so-called Fokker–Planck equation

$$\dot{f}(n,t) = \frac{\partial}{\partial n} \left\{ [r(n) - g(n)] + \frac{1}{2} \frac{\partial}{\partial n} [r(n) + g(n)] \right\} f(n,t). \qquad (7.19)$$

The first and second moments, $r(n) - g(n)$ and $\frac{1}{2}[r(n) + g(n)]$, are known as the drift and diffusion coefficients. These coefficients will, in general, depend upon n. However, for small fluctuations about the stationary state, determined by

$$r(\bar{n}) = g(\bar{n}),$$

we can linearize the Fokker–Planck equation (7.19) by introducing $\Delta n = n - \bar{n}$ and retaining lowest order terms. In this way we obtain

$$\dot{f}(\Delta n,t) = [r'(\bar{n}) - g'(\bar{n})] \frac{\partial}{\partial \Delta n} [\Delta n \, f(\Delta n,t)] + D(\bar{n}) \frac{\partial^2}{\partial \Delta n^2} f(\Delta n,t),$$

where the prime stands for differentiation with respect to $\bar{n}$ and D is the diffusion coefficient:

$$D(\bar{n}) = \frac{1}{2}[r(\bar{n}) + g(\bar{n})]. \qquad (7.20)$$

Our major interest lies in the determination of the invariant, stationary distribution which is a solution to the time independent, linearized Fokker–Planck equation

$$-[r'(\bar{n}) - g'(\bar{n})] \Delta n = D \frac{\partial}{\partial \Delta n} \ln f(\Delta n). \qquad (7.21)$$

This gives the condition for dynamical equilibrium in the continuous, diffusion limit between what is formally analogous to an osmotic pressure force, proportional to $\partial \ln f / \partial \Delta n$, and an (virtual) external force proportional to Δn.

In this analogy we may think of Δn as some spatial coordinate and the probability density $f(\Delta n)$ is just the number density divided by the total number of particles.

As an illustration, consider the Einstein radiation mechanism which involves the negative binomial distribution. The dynamic equilibrium condition (7.21) is explicitly given by

$$-\frac{\gamma}{\bar{n}}\Delta n = D\frac{\partial}{\partial \Delta n}\ln f(\Delta n), \tag{7.22}$$

where the diffusion coefficient is

$$D = \frac{\gamma}{1 - \beta/\alpha} = \frac{\gamma}{1 - e^{-h\nu/T}}. \tag{7.23}$$

This expression for the diffusion coefficient is a prototype of a "fluctuation–dissipation" relation. The fluctuation, which is measured by the correlation function

$$\overline{\Delta n(t)\,\Delta n(t + \Delta t)}/2\,\Delta t = D$$

over the small time interval Δt, is dissipated by the resistance due to friction, represented by the viscosity coefficient, γ, with dimensions of frequency. In the low frequency or equipartition limit, the diffusion coefficient (7.23) becomes

$$D = \frac{\gamma T}{h\nu} \qquad \left(\nu \ll \frac{T}{h}\right). \tag{7.24}$$

This limiting form of the diffusion coefficient can be converted into the well known Einstein formula by considering velocity fluctuations.

In order to transfer to velocity space, we multiply (7.24) by $h\nu/M$ to give

$$\mathcal{D} = \frac{\gamma T}{M} \qquad \left(\nu \ll \frac{T}{h}\right).$$

This is a prototype of Einstein's formula in the case of gases since it is the velocity rather than the displacement that is the random variable. In the case of gases, the diffusion coefficient is directly proportional to the viscosity rather than being inversely proportional to it as it would be for liquids.

The dynamic equilibrium condition (7.22) allows us to solve for the stationary probability density; integrating we obtain

$$f(\Delta n) = \left\{\frac{1 - \exp(-h\nu/T)}{2\pi\bar{n}}\right\}^{1/2} \exp\left\{-\frac{(\Delta n)^2[1 - \exp(-h\nu/T)]}{2\bar{n}}\right\}, \tag{7.25}$$

which is a normal error law. We can now appreciate that the mean square fluctuation in the particle number given by (7.25) is

$$\overline{(\Delta n)^2} = \frac{\bar{n}}{1 - e^{-h\nu/T}} = \bar{n} + \frac{\bar{n}^2}{m} = \sigma^2.$$

It is identical to the dispersion of the negative binomial distribution. In the low frequency limit, $\overline{(\Delta n)^2} = \bar{n}T/h\nu = \bar{n}^2/m$, which is the wave contribution to the dispersion that is characteristic of the Rayleigh–Jeans limit of black radiation.

We may now transfer our attention to changes in the velocity of a Brownian particle, occurring in a small time interval Δt, that are caused by radiative damping, $-RvM\,\Delta t$, where R is the resistance coefficient, and by the erratic motion of electric charges in the radiation field. The former tends to retard motion and is the systematic factor while the latter is responsible for the residual acceleration and is the random factor which can only be characterized statistically in terms of the correlation function

$$\overline{v(t)v(t+\Delta t)}/2\Delta t = \mathcal{D}.$$

The velocity space diffusion coefficients are

$$\mathcal{D} = \frac{h\nu}{M}D = \frac{\gamma h\nu}{M\left(1 - e^{-h\nu/T}\right)} \tag{7.26}$$

for the Einstein mechanism and

$$\mathcal{D} = \frac{\alpha m h\nu}{M\left(e^{h\nu/T} - 1\right)} \tag{7.27}$$

for the alternative mechanism. According to the fluctuation–dissipation theorem, $R = \gamma$ for the Einstein mechanism whereas $R = \alpha m$ for the alternative mechanism. This implies that the rate term, αm, in the alternative mechanism has the same role of dissipating energy as the coefficient of spontaneous emission, γ, in the Einstein radiation mechanism.

In a state of dynamic equilibrium, the radiative damping force per unit mass balances exactly the "osmotic" pressure force due to the radiation pressure on the particle:

$$-Rv = \mathcal{D}\frac{\partial}{\partial v}\ln f(v). \tag{7.28}$$

Integrating (7.28) gives

$$f(v) = \sqrt{\frac{R}{2\pi\mathcal{D}}}\exp\left\{-\frac{Rv^2}{2\mathcal{D}}\right\} \tag{7.29}$$

for the stationary probability density. But this distribution must not contain any information on how the asymptotic regime was reached for all physical systems "forget" their past with the passage of time. In order for this to be true, $\mathcal{D} \propto R$, which is precisely the fluctuation–dissipation theorem.

Consequently, we have the stationary probability density

$$f(v) = \left\{\frac{M[1 - \exp(-h\nu/T)]}{2\pi h\nu}\right\}^{1/2}\exp\left\{-\frac{Mv^2}{2h\nu}\left[1 - \exp\left(-\frac{h\nu}{T}\right)\right]\right\} \tag{7.30}$$

in the Einstein mechanism whereas

$$f(v) = \left\{ \frac{M[\exp(h\nu/T) - 1]}{2\pi h\nu} \right\}^{1/2} \exp\left\{ -\frac{Mv^2}{2h\nu} \left[\exp\left(\frac{h\nu}{T}\right) - 1 \right] \right\} \quad (7.31)$$

in the alternative radiation mechanism. These are Maxwellian type velocity distributions of a Brownian particle which is subjected to both a radiation damping and an erratic radiation pressure in a state of dynamical equilibrium. In the low frequency limit, both reduce to the usual Maxwellian distribution (7.18) with a modulus Θ proportional to the temperature.

Let $\bar{T}$ be the average kinetic energy of an oscillator which is just half the total energy. The two velocity distributions predict that the average kinetic energy of the Brownian particle and the average kinetic energy of the oscillator are related by

$$\tfrac{1}{2}M\overline{v^2} = \bar{T} \pm \tfrac{1}{2}h\nu, \quad (7.32)$$

where the plus and minus signs denote the Einstein and alternative radiation mechanisms, respectively. No matter which distribution we choose, we come out with the inevitable conclusion that, in general, *the law of equipartition of energy does not hold for translation motion of a particle interacting with a thermal radiation field in a state of dynamical equilibrium.* Only when equipartition holds for the oscillator will equipartition hold for the translational motion of the Brownian particle and that occurs in the low frequency limit where $T \gg h\nu$. We now have to find some way of discriminating between (7.30) and (7.31).

The distinction between the two velocity distributions is due to the absence or presence of a zero-point term in the expression for the average number of particles or energy. One way to distinguish between the two velocity distributions is to consider the case of Doppler broadening.

Atoms in a gas at temperature T have a spread in their velocities which through the Doppler effect produce a corresponding distribution in the frequencies at which they absorb or emit light. If light is emitted in one direction, then the line will be shifted by an amount

$$\Delta\nu = \nu_0 v/c,$$

where ν_0 is the frequency that the atom would absorb at if it were at rest before and after absorption and c is the velocity of light. Since the particle velocity v is governed by the Maxwell distribution, the probability that ν lies in the frequency interval $d\nu$ will be given by

$$f(\nu)\,d\nu = \frac{1}{\sqrt{2\pi\overline{(\Delta\nu)^2}}} \exp\left[-\frac{(\Delta\nu)^2}{2\overline{(\Delta\nu)^2}} \right] d\nu, \quad (7.33)$$

which is proportional to the intensity of the line. It is usually assumed that the classical equipartition result applies and hence

$$\overline{(\Delta\nu)^2} = (\nu_0/c)^2 T/M.$$

Consequently, according to (7.33), the full width of the Doppler-broadened line at half-maximum is [3]

$$\delta_C = \nu_0 \left(\frac{2T}{Mc^2} \ln 2 \right)^{1/2}. \tag{7.34}$$

If we assume that the Einstein radiation mechanism is at work and use the velocity distribution (7.30) to compute the line breadth, we find

$$\delta_{NB} = \nu_0 \left\{ \left(\frac{2h\nu}{Mc^2} \right) \ln 2 \cdot \frac{1}{1 - e^{-h\nu/T}} \right\}^{1/2}. \tag{7.35}$$

Alternatively, the velocity distribution (7.31) predicts a line breadth given by

$$\delta_P = \nu_0 \left\{ \left(\frac{2h\nu}{Mc^2} \right) \ln 2 \cdot \frac{1}{e^{h\nu/T} - 1} \right\}^{1/2}. \tag{7.36}$$

The line breadths at half-maximum predicted by the three Maxwellian distributions are reported in Table 7.1 for the typical values $h\nu/2Mc^2 \approx 10^{-9}$ and $v/c \approx 10^{-5}$ which for a rest energy of $Mc^2 \approx 1.5 \times 10^{-10}$ J corresponds to a wavelength of visible light between red and orange. It is evident that (7.31) cannot account for Doppler line broadening in this region of the spectrum. In the low frequency region where $\nu \ll T/h$, all three line breadths would coincide with the classical result based on the equipartition of energy.

Table 7.1 shows that in the high frequency region, the line breadth predicted by (7.30) becomes temperature independent which, at room temperature, occurs in the lower part of the visible region. Moreover, $\delta_{NB} \propto \sqrt{\nu}$ independent of T while $\delta_C \propto \sqrt{T}$ and is independent of the frequency ν of the external light source. At still higher frequencies the line breadth predicted by Einstein's radiation mechanism would be an order of magnitude greater than the classical prediction. This is to be expected because we are going further from the "classical" region, corresponding to the Rayleigh–Jeans limit, and not the Wien limit for which Wien's formula was derived from the classical Maxwell distribution (7.18).

Table 7.1 Line Breadths at Half-Maximum
Predicted by the Three Distributions

T (K)	λ (nm)	$h\nu_0/T$	δ_C/ν_0	δ_P/ν_0	δ_{NB}/ν_0
300	666	72	6.2×10^{-6}	10^{-20}	5.3×10^{-5}
1600	666	13.5	1.4×10^{-5}	6.2×10^{-8}	5.3×10^{-5}

[3]The peak of the Gaussian line is at $\nu = \nu_0$ and the line has half its maximum intensity at the frequencies that satisfy

$$\tfrac{1}{2} = \exp \left\{ -Mc^2(\nu - \nu_0)^2/2\nu_0^2 T \right\}.$$

Only in the wavelength limit are the two radiation mechanisms really comparable. Our alternative radiation mechanism cannot be continued into the visible part of the spectrum so that it does look like Einstein's radiation mechanism can account for all regions of the spectrum. But we have seen in §6.2 that radiation obeys a Clapeyron equation in which the pressure is independent of the volume. This indicates some type of equilibrium because it would be the equivalence of the chemical potentials of the two phases which would eliminate the volume dependence upon the pressure. Could it be possible that the stationary probability distributions, from which we derived the radiation mechanisms, really enter into a phase equilibrium with one another? If this were true, then the frequency at which our alternative radiation mechanism breaks down could be identified as a critical point beyond which this phase becomes unstable and breaks up. Let us now consider this possibility in greater detail.

7.3 Black Radiation as a First-Order Phase Transition

In §7.1 we proposed that the concept of a phase equilibrium can be applied to two probability distributions that belong to the same class. Here we show that it may explain the transition between thermal and nonthermal radiation. The analogy between the two stems from the fact that the Clapeyron equation, which characterizes a first-order phase equilibrium, applies to any equation of state in which the pressure is a function of the temperature independent of the volume. Black radiation, as we have seen, also gives an expression for the pressure which is independent of the temperature because the chemical potential of thermal radiation vanishes.

For the sake of concreteness, we may consider the phase equilibrium that is established between saturated vapor and liquid. As a microscopic picture of evaporation, we may consider the formation of an embryonic gas bubble which, if allowed to grow to macroscopic dimensions, would result in the breakdown of the liquid. The vapor pressure inside the bubble exceeds the surrounding pressure of the fluid by the surface, or capillary, pressure which is inversely proportional to the radius of the bubble. For variations in the radius of the bubble beyond a critical value, the mechanical equilibrium between the pressure of the bubble and the pressure of the surrounding liquid becomes unstable with the consequence that the liquid phase disappears.

Thermal radiation has its intensity and spectrum determined solely by the temperature of the emitting body. In contrast, nonthermal radiation which is characterized by its high intensity cannot be accounted for solely in terms of the temperature of the emitting body. The spectrum of thermal radiation may contain photons of any energy while a threshold in the photon energy due to an energy gap in the distribution of states is the earmark of nonthermal radiation. It has been common practice to account for nonthermal radiation by introducing an "effective" temperature, but more recently, it has been recognized that the correct way to treat nonthermal radiation is to introduce a chemical potential.

The cause of the transition between thermal and nonthermal radiation will be shown to be due to the "zero-point" energy in an exactly analogous way that the capillary pressure is implicated in transition between liquid and vapor. The critical photon frequency will be seen to be analogous to the critical value of the radius of the embyronic bubble. For photon energies greater than the critical value, the thermal radiation field tends to "break up" in the same way that a *negative* pressure tends to break-up a liquid.

As we have seen on numerous occasions [cf. §2.5.1], the radiation pressure of black radiation is

$$
\begin{aligned}
P_\varepsilon &= -\frac{mT}{V} \ln\left(1 - e^{(\varepsilon - \mu)/T}\right) \\
&= -\frac{mT}{V} \ln\left(e^{(\varepsilon - \mu)/T} - 1\right) + \frac{m(\varepsilon - \mu)}{V} \\
&= P_\varepsilon' + \frac{m(\varepsilon - \mu)}{V},
\end{aligned}
\tag{7.37}
$$

where we have introduced the chemical potential in order to make the transition from thermal to nonthermal radiation. The zero-point energy, which depends on the dimensions of the cavity, contributes to the radiation pressure in an analogous way that the capillary pressure, $2\sigma/r$, where σ is the surface tension and r the radius of the gas bubble, balances the external pressure, P', with the saturated vapor pressure P in a state of mechanical equilibrium:

$$
P = P' + \frac{2\sigma}{r}.
\tag{7.38}
$$

The analogy between the radiation pressure, (7.37), and the pressure of a saturated vapor in mechanical equilibrium with the external fluid pressure is summarized in the following:

Phase Transition		*Black Radiation*
Liquid–gas transition	$\longleftrightarrow$	Radiative transition
Gaseous phase	$\longleftrightarrow$	Nonthermal radiation
Liquid phase	$\longleftrightarrow$	Thermal radiation
Capillary pressure	$\longleftrightarrow$	Zero-point energy
Critical bubble radius	$\longleftrightarrow$	Critical photon energy

What is intriguing about the decomposition of the radiation pressure, (7.37), is that whereas P_ε is the radiation pressure of the negative binomial distribution, P_ε' is the radiation pressure of the Pascal distribution. The pressure,

$$
P_\varepsilon' = -\frac{mT}{V} \ln\left(e^{(\varepsilon - \mu')/T} - 1\right),
$$

is derived from the norming constant, or partition function,

$$\mathcal{Z}'_\varepsilon = \left(e^{(\varepsilon - \mu')/T} - 1\right)^{-m} = \sum_{n=m}^\infty \binom{n-1}{m-1} e^{-n(\varepsilon - \mu')/T},$$

according to the relation $P'_\varepsilon = T \ln \mathcal{Z}'_\varepsilon / V.$[4]

At sufficiently high photon energies, P'_ε can become negative; specifically, for $\bar{n}'_\varepsilon / m < 2$ it is negative, where

$$\bar{n}'_\varepsilon = \frac{m}{1 - e^{-(\varepsilon - \mu')/T}} = \frac{m}{e^{(\varepsilon - \mu')/T} - 1} + m = \bar{n}_\varepsilon + m,$$

while for $\bar{n}'_\varepsilon / m > 2$, it is positive. At the critical value, $\bar{n}'_\varepsilon / m = 2$, the radiation pressure vanishes. This critical value coincides with the state where the *a priori* probabilities in the Pascal distribution,

$$f(n) = \binom{n-1}{m-1} p^m q^{n-m},$$

are equal. Since the maximum likelihood values of the *a priori* probabilities are

$$q = e^{-(\varepsilon - \mu')/T} \qquad \text{and} \qquad p = 1 - e^{-(\varepsilon - \mu')/T},$$

the critical frequency or energy is

$$\varepsilon^* = \mu' + T \ln 2. \tag{7.39}$$

At this critical state, the entropy reduces to $S' = \bar{n}'_\varepsilon \ln 2$ which, however, is not the state of maximum disorder, since there are states of higher entropy than the $2^{\bar{n}'_\varepsilon}$ equally likely microscopic complexions.

In order to demonstrate that what we have is indeed a first-order phase transition, we must show that the radiation pressure satisfies the Clapeyron equation. Then, we can associate the two probability distributions with two phases. The liquid phase is represented by the Pascal distribution while the saturated vapor phase corresponds to the negative binomial distribution. Once the critical point (7.39) is surpassed, the pressure of the Pascal distribution becomes negative, which tends to break up the liquid phase so that the distribution, itself, is no longer stable.

The pair of Gibbs–Duhem equations, from which the Clapeyron equation is derived, are

$$\bar{n}_\varepsilon \, d\mu = -S \, dT + V \, dP_\varepsilon$$

and

$$\bar{n}'_\varepsilon \, d\mu' = -S' \, dT + V \, dP'_\varepsilon.$$

Although the entropies of the two phases are numerically equal, the "molar" entropies or the entropies per particle, $S_{\mathrm{m}} = S / \bar{n}_\varepsilon$ and $S'_{\mathrm{m}} = S' / \bar{n}'_\varepsilon$, are not.

---

[4]For an alternative derivation of the radiation pressure from the validity of the Euler relation in each frequency interval consult §2.5.2.

The condition for phase equilibrium is $\mu = \mu'$, and in order to maintain such an equilibrium in the face of variations in the temperature and pressures, we must have $d\mu = d\mu'$. From the later condition we obtain the Clapeyron equation:

$$V_{\mathrm{m}} \, dP_{\varepsilon} - V'_{\mathrm{m}} \, dP'_{\varepsilon} = \frac{\Delta H_{\mathrm{m}}}{T} \, dT, \tag{7.40}$$

where the molar heat of evaporation is

$$
\begin{aligned}
\Delta H_{\mathrm{m}} &= T(S_{\mathrm{m}} - S'_{\mathrm{m}}) \\
&= T \left(1 - e^{-(\varepsilon - \mu)/T} \right) \ln \left(e^{(\varepsilon - \mu)/T} - 1 \right) \\
&\quad - T \left(e^{(\varepsilon - \mu)/T} - 1 \right) \ln \left(1 - e^{-(\varepsilon - \mu)/T} \right)
\end{aligned}
$$

and the molar volume difference is

$$\Delta V_{\mathrm{m}} = 4 \frac{V}{m} \sinh^2 \left(\frac{\varepsilon - \mu}{2T} \right).$$

Since the pressures are related through the condition of mechanical equilibrium (7.37), we can eliminate P'_{ε} in favor of P_{ε}. We then obtain the generalized Clapeyron equation:

$$\frac{dP_{\varepsilon}}{dT} = \frac{\Delta H_{\mathrm{m}} + m\mu/\bar{n}'_{\varepsilon}}{T \Delta V_{\mathrm{m}}}, \tag{7.41}$$

where we have used the fact that we are dealing with a Stefan gas for which $\mu/T = \mathrm{const}$. We must now justify the presence of the additional term in this equation.

Let us recall from §6.2 that Lord Rayleigh used the equation $dP = h/T$ to derive Stefan's law. Since the modes of the electromagnetic field do not interact with one another, this equation should hold per mode[5]:

$$\frac{dP_{\varepsilon}}{dT} = \frac{h_{\varepsilon}}{T}, \tag{7.42}$$

where the enthalpy density of mode ε is $h_{\varepsilon} = \bar{n}_{\varepsilon} \varepsilon / V + P_{\varepsilon}$. It is easy to see that (7.37) satisfies this equation provided that

$$
\begin{aligned}
d \left(\frac{\mu}{T} \right) &= \frac{1}{T} \left(\frac{\partial \mu}{\partial T} \right)_{p_{\varepsilon}} dT + \frac{1}{T} \left(\frac{\partial \mu}{\partial p_{\varepsilon}} \right)_T dp_{\varepsilon} - \frac{\mu}{T^2} \, dT \\
&= -\frac{S_{\mathrm{m}}}{T} \, dT + \frac{V_{\mathrm{m}}}{T} \, dp_{\varepsilon} - \frac{\mu}{T^2} \, dT = 0.
\end{aligned}
$$

We recall from §6.3 that this is the condition imposed upon the chemical potential of an ideal gas which does not conserve the number of particles (i.e., a Stefan gas). Introducing the expression for the entropy per mode, $S = (\bar{n}_{\varepsilon} \varepsilon + P_{\varepsilon} V - \mu \bar{n}_{\varepsilon})/T$, it is readily seen that the terms containing the chemical

---

[5] We cannot, however, integrate this equation because the equation of state that relates the pressure to the energy density does not hold per mode.

potential cancel and it reduces to (7.42). What is not so evident is that (7.42) is identical to (7.41). Observe that $S/V = S_m/V_m$, which permits us to write (7.41) as

$$
\begin{aligned}
\frac{dP_\varepsilon}{dT} &= \frac{S}{V} + \frac{m}{\bar{n}'_\varepsilon} \frac{\mu}{T} \\
&= \frac{h_\varepsilon}{T} - \frac{\mu \bar{n}_\varepsilon}{VT} + \frac{m\mu}{VT} \frac{\bar{n}_\varepsilon}{\bar{n}'_\varepsilon - \bar{n}_\varepsilon}.
\end{aligned}
$$

The last two terms vanish on account of the relation $\bar{n}'_\varepsilon = \bar{n}_\varepsilon + m$, and this gives us back Lord Rayleigh's equation (7.42). Thus, we can write (7.41) as

$$
\frac{dP_\varepsilon}{dT} = \frac{\Delta S_m}{\Delta V_m} + \frac{\mu \bar{n}_\varepsilon}{VT}, \tag{7.43}
$$

where the second term arises from the zero-point energy.

When $\varepsilon - \mu \ll T$, the latent heat is negligibly small so that there is essentially only one phase present. At the critical point (7.39), the external pressure P'_ε of the liquid vanishes. This may be taken to define the threshold frequency that separates thermal from nonthermal radiation. At higher frequencies or larger energies, the liquid would be broken apart and forced to boil.

In the region $\varepsilon - \mu \gg T$, the molar enthalpy of evaporation $\Delta H_m \approx (\varepsilon - \mu)$ and the molar volume change $\Delta V_m \approx (V/m)e^{(\varepsilon-\mu)/T}$. The Clapeyron equation (7.43) thus reduces to

$$
\frac{dP_\varepsilon}{dT} = \frac{m\varepsilon}{VT} e^{-(\varepsilon-\mu)/T},
$$

which can be immediately integrated to give

$$
P_\varepsilon = \frac{m\varepsilon z}{V} \int \frac{e^{-x}}{x} \, dx
$$

on account of the fact that the ratio μ/T is constant. The integral is equal to the negative of the sum of the exponential integral and Euler's constant. For large values of x, the exponential integral is approximately e^{-x}/x so that the pressure is

$$
P_\varepsilon = \frac{mT}{V} e^{-\Delta H_m/T}, \qquad \varepsilon - \mu \gg T,
$$

which is the expression we would get from the radiation pressure (7.37) in the same limit. It tells us that we are in the Wien region in which the particle nature of light is manifested. Moreover, it has the same form as the solution to the Clapeyron equation in the case of large overheating.

In the high frequency limit, spontaneous emission predominates over stimulated emission. In fact, we recall from §2.4 that stimulated emission is not predicted by classical theory and Einstein found it necessary to introduce it in order to obtain a detailed balancing condition which yielded Planck's law. But what would happen if we could amply the rate of stimulated emission even beyond that permitted by inequality (7.14)?

7.4   Beyond Einstein's Radiation Theory: A Lasing Mechanism

The key to generalizing Einstein's radiation mechanism lies precisely in the ability to invert inequality (7.14), which is related to population inversion. However, we have seen that this inequality is ensured by the second law, and according to (7.15), it would formally correspond to a negative temperature. Certainly, Einstein's radiation mechanism does not cover such an eventuality since Einstein assumed that the atoms obey classical statistics.

If we solve the average equation of motion

$$\dot{\bar{n}}(t) = \gamma - (\alpha - \beta)\bar{n}(t) \tag{7.44}$$

subject to the initial condition $\bar{n}(0)$, we obtain

$$\bar{n}(t) = \bar{n}_0 e^{-t/\tau} + \bar{n}^s \left(1 - e^{-t/\tau}\right), \tag{7.45}$$

where $\tau = (\alpha - \beta)^{-1}$ is the relaxation time and $\bar{n}^s$ is the stationary, detailed balance solution, $\bar{n}^s = \gamma/(\alpha-\beta)$. This would mean that inverting the inequality in (7.14) would lead to an exponential growth which could not be continued indefinitely—if theory is to correspond to fact. When a perturbation grows without limits, it must be concluded that this is only valid during some finite interval of time, after which there should be some nonlinear bound to the unstable growth of the perturbation.

Only linear terms in the beam intensity have been taken into consideration in Einstein's expression for the rate of absorption. Higher order terms in the absorption rate correspond to the excitation of an atom by a process involving the absorption of two or more photons. In the simplest case involving two photons, we would add the term κn^2 to the absorption rate, which would mean replacing the average equation of motion (7.44) by

$$\dot{\bar{n}}(t) = \gamma - (\alpha - \beta)\bar{n}(t) - \kappa\bar{n}^2(t), \tag{7.46}$$

where κ is the two-photon absorption coefficient, assumed to be independent of the beam intensity.

There are two limiting stationary solutions to (7.46) that have a physical interest: in the limit as $\kappa \to 0$, the stationary mean photon number is given by the Bose–Einstein distribution (7.17) which we can write in the equivalent form as

$$\bar{n} = \frac{\gamma}{\alpha - \beta} \qquad (\alpha > \beta), \tag{7.47}$$

and in the limit as $\gamma \to 0$ for which we have

$$\bar{n} = \frac{\alpha - \beta}{\kappa} \qquad (\alpha < \beta). \tag{7.48}$$

In direct contrast to (7.47), the stationary solution (7.48) is maintained by a stimulated emission which prevails not only over spontaneous emission but also over single-photon absorption. It is in this case that a highly ordered situation arises even though the electrons may be excited by completely uncorrelated external pumping.

We now have to determine the stationary probability distribution corresponding to such a highly ordered state. This means solving the master equation

$$\dot{f}(n,t) = \left\{ (\mathbf{E} - 1)(\alpha + \kappa n)n + (\mathbf{E}^{-1} - 1)(\beta n + \gamma) \right\} f(n,t)$$

for its stationary solution. This can be done by iteration from the dynamical equilibrium condition

$$(\alpha + \kappa n)n f(n) = \mathbf{E}^{-1}[\beta n + \gamma] f(n).$$

We then obtain

$$f(n) = \frac{(\beta(n-1) + \gamma)!}{n!\,(\alpha + \kappa n)!} f(0), \tag{7.49}$$

where the normalization constant

$$f^{-1}(0) = \sum_{n=0}^{\infty} \frac{(\beta(n-1) + \gamma)!}{n!\,(\alpha + \kappa n)!}.$$

The stationary probability distribution (7.49) coincides with two well known distributions that belong to the exponential family in the limits as $\kappa \to 0$ and $\gamma \to 0$.

In the limit as $\kappa \to 0$, (7.49) reduces to

$$f(n) = \frac{1}{\beta} \frac{(m+n-1)!}{n!} \left(\frac{\beta}{\alpha} \right)^n f(0), \tag{7.50}$$

where we have set $\gamma = \beta m$. The normalization constant is easily found to be

$$f^{-1}(0) = \frac{(m-1)!}{\beta} \sum_{n=0}^{\infty} \binom{m+n-1}{n} q^n = \frac{(m-1)!}{\beta p^m},$$

which shows that (7.50) is none other than the negative binomial distribution.

In the limit as $\gamma \to 0$, the number of photons in the cavity mode becomes much greater than unity and the stationary probability distribution (7.49) is approximated by

$$f(n) = \frac{\beta^n}{(\alpha + \kappa n)!} f(0),$$

where normalization demands that

$$f^{-1}(0) = \sum_{n=0}^{\infty} \frac{\beta^n}{(\alpha + \kappa n)!} = \left(\frac{\beta}{\kappa} \right)^{-\alpha/\kappa} e^{\beta/\kappa}.$$

Consequently, the stationary distribution is the Poisson distribution

$$f(n) = \frac{(\bar{n} + \alpha/\kappa)^{n + \alpha/\kappa}}{(n + \alpha/\kappa)!} e^{-(\bar{n} + \alpha/\kappa)} \tag{7.51}$$

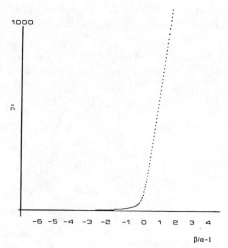

Figure 7.1: Mean photon number $\bar{n}$ as a function of the pumping rate which is proportional to $\beta/\alpha - 1$. What is not shown is the leveling off of the curve at higher values of the pumping rate.

provided the number of photons in the cavity mode is large compared to unity. The condition $\bar{n} > \alpha/\kappa$ corresponds to pumping rates in which the active atomic excited states approach *saturation*. For even greater values, $\bar{n} \gg \alpha/\kappa$, the Poisson distribution (7.51) reduces approximately to

$$f(n) = \frac{\bar{n}^n}{n!}\, e^{-\bar{n}}, \tag{7.52}$$

which is the well known photon distribution for coherent light.

The stationary photon probability distribution therefore transforms from the negative binomial distribution, (7.50), for single-mode *chaotic* light below threshold ($\beta < \alpha$) into a Poisson distribution for *coherent* light above threshold ($\beta > \alpha$). Let us recall that dispersion in the particle number of the negative binomial distribution contains both the "particle" and "wave" contributions whereas in the Poisson limit there remains only the particle contribution.

The analysis is confirmed in Fig. 7.1, which shows the mean photon number $\bar{n}$ as a function of the pump parameter, $\beta/\alpha - 1$. Note the extremely sharp rise in $\bar{n}$ at the laser threshold $\alpha = \beta$. The sharp rise shown in Fig. 7.1 tapers off at higher rates of pumping.

The entropy of the Poisson distribution, (7.51), is

$$S(\bar{n}) = \bar{n} - \left(\bar{n} + \frac{\alpha}{\kappa}\right) \ln\left(\frac{\bar{n} + \alpha/\kappa}{C}\right), \tag{7.53}$$

where $C\, d\nu$ is the number of quantum states of photons with frequencies between ν and $\nu + d\nu$.[6] The inverse of the two-photon absorption coefficient

---

[6]This has to be written in by hand in order to render the entropy extensive, but as we

is also extensive. Taking the derivative of the statistical entropy (7.53) and setting it equal to its thermodynamic counterpart give

$$-\ln\left(\frac{\bar{n} + \alpha/\kappa}{C}\right) = \frac{\varepsilon - \mu}{T}, \tag{7.54}$$

where μ is the chemical potential of radiation that we discussed at the end of §2.4.

A nonvanishing photon chemical potential is necessary for a radiative balance in the steady state while in the case of black radiation all electronic states possess a common chemical potential requiring that their difference $\mu = 0$. The condition for population inversion is $\mu > h\nu$, where ν is the photon frequency necessary to excite an electron from the lower to the upper level. At threshold, $\alpha = \beta$, while in the steady state above threshold, (7.54) and (7.48) combine to give

$$\frac{\beta}{\kappa} = Ce^{(\mu - h\nu)/T}.$$

When this is introduced into (7.48), we get

$$\bar{n} = \left(\frac{\beta - \alpha}{\beta}\right)Ce^{(\mu - h\nu)/T}. \tag{7.55}$$

Far above threshold, where $\beta/\alpha \gg 1$, the average number of photons $\bar{n} = Ce^{(\mu - h\nu)/T}$. If it weren't for the presence of the chemical potential, this would appear to be the classical Wien expression for the average number of photons in the freqency interval $d\nu$. An integration over all frequencies yields the total number of photons as $N = 16\pi VT^3 e^{\mu/T}/(hc)^3$. And since the radiation density has the same functional dependency as in Wien's law, the wavelength at which the energy density reaches its maximum value is $\lambda_m = hc/5T$. Introducing this expression into the total number of photons gives $N = 16\pi Ve^{\mu/T}/125\lambda_m^3$.

We can now appreciate that the condition for lasing is equivalent to

$$\lambda_m^3 \frac{N}{V} \gg 1.$$

If λ_m were replaced by the thermal wavelength, this would be the inequality for the appearance of "degenerate," or nonclassical, phenomena that occur at low temperatures or at high densities [cf. introductory remarks in §3.4]. The inequality which the chemical potential is required to satisfy for lasing is therefore consistent with the well known condition of requiring an extremely large photon density. This is ordinarily achieved by placing the lasing material into a resonator equipped with mirrors that provide a positive feedback by reflecting the emitted photons back into it.

---

recall from §3.4.2, it cancels out in Gauss' error law leading to the Poisson distribution. Let us also recall that since the Poisson distribution is the limiting form of the negative binomial distribution in which the particles are indistinguishable, it too considers the particles to be indistinguishable. It is precisely the term $C^{\bar{n}}$, for $\bar{n} \gg \alpha/\kappa$, or the number of ways that $\bar{n}$ particles can be placed in C cells which makes them appear distinguishable.

7.5   The Bose–Einstein Condensation Revisited

As we have mentioned in the introduction, Einstein was spurred by Bose's analysis into the almost inevitable conclusion that

> if it is justified to conceive of radiation as a quantum gas, then the analogy between a quantum gas and a molecular gas must be a complete one.

Unlike Bose's treatment, Einstein introduced a Lagrange multiplier for the constraint that the total number of particles be a constant. This led to a modified version of Planck's formula,

$$\bar{n}_\varepsilon = \frac{m}{z^{-1}e^{\varepsilon/T} - 1}, \tag{7.56}$$

for nonrelativistic particles. The energy, ε, is now proportional to the square of the momentum whereas in the relativistic case it is linear in the momentum. Einstein called the new parameter appearing in (7.56) the "degeneracy" parameter. Thermodynamically, it known as the fugacity $z = e^{\mu/T}$, which we introduced in §6.3. Since $z < 1$, it means that the chemical potential of a Bose–Einstein gas must be negative.

   In the second of the trio of papers that he published on the new statistics, between 1924 and 1925, Einstein asked the question of what happens as $\mu \to 0^-$. Integrating (7.56) over all energies and setting $z = 1$, Einstein obtained

$$T_0 = \frac{h^2}{2\pi M} \left[\frac{V g_{3/2}(1)}{N} \right]^{-2/3}, \tag{7.57}$$

where M is the mass of the particle and

$$
\begin{aligned}
g_n(z) &= \frac{1}{\Gamma(n)} \int_0^\infty \frac{x^{n-1}}{z^{-1}e^x - 1} \, dx \\
&= \frac{1}{\Gamma(n)} \int_0^\infty x^{n-1} \sum_{\ell=1}^\infty \left(z e^{-x} \right)^\ell \, dx \\
&= \sum_{\ell=1}^\infty \frac{z^\ell}{\ell^n}
\end{aligned}
$$

is a Bose–Einstein integral. Einstein concluded that, as the temperature drops below T_0 (for a given density!),

> a number of molecules steadily growing with increasing density goes over in the first quantum state (which has zero kinetic energy) while the remaining molecules distribute themselves according to the parameter $[z] = 1 \ldots$. A separation is effected; one part condenses, the rest remains a "saturated ideal gas".

But for $T < T_0$, the condition

$$\frac{N}{V} = \frac{g_{3/2}(z)}{\lambda^3}, \tag{7.58}$$

which implicitly determines that the chemical potential μ as a function of its temperature T and density N/V has no negative solutions! And we know that the chemical potential of a Bose–Einstein gas must be negative at all temperatures.

The usual way around this is to question the expression for the number of states in any given energy interval: since $m \propto \sqrt{\varepsilon}$, the density of states attributes a zero weight to the ground state where $\varepsilon = 0$. If the term representing the ground state is extracted from the integral and m is set equal to unity, it will tend to infinity in the limit $\mu \to 0^-$. The only way out of this dilemma is to let μ tend to some finite value so that this term can take the desired finite value. But if more and more particles occupy the first quantum state as the temperature is lowered, $z \sim N/(N+1)$, which is always effectively unity. Moreover, if z is effectively unity, then $N = z/(1-z)$ is infinite, and although we started with a fixed and finite number of particles—which is the original reason why z is introduced altogether—we wind up with an infinite number of particles!

Today, the Bose–Einstein condensation is generally recognized as a first-order phase transition, although Einstein did not call the condensation phenomenon a phase transition in his 1925 paper. Hence, a Clapeyron equation must be satisfied in which there is a latent heat. Yet, the "condensation" occurs in momentum space, and since no condensation actually occurs, the "condensed" phase has zero specific volume and entropy. Hence, the analogy with a first-order phase transition is illusive since the Clapeyron equation is really the Gibbs–Duhem equation in disguise.

According to Fritz London, the Bose–Einstein condensation had "the reputation of having only a purely imaginary character" until he proposed, in 1938, that the transition to superfluidity in ^{4}He was such a transition. The critical value of the temperature is 3.14 K, which is different from the experimentally determined value of 2.17 K. The difference has been attributed to weak interactions of an almost ideal Bose–Einstein gas. But, as we have mentioned in the introduction, the interactions in ^{4}He are stronger than would be appropriate for a Bose–Einstein condensation theory. In any event, the physical interactions should be incorporated into the statistics from the very start and not appended on as an afterthought.

There is also doubt concerning the fraction of the total number of particles that end up in the ground state. It has been estimated that even at $T = 0$, only 8% of the atoms are condensed into the lowest state. According to Lars Onsager, this background of condensed particles may be crucial for superfluidity of the entire system. It is generally accepted that the occupation of the zero momentum state may, in fact, be only a few percent even at absolute zero, yet, somehow this condensate dominates over the entire system by making it behave as a superfluid. In the picturesque terminology used by Herbert Frölich, the condensate acts as a "pilot-wave" dragging along the rest of the fluid. At finite temperatures, the superfluid, or liquid helium II, is known to have some entropy which is different from the entropy of the normal fluid, or helium I. Thus treating the two as separate phases will automatically lead to a latent heat and a discontinuity in the molar volume that justify considering helium I and helium II to be in a phase equilibrium.

Instead of extracting the ground state from the Bose–Einstein integral, we would rather view it as a separate phase which can form a phase equilibrium with the rest of the fluid which we will treat as an ideal Bose–Einstein gas. Since there is no entropy of a single state, which is the conclusion we reached in §7.2.1, we consider the condensate to be smeared out over a range of low-lying states in momentum space as in London's original formulation. When atoms of helium I, which are distributed over the higher-lying states of momentum space, enter that region of momentum space, they undergo a kind of allotropic modication which can be described as a separate phase.

The statistics of helium I, treated as an ideal Bose gas, are derived from the negative binomial distribution

$$f^{\mathrm{I}}(n) = \binom{m + n - 1}{n} p^m q^n,$$

while those of helium II would result from setting $m = 1$. This causes the negative binomial distribution to degenerate into the geometric distribution,

$$f^{\mathrm{II}}(n) = p q^n.$$

Probability distributions which belong to the same class differ in terms of the allocation problems they describe. For instance, the distribution of n particles among m cells with empty cells being admissible is described by the negative binomial distribution. The constraint that no cell should be empty is descibed by the Pascal distribution and that where only one cell is present is described by the geometric distribution. If the separation of phases is regarded as a constraint on how the cells are to be occupied or how many of them there are, then the probability distributions describing these events should be able to enter into an equilibrium with one another just as material phases do. The thermodynamic properties of these probability distributions are determined by their entropies which are derived from casting them as error laws. Once the entropies have been determined, all the other thermodynamic properties of the system follow.

The entropy of the negative binomial distribution is the entropy of "mixing":

$$S^{\mathrm{I}}(\bar{n}^{\mathrm{I}}) = \bar{n}^{\mathrm{I}} \ln\left(\frac{\bar{n}^{\mathrm{I}} + m}{\bar{n}^{\mathrm{I}}}\right) + m \ln\left(\frac{\bar{n}^{\mathrm{I}} + m}{m}\right).$$

For $m = 1$, the statistical entropy would be thermodynamically inadmissible since it would lose its property of extensivity. A statistical weight of unity would apply to the state of zero momentum, but we cannot define an entropy of a single state. In fact, the uncertainty relation would prevent us from ever being sure that a particle would be in such a state for it would amount to a complete uncertainty in its position which is usually stated as "the condensate has no volume."

If all the particles dropped into the ground state below the transition temperature, there would be neither an entropy nor a pressure of such a condensed phase. It would certainly not be a thermodynamic phase which would be able to come into equilibrium with any other phase. Moreover, the average number of particles in the ground state would no longer be an extensive quantity since it would be independent of the size of the system. By enlarging the domain of momentum space to a small but finite group of low-lying momentum states, our knowledge of the precise value of the momenta of the particles is diminished and at the same time we can say something about where they would be expected to be located in space.

We shall therefore argue that the Bose–Einstein condensation takes place in configuration space as well as momentum space having a well defined entropy, volume, and pressure. The pressure is the result of transitions between the zero momentum state and neighboring states with small but finite momenta.

With the aid of the second law, we find the average number of particles in liquid helium I to be

$$\bar{n}_\varepsilon^{\mathrm{I}} = \frac{m(\varepsilon)}{z_{\mathrm{I}}^{-1} e^{\varepsilon/T} - 1}, \tag{7.59}$$

while if we set $\varepsilon = 0$ and $m(0) = 1$, the average number of particles in the ground state would be given by

$$\bar{n}^{\mathrm{II}} = \frac{z_{\mathrm{II}}}{1 - z_{\mathrm{II}}}, \tag{7.60}$$

where the z_i are the fugacities in the two phases. But (7.60) is not extensive nor is the entropy from which it was derived. Hence, (7.60) cannot describe a macroscopic phase.

As the energy approaches zero, the discreteness of the translational levels will make itself felt. Since they are completely unlocalized in space, there is an inherent indeterminacy in being able to predict, with absolute certainty, the number of particles that will be found in the state of zero momentum. By collecting states with momenta less than a certain small value, $\mathbf{p}_0$, whose precise value will not be required and denoting this as the superfluid phase, we are able to increase the precision of measuring the positions of the particles at the expense of knowing their momenta. In other words, the particles will now be seen to occupy a volume and have a finite and extensive entropy.

The London mean energy is determined by

$$\varepsilon_0 \equiv \frac{\int_0^\varepsilon x m(x)\, dx}{\int_0^\varepsilon m(x)\, dx} = \frac{3}{5}\varepsilon,$$

where $\varepsilon = p_0^2/2M = hk_0^2/2MV^{2/3}$ and k_0^2 is the sum of the square of three integral quantum numbers. The number of states having an energy less than or equal to the mean energy is

$$m_0 = \frac{4\pi}{3} \left(\frac{10M\varepsilon_0}{3h^2} \right)^{3/2} V.$$

For $k_0 \approx 10$, $M = 6.65 \times 10^{-24}$ gm and $V = 27.6$ cm³, which are the mass and molar volume of liquid helium, the mean energy is $\varepsilon_0 \approx 10^{-29}$ ergs, and the density of states is $m_0/V \approx 1.5 \times 10^2$.

Therefore, a better approximation to $\bar{n}^{II}$ than (7.60) is

$$\bar{n}^{II}(T) = \frac{m_0}{z_{II}^{-1} e^{\varepsilon_0/T} - 1}. \tag{7.61}$$

This leads to an entropy of mixing

$$S^{II}(\bar{n}^{II}) = \bar{n}^{II} \ln \left(\frac{\bar{n}^{II} + m_0}{\bar{n}^{II}} \right), + m_0 \ln \left(\frac{\bar{n}^{II} + m_0}{m_0} \right)$$

which is extensive, and since it is a function of the volume, the condensate has a well defined pressure. In order that $\bar{n}^{II} > 0$, the difference $\varepsilon_0 - \mu^{II} > 0$. But since ε_0 is always such a small quantity, we must require $-\mu^{II}$ to be *finite*.

Comparing the terms in statistical entropies with their counterparts in the Euler relation, we obtain the pressures of the two phases as

$$P_\varepsilon^I = -\frac{m T}{V} \ln \left(1 - z_I e^{-\varepsilon/T} \right)$$

and

$$P^{II} = -\frac{m_0 T}{V} \ln \left(1 - z_{II} e^{-\varepsilon_0/T} \right).$$

Since $m_0 \propto V$, the pressure, P^{II}, of liquid helium II will not vanish in the thermodynamic limit as $V \to \infty$. If a phase equilibrium does exist, we will require the pressures of the two phases to satisfy a Clapeyron equation.

In order to establish a phase equilibrium at a common temperature T, it is necessary that

$$\mu^I \left(P^I, T \right) = \mu^{II} \left(P^{II}, T \right).$$

However, such a phase equilibrium will never be *complete* since the pressures of the two phases are unequal. Varying the temperature and pressures of the two phases,

$$\bar{n}_\varepsilon^I \, d\mu^I = -S^I \, dT + V \, dP_\varepsilon^I$$

and

$$\bar{n}^{II} \, d\mu^{II} = -S^{II} \, dT + V \, dP^{II},$$

the condition for the maintenance of equilibrium,

$$d\mu^I = d\mu^{II},$$

results in

$$V_m^I \, dP_\varepsilon^I - V_m^{II} \, dP^{II} = \Delta S \, dT. \tag{7.62}$$

This is the same Clapeyron equation we derived in (7.40) for a system comprising two degrees-of-freedom at different pressures. The latent heat of vaporization is

$$\begin{aligned}
T\Delta S &= T\left(S_m^I - S_m^{II}\right) \\
&= \varepsilon - \varepsilon_0 - T\left(z^{-1}e^{\varepsilon/T} - 1\right)\ln\left(1 - ze^{-\varepsilon/T}\right) \\
&\quad + T\left(z^{-1}e^{\varepsilon_0/T} - 1\right)\ln\left(1 - ze^{-\varepsilon_0/T}\right),
\end{aligned}$$

where the "molar" entropies, which are actually the entropies per particle, are denoted by $S_m^i = S^i/\bar{n}^i$ and V_m^i are the "molar" volumes, $V/\bar{n}^i$ or, more correctly, the volume per particle. However, unlike the Clapeyron equation (7.40), which establishes a phase equilibrium in the energy interval from ε to $\varepsilon + d\varepsilon$, there is no relation between the two pressures P_ε^I and P^{II} so that (7.62) cannot be reduced to the usual Clapeyron equation.

Although both the Clapeyron equations (7.40) and (7.62) are *local* relations, the former ensures an equilibrium between two phases in the same energy interval while the latter establishes an equilibrium between states in any given energy interval and the states whose energies are not superior to ε_0 that comprise the superfluid. A *global* Clapeyron equation can be obtained simply by integrating the entropy and pressure of the normal fluid over all energies. Certainly, the ground state will be excluded by the expression for the density of states which attributes zero weight to this state. The integral will, however, contain the low-lying momentum states. Nevertheless, since $V/\lambda^3 \gg 1$, [7] the contribution of these states to the rest of the continuum will be negligible. However, this does not mean that the molar quantities pertaining to phase II will be negligible with respect to those of phase I since the ratio V/λ^3 cancels out between numerator and denominator in the expression for the molar quantities.

With this understanding, we attribute a molar entropy and pressure

$$\hat{S}_m^I = \frac{5}{2}\frac{g_{5/2}(z)}{g_{3/2}(z)} - \ln z$$

and

$$P^I = \frac{T}{\lambda^3}g_{5/2}(z),$$

respectively, to the normal fluid. It is now easy to see that the global Clapeyron equation,

$$V_m^I \, dP^I - V_m^{II} \, dP^{II} = \frac{\Delta H}{T}\, dT,$$

---

[7]Even at the critical temperature of 2.19 K, $V/\lambda^3 = 1.35 \times 10^{23}(= N/4.5)$, where we have substituted the data for liquid ^{4}He: $M = 6.65 \times 10^{-24}$ g and $V = 27.6$ cm^3/mol.

is satisfied where

$$\Delta H = T \left(\hat{S}_m^I - S_m^{II} \right)$$

$$= T \left\{ \frac{5}{2} \frac{g_{5/2}(z)}{g_{3/2}(z)} + \left(z^{-1} e^{\varepsilon_0/T} - 1 \right) \ln \left(1 - z e^{-\varepsilon_0/T} - 1 \right) \right\} - \varepsilon_0$$

is the integrated latent heat of vaporization.

A proposal made long ago was to consider the λ-transition in liquid helium as a second-order phase transition since there is no latent heat in the transition from helium I to helium II.[8] The coexistence of helium I and helium II has shown that there is a latent heat between these two phases. The fact that the latent heat vanishes along the λ-curve, connecting the vapor pressure curve with the melting pressure curve, does not lend itself to the usual classification scheme of phase transitions that was proposed by Ehrenfest in 1933. Although we shall continue to refer to it as a λ-transition, because of the deep-rooted usage of the term, we shall nevertheless bear in mind that what we are actually talking about is a hybrid between a first- and a second-order phase transition. There is no change in symmetry at the transition point, as there would be in a second-order phase transition, and yet, there are no discontinuities in the first-order derivatives of the free energy. The latent heat and discontinuity in the molar volume vanish at the critical point but are nonzero on either side of it. It is precisely the latter, and with it the disappearance of the distinction between the two phases, that makes it look like a critical point of a first-order phase transition.

Therefore, if the phase transition is a hybrid between first- and second-order, both the distinction between the phases should disappear,

$$V_m^I = V_m^{II}, \tag{7.63}$$

and, at the same time, no heat should be emitted or absorbed so that

$$\hat{S}^I = S^{II}. \tag{7.64}$$

In other words, at the critical point the two phases become indistinguishable and the first-order transition, which is characterized by discontinuities in the *molar* potentials, degenerates into a second-order transition. However, unlike a critical point of a first-order phase transition there are discontinuities in the molar potentials on *both* sides of the transition point. We have no indication that the condensed phase becomes unstable for temperatures above T_λ as it would if the transition were first-order.

Now, the global Clapeyron relation is compatible with both (7.63) and (7.64) and imposes yet a further condition:

$$P^I - P^{II} = \text{const.} \tag{7.65}$$

---

[8]This latent heat should be kept distinct from the latent heat between liquid helium and its vapor.

At the transition point, the system will therefore lose one degree-of-freedom. Not only does (7.63) and (7.64) imply (7.65) but the converse would imply that the derivative dP/dT is indeterminate.

Condition (7.63) requires

$$\bar{n}^{\mathrm{II}} = \frac{N}{2} = \bar{n}^{\mathrm{I}} \tag{7.66}$$

on account of the fact that the total number of particles $N = \bar{n}^{\mathrm{I}} + \bar{n}^{\mathrm{II}}$ is constant. From (7.66) we can determine the critical value of the fugacity together with the λ-temperature, T_λ. The first equality in (7.66) gives the expression $z_\lambda = N e^{\varepsilon_0/T_\lambda}/(N + 2m_0)$ for the critical value of the fugacity while the second equality gives the expression

$$T_\lambda = \frac{h^2}{2\pi M} \left[\frac{2V}{N} g_{3/2}(z_\lambda) \right]^{2/3} \tag{7.67}$$

for the λ-temperature. Since $N \gg 2m_0$ and $\varepsilon_0/T_\lambda \ll 1$, $z_\lambda \approx 1$ and (7.67) becomes comparable to Einstein's expression (7.57). Instead of the critical value $T_0 = 3.14$ K we now obtain a λ-temperature $T_\lambda = 1.98$ K due to the fact that the density has been cut in half. Actually, the true λ-temperature will be slightly greater than this value because $g_n(z)$ is a bounded, positive, and monotonically increasing function of its argument for all values of $z \in [0, 1]$. Furthermore, we observe that (7.67) is independent of the explicit form of $\bar{n}^{\mathrm{II}}$. Nevertheless, it does effect the critical value of the fugacity, z_λ.

It is advantageous, at this point, to contrast our approach with the conventional formulation of the Bose–Einstein condensation that can be found in almost any textbook on statistical mechanics. In conventional formulation, the critical point is *defined* by the condition that $\mu = 0$. For a fixed number of particles and volume this defines a unique temperature T_0 which is given by (7.57). It is then argued that the chemical potential remains essentially zero for temperatures $T < T_0$. This causes strange things to happen below the critical temperature: the pressure becomes independent of the volume and the number of particles varies with the temperature! Hence, a changeover has been made from a system in which the particles are conserved into one in which they are not!

The formula

$$N = \frac{2}{\sqrt{\pi}} \frac{V}{\lambda^3} \int_0^\infty \frac{\sqrt{x}}{z^{-1}e^x - 1}\, dx, \tag{7.68}$$

where the variable of integration is $x = \varepsilon/T$, determines implicitly the chemical potential as a function of the temperature T and the density N/V. Setting $\mu = 0$ singles out a particular temperature, T_0, in terms of the density. For $T < T_0$, Eq. (7.68) has no negative solutions which for a Bose system must be negative for *all* temperatures. In order to circumvent this problem, it is claimed that it was not really legitimate to replace the sum

$$N = \sum_{\mathbf{p}} \frac{1}{z^{-1}e^{\varepsilon_{\mathbf{p}}/T} - 1},$$

where the sum extends over all momentum states $\mathbf{p}$, by an integral in order to arrive at (7.68), since it attributes zero weight to the ground state. The ground state, $\mathbf{p} = 0$, must be removed from the sum before it is converted into an integral so that the total number of particles is the sum of those in the ground state, $N_{\varepsilon=0}$, and the remaining particles in the excited states, $N_{\varepsilon>0}$:

$$\begin{aligned} N &= N_{\varepsilon=0} + N_{\varepsilon>0} \\ &= \frac{1}{z^{-1} - 1} + \frac{2}{\sqrt{\pi}} \frac{V}{\lambda^3} \int_0^\infty \frac{\sqrt{x}}{z^{-1}e^x - 1} \, dx, \end{aligned} \qquad (7.69)$$

where we see that the total number of particles in (7.68) are those with energies $\varepsilon > 0$. But with $\mu = 0$, the number of particles in the ground state is infinite! This is what we meant by going from a system of a finite number of particles to one of an infinite number of particles.

Therefore, the chemical potential cannot go to zero but to some small but finite value which will give the first term in (7.69) the desired finite value. Not only does this change the definition of the critical point but, moreover, it is contradictory to the fact that (7.68) is an equation of state which determines the chemical potential as a function of the temperature and density. Since the chemical potential has a nonvanishing value for the ground state, it must also be nonvanishing for the excited states as well. This means that (7.68) is now to be interpreted as an equation for $N_{\varepsilon>0}$ as a function of the temperature and an arbitrary small value of the chemical potential. But this is contradictory to our interpretation of this formula prior to the critical point where it was the chemical potential which was a function of the temperature and density!

Inverting this relation to give $N_{\varepsilon>0}$ as a function of the temperature and an arbitrary value of the chemical potential is *ad hoc* to say the least. A theory should not be bent so that it conforms to observation. And if we retain our original interpretation of (7.68) as an equation of state for the chemical potential, then there is no way in which $N_{\varepsilon>0}$ or $N_{\varepsilon=0}$ can vary. It is for this very reason that we introduced the notion of a phase equilibrium between helium I and helium II.

The requirement that the entropy be continuous at the transition point, (7.64), provides the relation

$$\frac{E^{\mathrm{I}}}{V} + P^{\mathrm{I}} = \tfrac{5}{2}P^{\mathrm{I}} = P^{\mathrm{II}} + \frac{E^{\mathrm{II}}}{V} \qquad (7.70)$$

between the pressures of the two phases. The first equality in (7.70) shows that we are treating liquid helium I as a gas of free, structureless particles. This is not as unrealistic as it would be for any other liquid since the viscosity of liquid helium I shows a *positive* temperature coefficient, the viscosity varying as $\sqrt{T}$ rather than that of a liquid where it would vary as $e^{-A/T}$ with A representing the activation energy that atoms must have in order to surmount the potential barriers that are normally found in liquids. The question arises as to what we are considering liquid helium II to be.

The energy $E^{\text{II}} = \bar{n}^{\text{II}}\varepsilon_0$ can be determined from the continuity of the entropy, (7.64). We remarked earlier that ε_0 should be the average energy of the low-lying translational states which can hardly be thought of as comprising a macroscopic phase. Numerical evaluation shows that $\varepsilon_0 \sim 10^{-16}$ ergs at the λ-point so that it must consist of excitations (rotons, phonons, etc.), other than particle excitations. It points to the conclusion that liquid helium II is not a single ground state or a group of low-lying states but other types of excitations are involved which have thermodynamic consequences.[9] For instance, the specific heat of liquid helium is proportional to T^3 (instead of the expected $T^{3/2}$ as it would be for an ideal gas), analogous to that of a phonon picture of a Debye solid at low temperatures. At any rate, it is this phase which accounts for the thermodynamic properties of liquid helium below the λ-point. If we are willing to assume that this phase can still be treated as an ideal gas at the λ-point, then $P^{\text{II}} = 2E^{\text{II}}/3V$ and $P^{\text{I}} = 2E^{\text{I}}/3V$ and (7.70) becomes the condition for the continuity of the energy,

$$E^{\text{I}} = E^{\text{II}},$$

at the λ-point.

This condition on the energy of phase II implies that it has grown into a full-fledged macroscopic phase at the critical point. It also points to the fact that our original model of this phase, as a conglomeration of geometric distributions, is inadequate to describe the actual excitations that are occurring. Other types of elementary excitations have contributed to the energy in addition to the excitation of low-lying energy levels. A more realistic description would take into account the statistical attractions by relaxing the invariancy of the *a priori* probabilities. However, these modifications will not affect our conclusions, which are entirely macroscopic and independent of the specific nature of phase II.

A basic feature of the λ-transition is the discontinuity in the specific heat curve, which is infinite, rather than being a finite jump, at the λ-point. The specific heat which possesses the discontinuity is the specific heat at saturation

$$C_{\text{sat}} = C_p - \alpha V T \left(\frac{dP}{dT}\right)_{\text{v.p.}},$$

where $\alpha = (1/V)(dV/dT)_P$ is the coefficient of thermal expansion and $(dP/dT)_{\text{v.p.}}$ is the slope of the vapor pressure curve. The characteristic λ-shape is shown in Fig. 7.2.

There is neither a discontinuity in the heat capacity at constant value, C_V, nor in the heat capacity at constant pressure, C_P. The divergence of C_{sat} has nothing whatsoever to do with a proposed cusp in C_V. An anomaly in

---

[9]There have been various attempts, with varying degrees of success, to modify the energy spectrum of an ideal Bose–Einstein gas in order to fit the experimental data. In the Landau model, an energy gap of $\sim 4T_\lambda$ seems to account for the experimental values quite well, which are still larger than our ε_0.

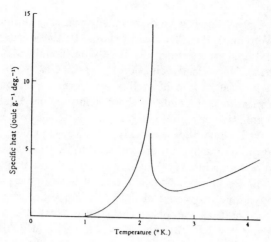

Figure 7.2: The specific heat of liquid helium under its saturated vapor pressure.

the coefficient of thermal expansion near the λ-point has been attributed as the cause of the anomaly in C_{sat}. In order to analyze such an anomaly in α, we would have to take physical interactions into consideration, as we would have to do if we want to gain a more detailed understanding of what actually comprises phase II of liquid helium. Yet, our analysis has shown that our representation of liquid helium II by the geometric distribution is too naïve, and one way of incorporating interactions is to forgo the invariancy of the *a priori* probabilities, as we have done in §3.3.2 and will come back to in §7.7.

7.6 Do Phase Transitions Higher than Second-Order Exist?

In the preceding sections we have seen that probability distributions can enter into phase equilibria just as material phases do. These involve first-order phase transitions, and the probability distributions associated with such phenomena belong to the family which has as members the negative binomial, Pascal, and geometric distributions. The latter has to be excluded on the grounds that the entropy of a single state has no meaning. The distributions possess different *molar* entropies, and when they enter into equilibrium signaled by the equivalence of their chemical potentials, there appears a discontinuity in the molar entropy. This is the hallmark of a first-order phase transition. We would now like to investigate what type of probability distributions can give rise to a second-order phase transition and whether there can exist phase transitions of still higher order.

Consider an array of polar molecules situated on a lattice and aligned in an "up" or "down" position. If n molecules on the m lattice sites point in the upward position, then it will give rise to an entropy of "mixing"

$$S(n) = n \ln \left(\frac{m-n}{n}\right) - m \ln \left(\frac{m-n}{m}\right). \tag{7.71}$$

The entropy achieves its maximum value subject to the constraints imposed at the most probable state $n = \bar{n}$. The constrained maximum yields the thermodynamic entropy, $S(\bar{n})$, which is related to the free energy according to

$$A(T, \bar{n}) = \bar{E}(\bar{n}) - TS(\bar{n}).$$

Since we are considering the case of a vanishing chemical potential, $A = -PV$. We do not assume that the average energy, $\bar{E}$, is necessarily a linear function of $\bar{n}$; it may happen that the energy needed to rotate a polar molecule in the lattice depends upon how many other molecules are pointing in either direction.

The equilibrium value of $\bar{n}$ can be obtained by minimizing the free energy. We then obtain the stationary condition

$$A' = \bar{E}' - T \ln \left(\frac{m-\bar{n}}{\bar{n}}\right) = 0,$$

where the prime means differentiation with respect to $\bar{n}$. Rearranging we get

$$\bar{n}(T) = \frac{m}{e^{\bar{E}'/T} + 1} \tag{7.72}$$

for the average number of particles as a function of the temperature. This type of "Fermi–Dirac distribution" was to be expected since the form of the entropy of mixing, (7.71), belongs to the binomial distribution.

Evaluating the entropy of mixing (7.71) at (7.72) results in

$$S(\bar{n}) = \frac{\bar{E}'}{T}\bar{n} + m \ln \left(1 + e^{-\bar{E}'/T}\right),$$

and inserting this expression into the free energy gives

$$A = \bar{E} - \bar{E}'\bar{n} - mT \ln \left(1 + e^{-\bar{E}'/T}\right). \tag{7.73}$$

For a vanishing chemical potential, $A = -PV$, and this will provide us with an analogy to the attractive term in van der Waals' equation of state. We leave the option open that $\bar{E}$ is not first-order homogeneous in $\bar{n}$ and show that it is responsible for critical behavior that would arise in a second-order phase transition.

Using the standard Bragg–Williams model of an order–disorder transition, we will show that the difference $\bar{E} - \bar{n}\bar{E}'$ is analogous to the attractive term in the van der Waals' equation of state, whose probabilistic origins will be discussed in the next section.

The entropy of mixing gives rise to the law of error

$$f(n) = \binom{m}{n} \left(\frac{1}{1 + e^{\bar{E}'/T}}\right)^n \left(\frac{1}{1 + e^{-\bar{E}'/T}}\right)^{m-n}. \qquad (7.74)$$

This is the maximum value of a binomial distribution whose *a priori* probabilities have been evaluated in terms of $\bar{n}$ by the method of maximum likelihood [cf. §3.3.2].

From the definition of the heat capacity and the expression for the entropy we obtain

$$C_V = T \left(\frac{\partial S}{\partial T}\right)_V = \bar{E}' \frac{d\bar{n}}{dT}, \qquad (7.75)$$

where

$$T \frac{d\bar{n}}{dT} = \frac{\bar{n}(m - \bar{n})\bar{E}'}{mT + \bar{n}(m - \bar{n})\bar{E}''}.$$

It is therefore apparent that in order for there to be a discontinuity in the specific heat, $\bar{E}$ must be a strictly *concave* function of $\bar{n}$.

The state $\bar{n} = m/2$ corresponds to the state of maximum disorder where

$$S_{\text{max}} = m \ln 2,$$

meaning that there are 2^m equiprobable configurations available to the polar molecules.

We now define the order parameter:

$$\lambda \equiv \frac{m - 2\bar{n}}{m}.$$

In order to determine the critical point, we cannot avail ourselves of expression (7.73) for the free energy since we have already used the fact that λ is a function of the temperature. Rather, we must return to the expression

$$A = \bar{E} - T\left\{S_{\text{max}} - \frac{m}{2}[(1 + \lambda)\ln(1 + \lambda) + (1 - \lambda)\ln(1 - \lambda)]\right\}$$

for the free energy and differentiate it with respect to λ holding the temperature constant. We then obtain

$$\left(\frac{\partial A}{\partial \lambda}\right)_T = m\left(T \tanh^{-1}\lambda - \tfrac{1}{2}\bar{E}'\right) = 0, \qquad (7.76)$$

which is a camouflaged way of writing the stationary condition (7.72). One solution to (7.76) is $\lambda_0 = 0$, which corresponds to the state of maximum disorder, $\bar{n} = m/2$. New solutions appear below the critical temperature, T_λ, which is determined from the bifurcation point condition

$$\left(\frac{\partial^2 A}{\partial \lambda^2}\right)_T = m\left\{\frac{m}{4}\bar{E}'' + \frac{T}{1 - \lambda^2}\right\} = 0. \qquad (7.77)$$

In the A, λ-plane there is a horizontal point of inflexion at $\lambda_0 = 0$ when the critical temperature,

$$T_\lambda = -\frac{m}{4}\bar{E}'',$$

is reached.

The condition for a bifurcation (7.77), which is a point of inflexion of the free energy as a function of the order parameter, implies a discontinuity in the specific heat, (7.75). Hence, any phase transition of higher order than second cannot occur since the bifurcation condition (7.77) always implies a discontinuity in the specific heat whereas higher order phase transitions would require its continuity with the appearance of a discontinuity in a still yet derivative of higher order. As an illustration, let us consider the classical Bragg–Williams model.

Developing the internal energy E in a truncated Taylor series about equilibrium gives

$$E(n) = E_0 + E'(n - \bar{n}) + \frac{\omega}{2}(n - \bar{n})^2$$

on the assumption that deviations from the equilibrium value E_0 are small. The parameter ω is assumed to be constant and independent of $\bar{n}$. Taking the average using the binomial distribution (7.74) results in

$$\bar{E}(\bar{n}) = \frac{\omega}{2}\frac{\bar{n}}{m}(m - \bar{n}) + \text{const.}$$

This is the Bragg–Williams expression which is obtained by making the approximation

$$\sum_{x=0}^{\bar{n}}(m - 2x) \approx \bar{n}(m - \bar{n}).$$

It is apparent that its validity depends upon the condition that $\bar{n} \gg 1$. Here, we have shown it to be a consequence of expanding the internal energy in a truncated series, implying small fluctuations, and averaging with respect to the binomial distribution.

Equipped with this expression for the average energy, the free energy (7.73) takes on the explicit form

$$A = \frac{\omega \bar{n}^2}{2 m} - mT \ln\left(1 + e^{-\omega\lambda/2T}\right). \qquad (7.78)$$

Since $A = -PV$, the first term is analogous to the attractive term in van der Waals' equation. In the usual case where the average energy is a linear function of the average number of molecules, the pressure would be given by

$$P = \frac{mT}{V}\ln\left(1 + e^{-\omega\lambda/2T}\right).$$

But since $\bar{E}$ is a nonlinear function, the entropy is

$$S(\bar{n}) = \frac{\bar{n}\bar{E}'}{T} + m\ln\left(1 + e^{-\omega\lambda/2T}\right)$$

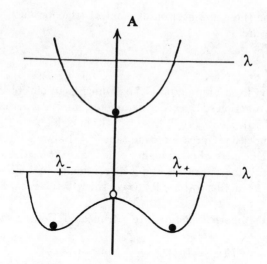

Figure 7.3: The free energy as a function of the order parameter above and below the transition point. The minimum transforms into a maximum with two new minima appearing at equal distances from the original equilibrium state.

which must be equivalent to the Euler relation:

$$S(\bar{n}) = \frac{\bar{E} + PV}{T}.$$

When the first term in (7.78) is subtracted from the average energy, it gives precisely the $\omega\lambda/2$ that is needed to form a proper error law.

The stationary states are determined from the extremum condition of the free energy (7.78) with respect to the order parameter. They are given by the implicit relation

$$\lambda = \tanh\left(\frac{T_\lambda}{T}\lambda\right), \tag{7.79}$$

where $T_\lambda = \omega/4$, which is known as the Curie temperature. In the A, λ-plane, there is a horizontal point of inflection at $\lambda_0 = 0$ when the critical temperature T_λ is reached. As the temperature is lowered below T_λ, the free energy passes from a minimum to a maximum at $\lambda_0 = 0$, as shown in Fig. 7.3, with two new minima occurring at points equidistant from the origin since the free energy is invariant under the transform $\lambda \to -\lambda$.

Provided the temperature satisfies the condition $1 - T/T_\lambda \ll 1$, approximate solutions to (7.79) are

$$\lambda_\pm^2 \simeq 3\left(\frac{T_\lambda - T}{T_\lambda}\right).$$

As the Curie temperature is approached from below, the heat capacity is

$$C_V(T \uparrow T_\lambda) = \lim_{z \downarrow 0} \frac{3m(1 - 3z)}{2(1 + z)} = \frac{3}{2}m,$$

where $z = 1 - T/T_\lambda$. Since $C_V(T \downarrow T_\lambda) = 0$, the heat capacity is discontinuous at the transition point.

At the state of maximum disorder, corresponding to $\lambda_0 = 0$, the binomial distribution (7.74) reduces to

$$f_0(n) = \binom{m}{n} 2^{-m}$$

for which the *a priori* probabilities are equal. Below the transition point, there is a splitting of this probability distribution into two new binomial distributions

$$f_\pm(n) = \binom{m}{n} \left(\frac{1}{1 + e^{\pm\alpha}}\right)^n \left(\frac{1}{1 + e^{\mp\alpha}}\right)^{m-n}, \qquad (7.80)$$

where $\alpha = 2T_\lambda|\lambda|/T$. The new stationary states,

$$\bar{n}_\pm = \frac{m}{e^{\pm\lambda} + 1},$$

differ from one another by an integral "zero-point" term, m. As we have seen in §7.2.1, the average number of particles belonging to the negative binomial and Pascal distributions also differ by an integral zero-point term. The corresponding phases coexist and can come to equilibrium; there is a latent heat precisely because the molar entropies are unequal. In the present case, there is a bifurcation of one solution to the stationary state equation into two new solutions as a characteristic parameter is varied. The original solution becomes unstable while the two new solutions are stable. This is characteristic of a Curie point rather than the critical point at which the distinction between the two phases disappears.

Interchanging the *a priori* probabilities in the two binomial distributions (7.80) interconverts the two distributions. Since the potential wells are of equal depth, it is commonly believed that the populations of both states will be equally probable. In fact, the two states should be different manifestations of the same phenomenon. But this would seem to go against the frugality that nature usually displays and, in fact, we will now show this not to be the case.

In order to accomplish our goal, we will employ a method developed by the American statistician Abraham Wald in the early forties called "sequential analysis" which can be described as a "decide as you go strategy." The method involves testing a simple hypothesis against a simple alternative.[10] We perform independent repetitions of a specified experiment and after experiment we form the likelihood ratio. A test is said to be a "sequential likelihood ratio test" if

---

[10]For a more detailed account, see the Bibliographic Notes.

there are fixed upper and lower limits to the likelihood function such that if this function exceeds either limit, one hypothesis is accepted and the other rejected.

In the present case, we make the hypothesis that the prior probability is $p = \bar{n}_+/m$ against the simple alternative $p = n_-/m$. The probability p is the probability that a polar molecule points in the upward direction. The difference between the *a priori* probabilities, $q - p$, induces a drift toward one of the two stationary states $\bar{n}_\pm$. If $q > p$, the drift will be two $\bar{n}_+$, while in the opposite case the drift will be toward $\bar{n}_-$. A characteristic feature of a second-order phase transition is that the two *a priori* probabilities are never equal so there will always be a drift to one or the other stationary state below the Curie point. This is surprising inasmuch as an external field is not necessary to destroy the symmetry between the two states in which one state becomes metastable.

The log-likelihood ratio is

$$\mathcal{L} = \ln \frac{f_+(n)}{f_-(n)} = m\alpha \left(1 - 2\frac{n}{m}\right),$$

which shows that $\mathcal{L}$ is, itself, a binomial random variable. Physically speaking, the log-likelihood ratio is a measure of the drift in one direction or the other. The stationary states $\lambda_\pm$ fix the limits on the log-likelihood ratio as

$$-\alpha m|\lambda| \leq \mathcal{L} \leq \alpha m|\lambda|.$$

Since, in general, these limits will not be integers, there will be an "excess" across the boundaries. Wald was only able to obtain approximate results by neglecting the excess across the boundaries.

Let $\pi_-(p)$ be the probability that the system enters state $\bar{n}_-$ when mp is the true portion of molecules pointing in the upward direction. Since the events $\mathcal{L}/m \geq \alpha|\lambda|$ and $\mathcal{L}/m \leq -\alpha|\lambda|$ are mutually exclusive, we have, for $p \neq q$, Wald's identity:

$$\pi_+ \overline{e^{\vartheta_0 \mathcal{L}/m}}^{\{\mathcal{L}/m \geq \alpha|\lambda|\}} + \pi_- \overline{e^{\vartheta_0 \mathcal{L}/m}}^{\{\mathcal{L}/m \leq -\alpha|\lambda|\}} = 1, \qquad (7.81)$$

where the first barred quantity is the conditional expectation value of $e^{\vartheta_0 \mathcal{L}}$ under the condition that $\mathcal{L} \geq \alpha m|\lambda|$ and ϑ_0 is the nonzero root of the equation

$$\sum_{n=0}^{m} \binom{m}{n} e^{\vartheta \mathcal{L}} p^n q^{m-n} = \left(qe^{\alpha\vartheta}\right)^m \sum_{n=0}^{m} \binom{m}{n} \left(\frac{p}{q} e^{-2\alpha\vartheta}\right)^n = 1.$$

If there is no drift, $p = q$, and there is clearly one root, $\vartheta = 0$. However, if $p \neq q$, there is a unique second real root to the equation

$$\left(qe^{\alpha\vartheta_0} + pe^{-\alpha\vartheta}\right)^m = 1$$

given by

$$e^{\alpha\vartheta} = \frac{p}{q}. \qquad (7.82)$$

From this we deduce that ϑ_0 will always have the opposite sign of the drift, $q - p$. We will now derive this result using Wald's approximations and show in this case that an exact result can be obtained.

Neglecting the excess across the boundaries and writing $\mathcal{L}/m \simeq \alpha|\lambda|$ when $\mathcal{L}/m \geq \alpha|\lambda|$ and $\mathcal{L}/m \simeq -\alpha|\lambda|$ when $\mathcal{L}/m \leq m\alpha|\lambda|$, Wald's identity (7.81) can be approximated by

$$\pi_+ e^{\vartheta_0 \alpha|\lambda|} + \pi_- e^{-\vartheta_0 m\alpha|\lambda|} \simeq 1.$$

Introducing the condition $\pi_+ + \pi_- = 1$ yields

$$\pi_- \simeq \frac{e^{\vartheta_0 \alpha|\lambda|} - 1}{e^{\vartheta_0 \alpha|\lambda|} - e^{-\vartheta_0 \alpha|\lambda|}}. \tag{7.83}$$

The log-likelihood ratio per "observation," $\mathcal{L}/m$, is a binomial random variable with expectation $\overline{\mathcal{L}}/m = \alpha(q - p)$, where $\bar{n}_+/m \leq p\bar{n}_-/m$. It takes on the value α with probability $1 - p$ and $-\alpha$ with probability p. We thus have

$$pe^{-\alpha\vartheta_0} + (1 - p)e^{\alpha\vartheta_0} = 1,$$

whose solution is precisely the exact result (7.82) which we found. Also observe that the sign of the expectation of the log-likelihood ratio is always opposite to that of ϑ_0.

Introducing (7.82) into (7.83) gives the absorption probabilities:

$$\pi_-(p) = \frac{(p/q)^{|\lambda|} - 1}{(p/q)^{|\lambda|} - (q/p)^{|\lambda|}}$$

and

$$\pi_+(p) = \frac{1 - (q/p)^{|\lambda|}}{(p/q)^{|\lambda|} - (q/p)^{|\lambda|}},$$

since $\pi_- + \pi_+ = 1$. If the *a priori* probability of finding a polar molecule pointing up is zero, $\pi_+(0) = 1$. In the opposite case, the system will enter the state $\bar{n}_-$ with probability 1. In the limiting case where $p = \bar{n}_+/m$,

$$\pi_+\left(p = \frac{\bar{n}_+}{m}\right) = \frac{e^{\alpha|\lambda|} - 1}{e^{\alpha|\lambda|} - e^{-\alpha|\lambda|}} > \pi_-.$$

In the other limiting case where $p = \bar{n}_-/m$,

$$\pi_-\left(p = \frac{\bar{n}_-}{m}\right) = \frac{e^{\alpha|\lambda|} - 1}{e^{\alpha|\lambda|} - e^{-\alpha|\lambda|}} > \pi_+.$$

Therefore, the probabilities of entering either stationary state depends upon the sign of the drift $q - p$. Since $p \neq q$, there will always be a tendency to evolve to one or the other stationary state. Although the two states are entirely equivalent from a macroscopic point of view, our analysis of a random walk with absorbing boundaries has shown that one of these two states will be preferred by the system as the temperature is lowered below the Curie temperature.

Finally, the expected number of "steps" to absorption or the number of "observations" that are required to reach a decision is

$$|\lambda| \frac{2 - (q/p)^{|\lambda|} + (p/q)^{|\lambda|}}{(p/q)^{|\lambda|} - (q/p)^{|\lambda|}}.$$

In the case of a vanishing drift this number is indeterminate while it tends to $|\lambda|$, the magnitude of the order parameter, as either p or q tend to 1. The latter implies that we know, with almost certainty, the direction in which the system is evolving.

7.7 Statistical Derivation of van der Waals' Equation

In the last section, we remarked on the analogy between the nonlinear term in the expression for the average energy and the attractive term in van der Waals' equation. In §3.3.2 we showed how statistics can simulate physical interactions by taking into account the fact that the presence of one particle in a given region will influence the probability of there being another in the same region. In other words, the *a priori* probabilities will vary from one observation to the next. The lack of ideality will naturally lead to probability distributions which do not belong to the exponential family of distributions which entail maximum entropy with respect to the constraints. However, in the limit of a large volume or parent population, we may approximate these distributions by exponential ones to which small corrective terms have been appended. This will permit us to determine the nonideal contributions to the entropy and, as an illustration, we shall give a probabilistic derivation of van der Waals' equation of state.

Let us recall from §3.3.2 that

$$f_{\mathrm{h}}(n) = \binom{n_0}{n} \binom{V_0/\sigma - n_0}{V/\sigma - n} \Big/ \binom{V_0/\sigma}{V/\sigma}$$

is the hypergeometric distribution for a hard sphere gas. In the limit as the excluded volume $\sigma \to 0$, the hypergeometric distribution transforms into the binomial distribution

$$f_{\mathrm{b}} = \binom{n_0}{n} \left(\frac{V}{V_0}\right)^n \left(\frac{V_0 - V}{V_0}\right)^{n_0 - n}.$$

This is the probability for finding n particles in the subvolume V; the *a priori* probability of finding a particle in the subvolume is independent of how many other particles there may already be in V. We recall that the maximum likelihood estimate of the subvolume V is given by the homogeneity condition:

$$\frac{n}{V} = \frac{n_0}{V_0}. \tag{7.84}$$

For large values V_0/σ, the hypergeometric distribution f_h is close to the binomial distribution f_b in the sense that

$$f_b \left(1 - \frac{n}{V}\sigma\right)^n \left(1 - \frac{n_0 - n}{V_0 - V}\sigma\right)^{n_0 - n} < f_h < f_b \left(1 - \frac{n}{V}\sigma\right)^{-n_0} . \tag{7.85}$$

A comparison of the two distributions, for large V_0/σ, can be formulated in terms of the testing of statistical hypotheses which we employed in the previous section. Denoting the lower limit in (7.85) by $f_h^<$, the log-likelihood ratio is given by

$$\begin{aligned}
\mathcal{L} &\equiv \ln\left(\frac{f_h^<}{f_b}\right) \\
&= n \ln\left(1 - \frac{n}{V}\sigma\right) + (n_0 - n)\ln\left(\frac{n_0 - n}{V_0 - V}\sigma\right).
\end{aligned}$$

Maximizing the log-likelihood ratio with respect to the parameter V results in the homogeneity condition (7.84). Imposing this condition on the log-likelihood ratio gives

$$\mathcal{L}_{\max} = n_0 \ln\left(1 - \frac{n_0}{V_0}\sigma\right), \tag{7.86}$$

since $\partial^2 \mathcal{L}/\partial V^2 < 0$.

The condition of homogeneity symmetrizes the upper and lower bounds of the ratio of the two probability distributions:

$$\left(1 - \frac{n_0}{V_0}\sigma\right)^{n_0} < \frac{f_h}{f_b} < \left(1 - \frac{n_0}{V_0}\sigma\right)^{-n_0}.$$

Alternatively, had we chosen the upper limit $f_h^>$ in (7.85) to compare with the binomial distribution, the logarithm of the likelihood ratio $f_b/f_h^>$ would have immediately yielded the maximum log-likelihood ratio (7.86) without having to maximize it. But we would not have known that it is, in fact, the maximum log-likelihood value.

We can now appreciate the maximum log-likelihood ratio (7.86) as the non-ideal gas contribution to the entropy. Adding this contribution to the entropy of an ideal gas, $S^0(V_0) = n_0 \ln V_0 +$ const., gives the total entropy

$$S(V_0) = n_0 \ln(V_0 - n_0\sigma) + \text{const.} \tag{7.87}$$

of configuration space. The nonideal contribution to the entropy in (7.87) will give rise to the repulsive term in van der Waals' equation of state.

In order to obtain the attractive contribution to van der Waals' equation, we are led to consider the remaining part of phase space which is occupied by a system with $2m_0$ degrees-of-freedom, just as we did in §5.2.1 in our discussion of the temperature ensemble. In the volume of phase space, demarcated by the energy E_0, we consider a subvolume where the phases are limited by an energy E. We want to determine the probability that there will be exactly $2m$ degrees-of-freedom in this subspace.

If we suppose that each pair of the degree-of-freedoms occupies a unit volume in phase space, with energy ε, then this probability will be proportional to the product of the number of ways in which m indistinguishable half degrees-of-freedom can be distributed over E/ε different "cells" and the number of ways that $m_0 - m$ indistinguishable objects can be distributed over $(E_0 - E)/\varepsilon$ cells with empty cells being admissible. Multiplying the product by the number of ways that m indifferent objects can be distributed into E/ε cells gives the probability

$$f_\mathrm{p} = \binom{-E/\varepsilon}{m} \binom{-(E_0 - E)/\varepsilon}{m_0 - m} \Big/ \binom{-E_0/\varepsilon}{m_0}.$$

This is the Pólya distribution, which we derived in §3.3.2.

For large values of E_0/ε, the Pólya distribution is close to the corresponding binomial distribution,

$$f_\mathrm{b} = \binom{m_0}{m} \left(\frac{E}{E_0}\right)^m \left(\frac{E_0 - E}{E}\right)^{m_0 - m},$$

in the sense that

$$f_\mathrm{b} \left(1 + \frac{m_0}{E_0}\varepsilon\right)^{-m_0} < f_\mathrm{p} < f_\mathrm{b} \left(1 + \frac{m}{E}\varepsilon\right)^m \left(1 + \frac{m_0 - m}{E_0 - E}\varepsilon\right)^{m_0 - m}. \qquad (7.88)$$

Calling the upper limit in (7.88) $f_\mathrm{p}^>$, we form the log-likelihood ratio

$$\begin{aligned} \mathcal{L} &= \ln\left(\frac{f_\mathrm{b}}{f_\mathrm{p}^>}\right) \\ &= -m \ln\left(1 + \frac{m}{E}\varepsilon\right) - (m_0 - m) \ln\left(1 + \frac{m_0 - m}{E_0 - E}\varepsilon\right). \end{aligned}$$

The maximum of the log-likelihood ratio with respect to E gives

$$\frac{E}{m} = \frac{E_0}{m_0},$$

which asserts that the degrees-of-freedom are distributed homogeneously throughout phase space. Introducing this maximum likelihood value back into the log-likelihood ratio results in

$$\mathcal{L}_{\max} = -m_0 \ln\left(1 + \frac{m_0}{E_0}\varepsilon\right). \qquad (7.89)$$

The maximum log-likelihood condition again symmetrizes the limits in (7.88):

$$\left(1 + \frac{m_0}{E_0}\varepsilon\right)^{-m_0} < \frac{f_\mathrm{p}}{f_\mathrm{b}} < \left(1 + \frac{m_0}{E_0}\varepsilon\right)^{m_0}.$$

And again we find a self-consistency condition: if we had chosen the lower limit in (7.88) to approximate the Pólya distribution, we would have seen that $\mathcal{L} = \ln(f_\mathrm{b}/f_\mathrm{p}^<)$ is identical to the maximum likelihood value (7.89).

The negative of (7.89) coincides with the nonideal entropy per particle divided by Boltzmann's constant. Adding this contribution to the ideal gas contribution $S^0(E_0) = m_0 n \ln E_0 +$ const. gives the total phase entropy

$$S(E_0) = m_0 n \ln(E_0 + m_0 \varepsilon). \tag{7.90}$$

With the aid of the second law,

$$\left(\frac{\partial S}{\partial E_0}\right)_V = \frac{m_0 n_0}{E_0 + m_0 \varepsilon} = \frac{1}{T}, \tag{7.91}$$

we obtain

$$E_0 = m_0 n_0 T \left(1 - \frac{\varepsilon}{n_0 T}\right) \simeq m_0 n_0 T e^{-\varepsilon/nT}, \tag{7.92}$$

on account of the fact that $\varepsilon/n_0 \ll T$.

Because $\varepsilon > 0$, the energy will be smaller than its equipartition value, $m_0 n_0 T$. This decrease in energy can be attributed to the potential energy of interaction, φ, between any one molecule and all the others in the system. It leads us to associate the energy, ε, with the potential energy, ϕ:

$$\frac{\varepsilon}{n_0} \leftrightarrow \frac{\phi}{2},$$

where the factor of $\frac{1}{2}$ is introduced because each interaction involves two molecules. Since the potential energy is proportional to the density, we can write

$$\varepsilon = \frac{1}{2} n_0 \varphi \equiv \frac{a n_0^2}{V_0},$$

where the coefficient of proportionality, a, is independent of the density.

Introducing this value of the unit energy of phase space, ε, into the equality in (7.92) gives the expression for the energy as

$$E_0 = m_0 n_0 T - \frac{a m_0 n_0^2}{V_0}. \tag{7.93}$$

We pause to observe that the specific heat is unaltered: the specific heat of a van der Waals' gas is no different than the specific heat of an ideal gas. There is also no difference if we use expression (7.93) to define the heat capacity, $C_V = (\partial E_0/\partial T)_V = m_0 n_0$, or the entropy expression (7.90), in which case we have $C_V = T(\partial S/\partial T)_V$, on account of the second law, (7.91).

The final step in the derivation of van der Waals' equation requires the construction of the free energy from (7.93) and (7.87):

$$A = m_0 n_0 T - \frac{a m_0 n_0^2}{V_0} + T n_0 \ln(V_0 - n_0 \sigma) + \text{const.}$$

Differentiating with respect to V_0 and changing sign gives the pressure as

$$P = -\frac{\partial A}{\partial V_0} = \frac{n_0 T}{V_0 - n_0 \sigma} - \frac{a m_0 n_0^2}{V_0^2}$$

or

$$\left(P + \frac{amn^2}{V^2}\right)(V - n\sigma) = nT, \tag{7.94}$$

on account of the condition of homogeneity, (7.84). Equation (7.94) is van der Waals' equation of state.

As a final point, it is worth pointing out that had we used the expressions for the entropy which are derived directly from the probability distributions themselves, we would have obtained concave functions of the form $-x \ln x$. These are the entropies of mixing that result from allocation problems. Identifying the maximum of the log-likelihood ratio with the negative of the entropy leads to concave functions of the form $\ln x$ which is the form found in thermodynamic relations. Consequently, the method of estimating nonexponential probability distributions in terms of their limiting forms of exponential distributions, in which the lowest order correction terms are maintained, has enabled us to transform from one class of concave functions, associated with the microscopic world, to the other class of concave functions, associated with the macroscopic world.

Bibliographic Notes

§7.1

The anecdotes about Einstein can be found in

- A. Pais, *Subtle is the Lord. . . The Science and the Life of Albert Einstein* (Oxford Univ. Press, Oxford, 1982), §23b, §23d.

The Einstein and Stern paper referred to is

- A. Einstein and O. Stern, "Einige Argumente für die Annahme einer molekularen Agitation beim absoluten Nullpunkt," *Ann. d. Phys.* **40**, 551-560 (1913).

In this article Einstein and Stern attempt to discriminate between the two expressions for the average energy of an oscillator

$$\bar{E} = \frac{h\nu}{e^{h\nu/T} - 1}$$

and

$$\bar{E} = \frac{h\nu}{e^{h\nu/T} - 1} + \frac{h\nu}{2}$$

from Eucken's measurements of the specific heat of hydrogen at low temperatures. The second expression comes from Planck's second theory, which corrects the asymptotic expression of the first expression for large T,

$$\bar{E} = T - \frac{h\nu}{2}, \qquad\qquad T \to \infty,$$

so that it would agree with the law of equipartition of energy.

Their procedure was to fix the energy and investigate the behavior of the frequency as the temperature approaches zero. They used diatomic hydrogen and set the average energy equal to the rotational energy. Then solving for T, they found that in the first case $\nu \to 0$ as $T \to 0$ while in the second case the frequency approached the limiting value $h/(2\pi)^2 I$, where I is the moment of inertia. Eucken's data seemed to prefer the second possibility as well as the existence of thermal agitation at absolute zero.

In this paper Einstein and Stern give still a third expression for the average energy,

$$\bar{E} = \frac{h\nu}{e^{h\nu/T} - 1} + h\nu,$$

with the cryptic passage cited in the text. The integral zero point is the topic of the next section.

A lively discussion of the history of liquid helium (including the heated debate between Tisza and Landau) can be found in

- S.G. Brush, *Statistical Physics and the Atomic Theory of Matter, from Boyle and Newton to Landau and Onsager* (Princeton Univ. Press, Princeton, NJ, 1983), §4.9.

§7.2

This section follows our papers,

- B.H. Lavenda and W. Figueiredo, "Mechanism of blackbody radiation," *Int. J. Theoret. Phys.* **28**, 391-406 (1989);

- B.H. Lavenda and W. Figueiredo, "On the law of equipartition of energy for translational motion of excited molecules in equilibrium with thermal radiation," *Z. Naturforsch.* **44A**, 273-277 (1989).

§7.3

The analogy between black radiation and a phase equilibrium was made in

- B.H. Lavenda and J. Dunning-Davies, "On the analogy between first-order phase transitions and blackbody radiation," *Found. Phys. Lett.* **2**, 251-259 (1989).

Mechanical models of phase transitions are described in

- J. Frenkel, *Kinetic Theory of Liquids* (Oxford Univ. Press, Oxford, 1946), Ch. VII.

§7.4

The idea of generalizing Einstein's radiation mechanism by taking into account two-photon absorption is given in

- B.H. Lavenda, "Beyond Einstein's radiation theory," *Found. Phys. Lett.* **1**, 333-341 (1988).

§7.5

This section is based on

- B.H. Lavenda and J. Dunning-Davies, *Bose–Einstein Condensation: A Two Phase Equilibrium at Different Pressures* (unpublished).

§7.6

This section follows

- B.H. Lavenda and J. Dunning-Davies, *On the Theory of Second-Order Phase Transitions* (unpublished).

Wald's sequential analysis of testing statistical hypotheses is given in

- A. Wald, *Sequential Analysis* (Wiley, New York, 1947).

It is also discussed in

- D. Blackwell and M.A. Girshick, *Theory of Games and Statistical Decisions* (Wiley, New York, 1954), §10.4

Good accounts of Wald's theory can be found in

- W. Feller, *An Introduction to Probability Theory and Its Applications*, Vol. II, 2nd ed. (Wiley, New York, 1971), §XVIII.2;

- D.R. Cox and H.R. Miller, *The Theory of Stochastic Processes* (Chapman and Hall, London,1965), §2.3.

Essentially, Wald's theory of sequential analysis is a generalized random walk with absorbing barriers. The basic idea is the following. Certain distributions are defective. If $L(x)$ is a defective distribution, then $L(\infty) < 1$ and the defect $1 - L(\infty)$ represents the probability of termination. To remedy this situation, Cramér introduced a number ϑ_0 such that

$$\int_0^\infty e^{\vartheta_0 x}\, dL(x) = 1,$$

implying that $e^{\vartheta_0 x} L'(x)$ is a proper probability density. We may now translate this into a random walk problem with absorbing barriers where the process terminates.

Consider a random walk on the entire interval $(-\infty, \infty)$ with a probability distribution $F(x)$ having a finite first moment $b \neq 0$. Suppose there exists a number ϑ_0 such that

$$\int_{-\infty}^{\infty} e^{-\vartheta_0 x}\, dF(x) = 1.$$

Since F is a proper probability distribution, $\vartheta_0 = 0$ is certainly one root. However, there may exist another solution. Consider the integral

$$G(\vartheta) = \int_{-\infty}^{\infty} e^{-\vartheta x}\, dF(x) = 1,$$

which exists for $0 < \vartheta < \vartheta_0$. Here $G(\vartheta)$ is a convex function, and if there is no other root, it will have a minimum at $\vartheta = \vartheta_0$. However, if there is a second real root, the minimum is displaced, in this case to the right of $\vartheta = 0$ since ϑ_0 has been assumed to be positive. On account of the convexity of $G(\vartheta)$, the drift $G'(0) = b$ will be negative.

We may apply the foregoing analysis to a random walk X_t, whose value x represents the position of the particle at time t located somewhere between the barriers placed at $-b$ and a where $a, b > 0$. Neglecting the excess across boundaries and writing $X_t \simeq a$ when $X_t \geq a$ and $X_t \simeq -b$ when $X_t \leq -b$, we have

$$\pi_- e^{-b\vartheta_0} + \pi_+ e^{a\vartheta_0} \simeq 1,$$

where the probabilities $\pi_- = \Pr(X \leq -b)$ and $\pi_+ = \Pr(X \geq a)$. This is precisely Wald's approximation.

§7.7

This section follows

- B.H. Lavenda, "Probabilistic derivation of van der Waals' equation," *Found. Phys. Lett.* **3**, 285-290 (1990).

Chapter 8

Kinetic Foundations of Gauss' Error Law

8.1 Introduction

The central theme of this book has been the use of probability theory in statistical physics. We have found that the different forms of statistics are derived from probability distributions which are laws of error for extensive variables leading to the average value as the most probable value of the quantity measured. It is these error laws, rather than Boltzmann's principle, that are of fundamental importance; Boltzmann's principle relating the entropy to the fluctuating component of these error laws follows as a consequence. Yet, we have had to identify the entropy in each particular case from a Markov process in the stationary limit that is characterized by an invariant probability distribution (i.e., one that does not change with time).

For a general derivation of the error law in terms of the entropy, we would like to make no reference to any particular kinetic process. Therefore we are led to consider the full time dependent transition probability and show in the asymptotic time limit that it coincides with the Gaussian error law

$$f(n) = A \exp\left\{ S(n) - S(\bar{n}) - \left(\frac{\partial S}{\partial \bar{n}}\right)_V (n - \bar{n}) \right\}, \tag{8.1}$$

where

$$\left(\frac{\partial S}{\partial \bar{n}}\right)_V = \sum_i \left(\frac{\partial S}{\partial X_i}\right)_{V,\bar{n}} \left(\frac{\partial X_i}{\partial \bar{n}}\right)_V$$

and the X_i denote the extensive independent variables. The concavity property of the entropy is what is responsible for (8.1) being a *bona fide* probability distribution. The entropy shows the monotonic tendency to increase; its derivative is a decreasing function of the extensive variable which remains, however, positive for all values.

Gauss' law looks at the situation in reverse: it considers deviations from the most probable state which defines equilibrium that is characterized by the average value of the extensive quantities. Deviations from this state lower the probability whose maximum value coincides with the state of thermodynamic equilibrium defined in terms of the average values of the extensive variables. Hence, we can already begin to appreciate that any kinetic derivation of Gauss' error law will involve the optimal path for the growth of a spontaneous fluctuation from equilibrium. Since any two nonequilibrium states through which the system will pass at successive instants in time will undoubtedly be correlated statistically, the transition probability is usually path dependent and cannot be expressed as a difference of functions of state. But precisely because we are considering spontaneous fluctuations arising from equilibrium, one of the two states coincides with equilibrium which is statistically uncorrelated with all nonequilibrium states, and hence the transition probability will be path independent so that it can be expressed in terms of differences of a function of state.

Alternatively, if we consider the reverse process of a system evolving toward equilibrium, we find that as the size of the fluctuation decreases, one path—which coincides with the solution to the deterministic rate equation—overwhelmingly maximizes the path probability and, again, we find no statistical correlations between nonequilibrium states. Statistical correlations can, however, be superimposed on the deterministic evolution by taking into account thermal disturbances that place us in the so-called diffusion limit [cf. §7.2.3]. The behavior of the system simulates that of a Brownian motion for which the optimal paths that overwhelmingly maximize the path probability disappear. Only in the asymptotic time limit, where the statistical correlations have had sufficient time to have worn off, do we get a path independent transition probability that is given by the error law (8.1). This establishes the kinetic origin of Gauss' principle.

8.2   Optimal Paths of Evolution

For our purposes it will be sufficient to consider the bivariate Poisson distribution

$$
\begin{aligned}
f(k,\ell) &= \Pr\{n(t+\Delta t) - n(t) = k - \ell\} \\
&= \frac{(g\,\Delta t)^k}{k!} e^{-g\Delta t} \cdot \frac{(r\,\Delta t)^\ell}{\ell!} e^{-r\Delta t},
\end{aligned}
\tag{8.2}
$$

where we recall that the intensity parameters, g and r, stand for the rates of generation and recombination, respectively.

In a certain sense, which will soon be made clear, the bivariate Poisson distribution (8.2) gives latent probabilities for the outcome of a pair of trials. Through the process of observation, we obtain values of k and ℓ which we can compare with their respective expectations, $g\,\Delta t$ and $r\,\Delta t$. As a result of these observations, we acquire new information so that the bivariate Poisson distribution (8.2) transforms into the "conjugate" distribution

$$f^{\#}(k, \ell; \chi) = \frac{e^{\chi(k-\ell)}}{\mathcal{Z}(\chi)} f(k, \ell), \tag{8.3}$$

where

$$\begin{aligned}
\mathcal{Z}(\chi) &= \sum_{k=0}^{\infty} \sum_{\ell=0}^{\infty} e^{\chi(k-\ell)} f(k, \ell) \\
&= \exp\left\{ g \, \Delta t \, (e^{\chi} - 1) + r \, \Delta t \left(e^{-\chi} - 1 \right) \right\}
\end{aligned}$$

is the moment generating function.

The probability distribution (8.3) is again a bivariate Poisson distribution but differs from (8.2) by having modified expectation values. This is made clear by observing that (8.3) is the conjugate bivariate distribution

$$f^{\#}(k, \ell; \chi) = \frac{(g \, \Delta t e^{\chi})^{k}}{k!} e^{-g \Delta t e^{\chi}} \cdot \frac{(r \, \Delta t e^{-\chi})^{\ell}}{\ell!} e^{-r \Delta t e^{-\chi}}.$$

The parameter χ, which is the conjugate of number difference $k - \ell$ in a small time interval Δt, is a measure of how closely the conjugate distribution $f^{\#}$ resembles the original distribution f. In view of our discussion in Chapter 4 of conjugate distributions, we see a general pattern being followed: we acquire new information about our system and modify the probabilities accordingly. A parameter is introduced that depends upon the "state of nature," or what we have referred to as the reservoirs of our system, and we want to estimate the optimal values of χ based on our observations. The parameter χ is intimately related to a nonequilibrium constraint imposed upon the system or, what is equivalent, a spontaneous fluctuation caused by random perturbations of a thermal origin. Because the process is Markov, we cannot distinguish between the two: the system does not recall how it got into its nonequilibrium state.

In order to estimate χ, we may turn to the method of maximum likelihood described in §4.3. The likelihood function is the ratio of the conjugate to the prior distribution, and its logarithm is

$$\mathcal{L}(\chi) = \chi(k - \ell) - \left[g \left(e^{\chi} - 1 \right) + r \left(e^{-\chi} - 1 \right) \right] \Delta t. \tag{8.4}$$

Taking the derivative of the log-likelihood function with respect to the parameter χ and setting it equal to zero give the likelihood equation

$$\frac{\partial \mathcal{L}}{\partial \chi} = (k - \ell) - \left[g e^{\chi} - r e^{-\chi} \right] \Delta t = 0.$$

Solving for the maximum likelihood value of the parameter χ, we obtain

$$\chi^{*} = \ln \left\{ \frac{(k - \ell) + \sqrt{(k - \ell)^2 + 4rg(\Delta t)^2}}{2g \, \Delta t} \right\}. \tag{8.5}$$

Introducing this into (8.4) gives the maximum log-likelihood function

$$\mathcal{L}(\chi^*) \approx \left[\frac{(k-\ell)}{\Delta t} - (g-r) \right] \chi^* \, \Delta t$$

for small χ^*.

Significant values of the maximum log-likelihood function imply large deviations from the expected result $g - r$. But this is just the deterministic rate so that we can identify $(k-\ell)/\Delta t$ with the velocity $\dot{n}$. Then, as a first approximation to the logarithm in the expression for χ^*, we may replace it by the mean of its upper and lower bounds, namely

$$\ln \left(\frac{x}{y} \right) \approx \frac{1}{2} \left(\frac{x-y}{x} + \frac{x-y}{y} \right) = \frac{x^2 - y^2}{2xy}.$$

The approximation is better the closer x/y is to unity. Introducing the change of notation together with this approximation into the maximum log-likelihood estimate results in

$$2\mathcal{L}(\chi^*) \approx \frac{[\dot{n} - (g-r)]^2}{g} \Delta t.$$

We may, at this point, wish to apply a chi-square test or criterion for goodness of fit. This test determines the probability that a given set of observations follows the normal law. We will come back to this approximation in our discussion of the diffusion limit. We can proceed in an entirely general way and show how the transition probability can be expressed in terms of a variational principle when a small parameter characterizing the intensity of the fluctuations tends to zero.

In the continuous time approximation, the conjugate distribution, (8.3), can be written in the canonical form

$$f^{\#}(n, n_0; \chi) = \int_{(n,n_0)} \exp \left\{ \int_{T_0}^{T} (\chi \, dn - \mathcal{H}(n,\chi) \, dt) \right\} f(dn), \qquad (8.6)$$

where the symbol $\int_{(n,n_0)}$ stands for the integral over all paths that connect the state n_0 to n. The function

$$\mathcal{H}(n, \chi) = g \left(e^{\chi} - 1 \right) + r \left(e^{-\chi} - 1 \right) \qquad (8.7)$$

will be seen to be analogous to a classical Hamiltonian which generates the canonical equations of motion.

Inverting (8.6), we can express our original distribution as a *conditional* expectation

$$f(n, n_0) = \overline{\left[\exp \left\{ \int_{T_0}^{T} -(\chi \, dn - \mathcal{H}(n,\chi) \, dt) \right\} \Big| n_0 < n(t) < n \right]}^{\#}$$

with respect to the distribution $f^{\#}$, keeping the initial and final states of transition fixed. According to Jensen's inequality, $\overline{e^{-x}} \geq e^{-\bar{x}}$, which says that the center of the mass weights, which lie along the convex curve e^{-x}, lies above or, at most, on the curve. On the strength of Jensen's inequality we have

$$f(n, n_0) \geq \exp\left\{-\overline{\left[\int_{T_0}^{T} (\chi\, dn - \mathcal{H}(n, \chi)\, dt)\bigg| n_0 < n(t) < n\right]^{\#}}\right\}.$$

Now we equate average with most probable values by replacing the condition expectation by the minimum value of the integral,

$$\left\{\int_{T_0}^{T} (\chi\, dn - \mathcal{H}(n, \chi)\, dt)\right\}_{\min}, \tag{8.8}$$

subject to the given endpoint conditions, since this will make the probability greatest.

It is important to bear in mind that (8.8) is *no* longer a functional of the fluctuating paths. It is a function only of the endpoints of transition. The function $n(t)$ appearing in the expression is a continuous and, at least, twice-differentiable function which makes the functional in brackets a minimum subject to the given endpoint conditions. The equivalence of the conditional expectation with the most probable value thus yields the asymptotic equivalence,

$$f(n, n_0) \asymp \exp\left\{-\left[\int_{T_0}^{T} (\chi\, dn - \mathcal{H}(n, \chi)\, dt)\right]_{\min}\right\}, \tag{8.9}$$

between the transition probability and the minimum of the so-called thermodynamic action.

This asymptotic equivalence is rigorously established, in general, by allowing a small parameter, characterizing the intensity of the fluctuations, to tend to zero or, in particular, in the strict Gaussian limit, where there are no conditions on the characteristic parameter. This is because we are already in the limit of small fluctuations. The Gaussian limit was dealt with by Lars Onsager and Stefan Machlup in 1953, while the asymptotics of the probabilities, given in terms of the Onsager–Machlup function, were only derived more recently by the Russian mathematicians A. D. Wentzel and M. I. Freidlin for non-Gaussian, Markov processes in what we have designated as the "thermodynamic limit."[1] This limit is achieved by allowing Boltzmann's constant, which characterizes the intensity of the fluctuations, to tend to zero. In the limit there will be two paths that make the term in brackets a minimum, analogous to the classical equation of motion which renders the action an extremum.

If we consider the term within the brackets of (8.8) an action, then we may define a Lagrangian according to

$$L(n, \dot{n}) = \max_{\chi}[\dot{n}\chi - \mathcal{H}(n, \chi)]. \tag{8.10}$$

---

[1]This limit should be kept completely distinct from the conventional notion of the thermodynamic limit which we defined in §7.1.

The maximum property is a consequence of the fact that the Hamiltonian $\mathcal{H}$ is a convex function of χ. The stationary condition,

$$\dot{n} = \frac{\partial \mathcal{H}}{\partial \chi} = ge^{\chi} - re^{-\chi}, \tag{8.11}$$

determines the equations of motion. Solving for χ, we obtain our maximum likelihood estimate, (8.5).

Now suppose that U is the (quasi-) potential of χ where U is the solution to the equation

$$\mathcal{H}\left(n, \frac{\partial U}{\partial n}\right) = 0. \tag{8.12}$$

We have referred to this as a "dissipation balance condition" because on the paths which satisfy this condition, the two dissipation functions are numerically equal to one another.[2] We will now show that if a solution to the dissipation balance condition (8.12) exists, then

$$U(n) - U(\bar{n}^{s})$$
$$= \left\{ \int_{T_0}^{T} (\chi \, dn - \mathcal{H}(n, \chi) \, dt) \, \middle| \, n(T_0) = \bar{n}^{s}, n(T_1) = n \right\}_{\min} \tag{8.13}$$

for $-\infty \leq T_0 < T_1 \leq \infty$. The equilibrium solution, $\bar{n}^{s}$, satisfies the detailed balance condition

$$g(\bar{n}^{s}) = r(\bar{n}^{s}).$$

For the Hamiltonian given by (8.7), we find there are two values,

$$\chi^{\dagger} = 0 \qquad \text{and} \qquad \chi^{\ddagger} = \ln\left(\frac{r}{g}\right), \tag{8.14}$$

which satisfy the dissipation balance condition, (8.12). In view of the equation of motion (8.11), the first value gives the deterministic rate equation.

---

[2] The dissipation functions are derived from the kinetic equation which we write as a phenomenological relation, $R\dot{n} = \chi$, where R is a generalized resistance. Multiplying both sides by $\dot{n}$, we get $2\Phi(\dot{n}) = \dot{S}$, where $\Phi(\dot{n}) = \frac{1}{2}R\dot{n}^2$ is the Rayleigh–Onsager dissipation function and $\dot{S} = \dot{n}\chi$ is the entropy production. Alternatively, we can write the phenomenological relation in the form $\dot{n} = L\chi$, where $L = R^{-1}$ is a generalized conductance. Multiplying both sides by χ, we get $\dot{S} = 2\Psi(\chi)$, where $\Psi(\chi) = \frac{1}{2}L\chi^2$ is another dissipation function which Landau and Lifshitz refer to as the generating function. Although this may cause some confusion with a moment generating function, the terminology is so widespread that we will adhere to it.

Whereas the dissipation function Ψ is a function of state—a function of n—the dissipation function Φ is a function of its rate of change—a function of $\dot{n}$. Since both are equal to the entropy production in an isolated system, they are numerically equal. However, irreversible thermodynamics does not take into consideration the thermal source of noise and this is responsible for their numerical equivalence. When thermal fluctuations are taken into account, this numerical equivalence no longer holds in general. However, the numerical equivalence of the dissipation functions can be achieved in particular instances which single out optimal paths of evolution. Necessarily the deterministic path belongs to this set.

Its solution maximizes overwhelmingly the transition probability, since both terms in the action vanish. However, it tells us nothing about the magnitude of the fluctuations from equilibrium and can be excluded by requiring that the thermodynamic force, $\partial U / \partial n \neq 0$ for $n \neq \bar{n}^s$.

The second value of χ in (8.14) gives the mirror image solution to the deterministic rate equation when it is substituted into (8.11). This also justifies our choice of endpoints in (8.13), for surely the aged system must have been at equilibrium at some distant time in the past. From (8.10) it follows that

$$
\begin{aligned}
\int_{T_0}^{T} L(n, \dot{n})\, dt &\geq \int_{T_0}^{T} \frac{\partial U}{\partial n} \dot{n}\, dt - \int_{T_0}^{T} \mathcal{H}\left(n, \frac{\partial U}{\partial n}\right) dt \\
&= U(n) - U(\bar{n}^s) \\
&= \int_{\bar{n}^s}^{n} \ln\left(\frac{r}{g}\right) dn,
\end{aligned}
$$

since the second integral vanishes according to (8.12). Writing this last equality as

$$
U(n) - U(\bar{n}^s) = \int_{T_0}^{T} (r - g) \ln\left(\frac{r}{g}\right) dt,
$$

we can appreciate that it is always positive, regardless of whether $n > \bar{n}^s$ or $n < \bar{n}^s$. It is for this reason that U has been referred to as a quasi-potential.

Observe that our initial condition has wiped out all statistical correlations between the endpoints of transition since one of these states coincides with thermodynamic equilibrium. Hence, the probability of a fluctuation from thermodynamic equilibrium can be expressed in terms of a difference in a function of state. Not knowing exactly when the system has reached the state of equilibrium, we have to take the limit as $T_0 \to -\infty$.

The monotonic increase in the quasi-potential can only be due to the same property possessed by the entropy, namely,

$$
U(n) - U(\bar{n}^s) = S(\bar{n}^s) - S(n) + \left(\frac{\partial S}{\partial \bar{n}^s}\right)_V (n - \bar{n}^s) \geq 0.
$$

Differentiating both sides with respect to n results in

$$
\frac{\partial U}{\partial n} = -\left\{ \frac{\partial S}{\partial n} - \left(\frac{\partial S}{\partial \bar{n}^s}\right)_V \right\} = \chi^{\ddagger}. \tag{8.15}
$$

This identifies $\chi^{\ddagger}$ as the *negative* of the thermodynamic force, or what has often been referred to as an affinity by Théophile De Donder and his followers.

In an analogous way that the thermodynamic force is a measure of the tendency of the system to *seek* equilibrium, $\chi^{\ddagger}$ is a measure of the system's ability to *deviate* from equilibrium. Hence, it is a measure of the strength of the fluctuations. By virtue of the relation between the quasi-potential and entropy we identify $U(\bar{n}^s) = W/T$, where W the grand-canonical potential that we defined in §3.6.

The initial condition can be used to discriminate between the two solutions (8.14) that represent optimal paths of evolution. The initial condition, $T_0 = -\infty$, excludes the deterministic solution to the rate equation and gives the asymptotic expression

$$
\begin{aligned}
f(n, \bar{n}^s) &\asymp \exp\left\{-\frac{1}{k}[U(n) - U(\bar{n}^s)]\right\} \\
&= \exp\left\{-\int_{\bar{n}^s}^{n} \chi^{\dagger} \, dn\right\} \\
&= \exp\left\{\frac{1}{k}\left[S(n) - S(\bar{n}^s) - \left(\frac{\partial S}{\partial \bar{n}^s}\right)_V (n - \bar{n}^s)\right]\right\} \quad (8.16)
\end{aligned}
$$

for the spontaneous growth of a fluctuation from equilibrium. We have reinstated Boltzmann's constant k in order to emphasize the fact that Gauss' principle is obtained as a logarithmic equivalence

$$
\ln f(n, \bar{n}^s) \sim \frac{1}{k}\left[S(n) - S(\bar{n}^s) - \left(\frac{\partial S}{\partial \bar{n}^s}\right)_V (n - \bar{n}^s)\right]
$$

in the "thermodynamic" limit as $k \downarrow 0$. In other words, Gauss' principle emerges from the full transition probability when the initial condition has been chosen such that the system must have surely been at equilibrium at some distant time in the past *and* in the thermodynamic limit as the characteristic parameter k, measuring the intensity of the fluctuations, tends to zero. The initial condition that the system was at equilibrium at some very distant time in the past has excluded the deterministic solution to the rate equation and wiped out all correlations since no state can be correlated statistically with the equilibrium state. We reiterate that for Gaussian fluctuations in which the rate equations are linear, the thermodynamic limit is not necessary since we are already in the small fluctuation limit.

8.3   Nonequilibrium Radiation

As an illustration of how the analysis of optimal paths of evolution can be applied in practice, we consider the phenomenon of nonequilibrium radiation. The radiation mechanism which we will use has been amply discussed in §2.4 and generalized in §7.4. For the sake of simplicity, we shall neglect the nonlinear effects related to two-photon absorption.

The radiation mechanism specifies a rate of recombination due to photon absorption $r = \alpha n$ while the rate of generation that results from the combined processes of stimulated and spontaneous emission is $g = \beta n + \gamma$. The difference between the average rates of absorption and emission is the average rate of change of the number of photons with time, viz.,

$$
\dot{\bar{n}}(t) = \gamma - (\alpha - \beta)\bar{n}(t).
$$

The general solution of this average equation of motion is

$$\bar{n}(t) = \bar{n}_0 e^{-t/\tau} + \bar{n}^s \left(1 - e^{-t/\tau}\right),$$

where $\bar{n}_0$ is the inital number of photons at time $t = 0$ and $\tau = (\alpha - \beta)^{-1}$ is the characteristic relaxation time. If all atoms are in their ground state at $t = 0$, $\bar{n}_0 = 0$. Let us also remember that in order for the stationary solution, $\bar{n}^s = \gamma/(\alpha - \beta)$, to be equivalent to Planck's law, we have to set $\alpha/\beta = e^{h\nu/T}$ and $m = \gamma/\beta$, which is the number of states in the given frequency interval.

With the rate expressions for the Einstein radiation mechanism, we find the optimal value of the force for deviations from equilibrium to be given by

$$
\begin{aligned}
-\chi^\ddagger &= \ln\left(\frac{\beta n + \gamma}{\alpha n}\right) = \ln\left(\frac{m + n}{n}\right) - \frac{h\nu}{T} \\
&= \frac{\partial S}{\partial n} - \left(\frac{\partial S}{\partial \bar{n}^s}\right)_V,
\end{aligned}
\tag{8.17}
$$

where the stochastic entropy is given by Boltzmann's principle,

$$S(n) = \ln\binom{m + n - 1}{n},$$

and provided there are enough photons and modes of the electromagnetic field present so as to justify Stirling's approximation. Integrating (8.17) between the limits $\bar{n}^s$ and n and introducing the result into the second line of (8.16) give Gauss' principle on the third line.

Solving (8.17) for n in terms of the optimal value of the force $\chi^\ddagger$ leads to

$$n = \frac{m}{e^{h\nu/T - \chi^\ddagger} - 1},\tag{8.18}$$

which can be considered as the nonequilibrium generalization of Planck's law. Expression (8.18) has the same form as the generalized Planck law for luminescent radiation where the chemical potential divided by T replaces the optimal value of the force, $\chi^\ddagger < 0$. The requirement that the optimal value of the force be less than zero, necessitating that the rate of emission should be greater than the rate of absorption, follows from considerations of the spectrum in the long wavelength limit. For in the region where $h\nu < -\chi^\ddagger$, the energy spectrum differs quite distinctly from the equilibrium Planck spectrum. Going to the asymptotic limit of very small frequencies, the energy density $\rho(\nu) = \alpha\nu^3/(e^{-\chi^\ddagger} - 1)$, instead of the Rayleigh–Jeans law, $\rho(\nu) = \alpha\nu^2 T$, where $\alpha = 8\pi/c^3$.

The system is still in thermal equilibrium because we can still talk about an equilibrium between rays of different frequencies, all having a common temperature T. Due to the tiny speck of carbon in the cavity whose perfectly reflecting walls are maintained at a uniform temperature T, there can be an energy exchange between the different frequency intervals in an analogous way that material can pass between different phases of the same substance when a phase equilibrium has been established [cf. §7.3]. The scattering of thermal radiation by electrons, referred to as Compton scattering, is known to modify the Planck spectrum. The energy imparted to the electrons increases their temperature so that the temperature of the electrons is always much greater than the radiation temperature.[3] The scattering process redistributes the photons among the different modes so that the system can no longer be considered to be in equilibrium. This is described by a modified Planck spectrum of the form (8.18).

Such processes of energy liberation may have been important in the early stages of the universe. That no such deviations have been observed in the relic-radiation spectrum makes it doubtful that such types of scattering processes were important in the early evolution of the universe. In any event, it places an upper limit on how large $-\chi^{\ddagger}$ can be.[4]

For small deviations from equilibrium, we can expand (8.18) in powers of $\chi^{\ddagger}$ and limit ourselves to linear terms. We then obtain

$$n = \bar{n} + \left(\frac{\partial n}{\partial \chi^{\ddagger}}\right)_{\chi^{\ddagger}=0} \chi^{\ddagger} \tag{8.19}$$

$$= \bar{n}^s + \left(\bar{n}^s + \frac{(\bar{n}^s)^2}{m}\right)\chi^{\ddagger}, \tag{8.20}$$

where the coefficient of $\chi^{\ddagger}$ is the dispersion in the number fluctuations, σ^2. We may use (8.20) to calculate the total radiant energy in a cavity of volume V.

If $\chi^{\ddagger}$ is frequency independent, we find

$$E = E_{\text{black}}\left[1 + \frac{\zeta(3)}{\zeta(4)}\chi^{\ddagger}\right] = E_{\text{black}}\left[1 - 1.11|\chi^{\ddagger}|\right],$$

where $E_{\text{black}} = 8\pi^5 VT^4/15(hc)^3$ and ζ is the Riemann zeta function. Hence, we will expect that there will be a *decrease* in the total radiant energy due to the external constraint since $\chi^{\ddagger} < 0$. Likewise, the total number of photons will be decreased by an amount

$$N = N_{\text{black}}\left[1 + \frac{\zeta(2)}{\zeta(3)}\chi^{\ddagger}\right] = N_{\text{black}}\left[1 - 1.37|\chi^{\ddagger}|\right].$$

---

[3]We saw in §7.2.3 that if electrons were to be in equilibrium with thermal radiation, then it entailed a modification of their Maxwellian distribution such that equipartition is not generally valid except in the long wavelength limit. If collisions between electrons predominate, then this process will thermalize their distribution leading to a Maxwellian distribution where equipartition applies. We can expect that there will be a difference in the temperature of the electrons and the thermal radiation.

[4]Data taken from the Far Infrared Absolute Spectrometer aboard the COBE (Cosmic Background Explorer) satellite show that present agreement of the long wavelength data with Planck's law sets an upper bound of 0.009 on $-\chi^{\dagger}$.

Finally, in the range where (8.20) is valid, the negative binomial distribution reduces to the normal distribution

$$f(n) = \frac{1}{\sqrt{2\pi\sigma^2}} \exp\left\{ -\frac{(n - \bar{n}^s)^2}{2\sigma^2} \right\},$$

and all the results of linear response theory follow. The normal form of the density is a hallmark that we are in the small fluctuation limit.

8.4　The Diffusion Limit

As we have already mentioned, the parameter χ is a measure of the intensity of the thermal fluctuations. In the limit of small χ, we arrive at the continuous diffusion limit which is of interest because explicit irreversible thermodynamic principles can be obtained.

Expanding the Hamiltonian in (8.7) to second-order in the small parameter χ gives

$$\mathcal{H}_{\text{diff}} = b\chi + \frac{D}{2}\chi^2, \tag{8.21}$$

where $b = g - r$ is the drift and $D = g + r$ is the diffusion coefficient. If we look back at our master equation approach of §7.2.3, we can appreciate that this is equivalent to the Fokker–Planck approximation, which breaks off the infinite series at second-order terms.

Differentiating (8.21) with respect to χ gives the equation of motion

$$\dot{n} = \frac{\partial \mathcal{H}_{\text{diff}}}{\partial \chi} = b + D\chi, \tag{8.22}$$

which has the form of a linear Langevin equation with χ playing the role of a fluctuating force. The optimal values of the force, χ, are obtained from the dissipation balance condition, (8.12). Upon setting (8.21) equal to zero, or what is equivalent,

$$\mathcal{H}_{\text{diff}} = \frac{1}{2D}\left(\dot{n}^2 - b^2\right) = 0, \tag{8.23}$$

we obtain the optimal values

$$\chi^\dagger = 0 \qquad \text{and} \qquad \chi^\ddagger = -2\frac{b}{D}. \tag{8.24}$$

In view of (8.23), these optimal values of the force correspond to

$$\dot{n} = b \qquad \text{and} \qquad \dot{n} = -b,$$

respectively.

The first value of the optimal force corresponds to the deterministic path, while the second corresponds to its mirror image in time. In the multidimensional case, the mirror-image symmetry in time is only preserved when $\chi^\ddagger$ is the gradient of a scalar potential. It is for this reason that we insisted that $\chi^\ddagger$ be derived from the quasi-potential, U. The connection between the non-vanishing optimal force $\chi^\ddagger$ given in expressions (8.14) and (8.24) is made by approximating the first expression, $\ln(r/g)$, by the average of its upper and lower bounds so that

$$\chi^\ddagger = \ln\left(\frac{r}{g}\right) \approx \frac{1}{2}(r - g)\frac{r + g}{rg} \approx -2\frac{b}{D},$$

which is precisely the second expression (8.24). The second of the two approximations comes from evaluating $(r + g)/rg$ at the equilibrium state, $\bar{n}^s$.

We have already limited ourselves to the domain of small fluctuations by breaking off the series expansion in (8.21) at second-order. In order that the two terms be of equal magnitude, the drift should be linearized to $b = [g'(\bar{n}^s) - r'(\bar{n}^s)]\Delta n$, where $\Delta n \equiv n - \bar{n}^s$. This requires the diffusion coefficient to be evaluated at $\bar{n}^s$, namely, $D = r(\bar{n}^s) + g(\bar{n}^s)$. However, the decisive property of diffusion processes is that their transition probability is, under certain regularity assumptions, determined uniquely by the drift and diffusion coefficient—whatever they may be, linear or nonlinear. This is an extraordinary property of diffusion processes inasmuch as the first two moments, b and D, are all that is required to specify the transition probability. Ordinarily, a knowledge of all the moments is equivalent to a knowledge of the distribution function.

Solving (8.22) for χ and introducing it together with (8.21) into (8.9) result in

$$f(n, n_0) \asymp \exp\left\{-\left[\int_{T_0}^{T} \frac{(\dot{n} - b)^2}{2D} \, dt\right]_{\min}\right\}. \tag{8.25}$$

It is clear that we are not dealing with optimal paths since (8.21) does not vanish. This is due to the presence of noise. If the drift is linear and the diffusion coefficient is constant, we can do better than an asymptotic equivalence. For then the normalization constant is a function only of the time interval, $T - T_0$, and we would get an exact expression. This is the Gaussian limit, where the means and the modes of the distribution are equal. The transition probability was originally proposed in this form by Onsager and Machlup almost 40 years ago for Gaussian–Markov processes. By expanding the exponent in (8.25) and identifying the expressions in terms of the dissipation functions and entropy production [cf. footnote 1], they hoped to characterize the transition probability in terms of the "principle of least dissipation of energy" in an analogous way that the probability density is characterized by the "principle of maximum entropy." This desideratum can already be found in Onsager's 1931 papers. However, the time was not ripe, and it took another 20 years for the mathematical theory of stochastic processes to develop to the point where Onsager could generalize the phenomenological relations to include a stochastic element. Apparently, it was no accident that the Onsager and Machlup papers appeared in the same year as J. L. Doob's book on stochastic processes.

However, expression (8.9) indicates that we can go beyond the purely linear range of Gaussian processes and show that an asymptotic equivalence is established as a small parameter measuring the intensity of the fluctuations tends to zero. This small parameter is Boltzmann's constant, k, which is incorporated in to the expression of the diffusion coefficient by Einstein's formula, $D = kT/R$, where R is a generalized resistance. More recently, Wentzell and Friedlin have shown how the non-Gaussian, Onsager–Machlup formulation can be converted into asymptotic limit theorems which form a part of large deviation theory. Large deviation theory considers a diffusion process in which the variance contains a small parameter, which we have identified thermodynamically as Boltzmann's constant, and studies the asymptotic limit as this parameter tends to zero. In doing so, upper and lower limits can be found for the logarithm of the transition probability in terms of the entropy.

Thermal noise, modeled as Brownian motion, has no effect upon optimal paths determined by the dissipation balance condition, (8.23). In the Gaussian limit, this is true without having to take the limit as Boltzmann's constant tends to zero. This is just another way of stating Onsager's regression hypothesis:

> The decay of a system from a given nonequilibrium state produced by a spontaneous fluctuation obeys, *on the average*, the (empirical) law for the decay from the same state back to equilibrium, when it has been produced by a constraint which is then suddenly removed.

In addition to the Markov assumption, Onsager's hypothesis relies on the fact that the process is Gaussian, for only then will the *linear* deterministic rate equation coincide with the average of the stochastic equation. Linearity of the stochastic equation implies that the process is Gaussian if the noise is "white." In this case there is no average effect of the fluctuating force. However, Onsager's hypothesis cannot be generalized to non-Gaussian fluctuations which are described by *nonlinear* stochastic equations, because the average of a function will not be equal to the function of the average in the presence of fluctuations. In this case, we are compelled to go to the asymptotic limit, where irreversible thermodynamic principles come into play.

In this limit we obtain not only the deterministic path, satisfying $\dot{n} = b$, but also the mirror image path which is a solution of $\dot{n} = -b$, provided we fix the initial condition at some very distant time in the past where the aged system must have certainly been at equilibrium. In general, the paths of a diffusion process are not optimal paths in the sense described above. However, we can determine the statistical correlations between any two nonequilibrium states through which the system passes at successive instants in time. In doing so, we will find a quasi-potential for diffusion processes.

We again resort to the method of statistical estimation. Our prior knowledge is summarized by the Wiener transition probability density[5]

$$f(n, n_0; \tau) = \frac{1}{\sqrt{2\pi D\tau}} \exp\left\{-\frac{(n-n_0)^2}{2D\tau}\right\} \tag{8.26}$$

subject to the usual delta function condition at the initial instant in time and we have set $\tau = T - T_0$ since the process is homogeneous. This is to say that without any additional information about the nature of the process, we assume that it is a "free" diffusion process. We now make an observation by noting the transition that occurs in a given time interval which is a time-of-flight method for measuring the velocity. Our original diffusion process, described by the stochastic differential equation

$$dn = dw, \tag{8.27}$$

where w is a Wiener process, is thus converted into a new process which is described by the stochastic differential equation

$$d\tilde{n} = b + dw. \tag{8.28}$$

Necessarily, it has the same initial condition, $n(T_0) = n_0$, as the free diffusion process.

The presence of the drift, b, induces a transformation of the free diffusion process into a "biased" diffusion process that is described by the conjugate distribution

$$F^{\#}(n, n_0; \tau) = \int_{(n,n_0)} \exp\left\{\int_{T_0}^{T} (\chi \, d\tilde{n} - \mathcal{H}_{\text{diff}} \, dt]\right\} F\{dn\},$$

where F is the distribution whose density is (8.26). Since F possesses a density and the conjugate distribution, $F^{\#}$, is absolutely continuous with respect to it,[6] it too will have a density. The probability density is

$$f^{\#}(n, n_0; \tau) = \overline{\left[\exp\left\{\int_{T_0}^{T} (\chi \, d\tilde{n} - \mathcal{H}_{\text{diff}} \, dt)\right\} \middle| n_0 < n(t) < n\right]}.$$

Introducing (8.28) and noting the expression for the Hamiltonian (8.21) give

$$f^{\#}(n, n_0; \tau) = \overline{\left[\exp\left\{\int_{T_0}^{T} (\chi \, dn - \frac{D}{2}\chi^2 \, dt)\right\} \middle| n_0 < n(t) < n\right]}. \tag{8.29}$$

The difference between ordinary and Brownian motion calculus is that upon taking the differential of a smooth function V, say, it is necessary to retain terms up to second-order in the power series and replace $(dn)^2$ by its conditional expectation, $D \, dt$.

---

[5]This expression was first derived by Einstein in his celebrated 1905 paper on Brownian motion.

[6]This is ensured by the fact that both processes possess the same diffusion coefficient.

This was first appreciated by Einstein, who noted that $\Delta n/\Delta t$ is of order $1/\sqrt{\Delta t}$ and as $\Delta t \to 0$, it becomes infinite. Although the paths of a Brownian particle are continuous, they are nevertheless not differentiable, for the closer we look at the motion of a Brownian particle, the more zig-zag it becomes.

Thus, if

$$\frac{1}{k}\frac{\partial V}{\partial n} = \chi, \tag{8.30}$$

then

$$\begin{aligned} dV &= \frac{\partial V}{\partial n}\,dn + \frac{1}{2}\frac{\partial^2 V}{\partial n^2}(dn)^2 \\ &= \frac{\partial V}{\partial n}\,dn + \frac{D}{2}\frac{\partial^2 V}{\partial n^2}\,dt, \end{aligned}$$

or equivalently,

$$V(n) - V(n_0) = k \int_{T_0}^{T} \left(\chi\,dn + \frac{D}{2}\chi'\right) dt, \tag{8.31}$$

where the prime stands for differentiation. Introducing this rule into our expression for the transition probability density, (8.29), results in

$$f^{\#}(n, n_0; \tau) = \exp\left\{\frac{1}{k}[V(n) - V(n_0)]\right\} \cdot \mathcal{K}(n, n_0; \tau), \tag{8.32}$$

where the kernel

$$\mathcal{K}(n, n_0; \chi) = \overline{\left[\exp\left\{-\int_{T_0}^{T} \mathcal{O}\,dt\right\}\bigg| n_0 < n(t) < n\right]} \tag{8.33}$$

is given in terms of the so-called Onsager–Machlup potential

$$\mathcal{O} = \frac{D}{2}\left\{\chi^2 + \chi'\right\}.$$

Now for Gaussian processes, χ' is a constant, and hence the last term is proportional to the time interval. It is noteworthy that this term provides the precise normalization constant for the transition probability density. As we mentioned earlier, in the non-Gaussian case all we can do is obtain an asymptotic equivalence as $k \downarrow 0$.

The kernel (8.33) is in a standard form, known as the Feynman–Kac formula, and it is well known that it can be "tamed" by solving the diffusion equation

$$\frac{\partial \mathcal{K}}{\partial \tau} = \frac{D}{2}\mathcal{K}'' - \mathcal{O}\mathcal{K}.$$

The solution can be expressed as a bilinear sum,

$$\mathcal{K}(n, n_0; \tau) = \sum_{j=0}^{\infty} \psi_j(n)\psi_j(n_0)\,e^{-\lambda_j \tau},$$

in terms of normalized eigenfunctions, ψ_i with a discrete spectrum of eigenvalues, λ_j. In this way, an integral over a space of functions has been domesticated by turning it into a purely classical quantity.

With the passage of time, all terms in the sum will decay and, in the end, we will be left with only the eigenvalue $\lambda_0 = 0$. The solution to

$$\psi_0'' - (\chi^2 + \chi')\psi_0 = 0$$

is clearly seen to be $\psi_0 \propto \exp(V/k)$. Consequently, in the asymptotic time limit we have

$$k \lim_{\tau \to \infty} \ln \mathcal{K}(n, n_0; \chi) = V(n) + V(n_0) + \text{const.} \tag{8.34}$$

In the same limit the transition probability density transforms into

$$\lim_{\tau \to \infty} f^{\#}(n, n_0; \tau) = f(n) = \exp\{2V(n)/k + \text{const.}\}, \tag{8.35}$$

which says that given a long enough time all physical systems tend to "forget" their past. The asymptotic form of the kernel (8.34) is responsible for wiping out all information regarding the initial state of the system. It is clear from (8.32) that the kernel is what accounts for the statistical correlations between nonequilibrium states that are not widely separated in time. The fact that the smallest eigenvalue is zero ensures that the correlations decay, *on the average*, in a completely monotonic way.

The direction of evolution is determined by the first factor in (8.32). For in order that there be an asymptotic equivalence between the invariant probability density, $f(n)$, apart from a normalizing factor, the quasi-potential must be given by

$$V(n) = \frac{1}{2}\left\{S(n) - \left(\frac{\partial S}{\partial \bar{n}}\right)_V\right\}. \tag{8.36}$$

Consequently, the force (8.30) is

$$\chi = \frac{1}{2k}\left\{\frac{\partial S}{\partial n} - \left(\frac{\partial S}{\partial \bar{n}}\right)_V\right\} = \frac{1}{2}\chi_{\text{thermo}}, \tag{8.37}$$

which is precisely $\frac{1}{2}$ of the thermodynamic force, χ_{thermo}, which is a measure of the tendency of the system to evolve to equilibrium. The reason for the one-half is the following.

According to (8.15) and the second solution in (8.24), the drift will be given by

$$b = \frac{D}{2k}\left\{\frac{\partial S}{\partial n} - \left(\frac{\partial S}{\partial \bar{n}}\right)_V\right\}. \tag{8.38}$$

Therefore, in view of the equation of motion (8.22), we obtain the phenomenological relation

$$\dot{n} = \frac{D}{k}\left\{\frac{\partial S}{\partial n} - \left(\frac{\partial S}{\partial \bar{n}}\right)_V\right\} = \frac{D}{k}\chi_{\text{thermo}}, \tag{8.39}$$

where the ratio D/k is the temperature times a phenomenological mobility coefficient, independent of k. In other words, $b = D\chi$ and each is proportional to one-half of the thermodynamic force. The phenomenological relation does not coincide with either the deterministic rate equation, $\dot{n} = b$, or its mirror image in time, $\dot{n} = -b$, and hence its solution is not an optimal path in the sense described above.

Introducing (8.36) into (8.35) gives

$$f(n) = \exp\left\{\frac{1}{k}\left[S(n) - \left(\frac{\partial S}{\partial \bar{n}}\right)_V n + \text{const.}\right]\right\}. \tag{8.40}$$

Recalling our discussion in §1.8, the correct normalizing factor of $f(n)$ is given by

$$f(\bar{n}) = \exp\left\{\frac{1}{k}\left[S(\bar{n}) - \left(\frac{\partial S}{\partial \bar{n}}\right)_V \bar{n} + \text{const.}\right]\right\}, \tag{8.41}$$

so that the ratio of the two is none other than Gauss' principle, (8.16).

8.5 Gauss' Principle Vindicated

We have thus run the full gamut and returned to our starting point: a comparison between Boltzmann's and Gauss' principles as we have applied it to thermodynamics. To accomplish our goal, we must recast Boltzmann's principle in the form

$$\frac{f(n)}{f(\bar{n})} = \frac{\Omega(n)}{\Omega(\bar{n})} = \exp\left\{\frac{1}{k}[S(n) - S(\bar{n})]\right\} \tag{8.42}$$

in order to relate the thermodynamic probability to an actual probability. In this section, we shall refer to (8.42) as Boltzmann's principle instead of the more generic form $S(n) = k\ln\Omega(n)$. We will show that while the latter retains a meaning precisely *because of* Gauss' principle, the former is devoid of any physical content.

For to argue in favor of Boltzmann's principle, (8.42), would mean that we would have to consider our system in contact with a heat reservoir at *infinite* temperature. The entropy of such a system would no longer show the tendency to increase, since the entropy curve is now parallel to the energy axis. We now want to show that this conclusion, which Fowler has termed as "rather clumsy," is a necessary consequence of the hypothesis of equal *a priori* probabilities.

As an illustration, consider the negative binomial distribution which we have implicated in the mechanism of black radiation. The invariant density is

$$f(n) \propto \binom{m+n-1}{n}\left(\frac{\bar{n}}{m+\bar{n}}\right)^n$$

$$= \exp\left\{\frac{1}{k}\left[S(n) - \left(\frac{\partial S}{\partial \bar{n}}\right)_V n\right]\right\}$$

provided Stirling's approximation holds. Now, in order to eliminate the *a priori* probability $\bar{n}/(m+\bar{n})$, we would have to let $T \to \infty$. This would be the only way to justify the hypothesis of equal *a priori* probabilities.

Furthermore, the normalization constant is easily worked out to be

$$\sum_{n=0}^{\infty} \binom{m+n-1}{n} \left(\frac{\bar{n}}{m+\bar{n}}\right)^n = \left(\frac{m+\bar{n}}{m}\right)^m$$

$$= \exp\left\{\frac{1}{k}\left[S(\bar{n}) - \left(\frac{\partial S}{\partial \bar{n}}\right)_V \bar{n}\right]\right\}.$$

This ensures that the ratio $f(n)/f(\bar{n})$ is a proper fraction, as it must be. Setting the invariant density equal to the negative binomial coefficient throws probability theory to the wind, and it does not employ the fundamental property of the *concavity* of the entropy when Boltzmann's principle is invoked.

In Boltzmann's formulation, the thermodynamic probability is to be maximized, which fixes the most probable state of the system by itself. Then setting the logarithm of the thermodynamic probability equal to the entropy relates the system to the outside world, and finally, the absolute temperature scale is introduced by the derivative $\partial S/\partial \bar{E} = 1/T$. But when we compare this procedure with an actual probability distribution, we come to the conclusion that we cannot define a finite temperature in the last step because it contradicts the initial hypothesis that the *a priori* probabilities are equal.

Boltzmann's principle, (8.42), expresses the fact that the entropy of the equilibrium state is maximum with respect to all those nonequilibrium states which are compatible with the constraints of constant total energy and particle number. In order to eliminate the linear term in the Taylor series expansion of the entropy, one must consider a composite system and use the closure conditions of the extensive variables [cf. §1.2].

Gauss' principle identifies the average as the most probable value, but the entropy of the average need not be the maximum entropy. Recall that Gauss' principle is based on the concavity property of the entropy so it is immaterial which entropy is greater than the other. If $S(n) > S(\bar{n})$, then we have

$$S(\bar{n}) + S'(\bar{n})(n - \bar{n}) > S(n), \tag{8.43}$$

by construction, whereas if $S(n) < S(\bar{n})$, then

$$S(n) - S(\bar{n}) - (n - \bar{n})S'(n)(n - \bar{n}) > 0 \tag{8.44}$$

by interchanging the endpoints. Either one is a valid definition of concavity; the essence of the second law is obtained by adding the two results [cf. §1.8],

$$(n - \bar{n})\left[S'(n) - S'(\bar{n})\right] < 0.$$

This affirms that the slope of a *strictly* concave function is monotonically decreasing.

Suppose the constraints fix the equilibrium value of the entropy at $S(\bar{n})$. This is a constrainted maximum, since necessarily $S'(\bar{n}) > 0$. For any other state we may consider $S(n)$ as the *microcanonical* entropy, as opposed to the *canonical* entropy, $S(\bar{n})$. The microcanonical entropy, $S(n)$, may be larger or smaller than the canonical entropy. When we say that the entropy is a maximum at $\bar{n}$, we mean that it is a maximum subject to the given constraints. The microcanonical entropy has no such extremum condition, except when it coincides with the canonical entropy. Now, when we look at Boltzmann's principle, (8.42), we are struck by the fact that in order for the left-hand side to be a proper probability distribution, we must necessarily have $S(n) \leq S(\bar{n})$ for all values of n, which means that $S(\bar{n})$ is an absolute maximum and not a constrained maximum. But this would imply that $S'(\bar{n}) = 0$, and as we have seen, this would imply $T = \infty$. At finite temperatures we cannot make any claim on the relative magnitude of the difference $S(n) - S(\bar{n})$, and consequently, Boltzmann's principle, (8.42), would appear to have no physical content.

Yet, Boltzmann's principle has been shown to apply to the case where there are equal *a priori* probabilities. In the case of the negative binomial distribution this has no meaning since $S'(\bar{n}) \neq 0$, except asymptotically. At first sight, it may appear that the entropy can have an unconstrained maximum in the binomial distribution. It would occur at $\bar{n} = m/2$ where the entropy, $S(\bar{n}) = m \ln 2$. But according to Fermi–Dirac statistics the occupation index $\bar{n}/m = 1/(e^{(\varepsilon-\mu)/T}+1)$, and if we set it equal to $\frac{1}{2}$, we come out with $e^{(\varepsilon-\mu)/T} = 1$, which can be satisfied asymptotically as $T \to \infty$ *or* when $\mu = \varepsilon_{\mathrm{f}}$, the Fermi energy, which is the energy of the highest occupied momentum state at $T = 0$. So, here again, the entropy cannot have an unconstrained maximum as it would be required if the hypothesis of *a priori* probabilities were valid. It can only occur asymptotically, at the extremes of the entropy versus energy curve, and not at finite temperatures. Therefore, we conclude again that Boltzmann's principle, (8.42), is invalid. It must be replaced by Gauss' principle (8.1) in all cases where processes are occurring at a *finite* temperature.

Bibliographic Notes

This chapter provides the bridge with our earlier work,

- B.H. Lavenda, *Nonequilibrium Statistical Thermodynamics* (Wiley, Chichester, 1985).

It is an extension of our article

- B.H. Lavenda and J. Dunning-Davies, "Kinetic derivation of Gauss' Law and its thermodynamic interpretation," *Z. Naturforsch. A* **45**, 873-878 (1990).

One of the problems that Onsager posed in

- L. Onsager, "Reciprocal relations in irreversible processes II," *Phys. Rev.* **38**, 2265-2279 (1931) was to find a relation analogous to Boltzmann's principle, which relates the state of maximum entropy with maximum probability, for the probability of transition. Since it involved rate processes, Onsager conjectured that the maximum probability of transition should be given in terms of the Rayleigh–Onsager dissipation function, written in a finite difference form. It took a little over two decades until Onsager, with his graduate student Machlup, in

- L. Onsager and S. Machlup, "Fluctuations and irreversible processes," *Phys. Rev.* **91**, 1505-1512 (1953),

to arrive at the correct formulation for Gaussian processes with perturbations of the white noise type. Undoubtedly, Onsager's reading of Doob's recently published book on stochastic processes,

- J.L. Doob, *Stochastic Processes* (Wiley, New York, 1953),

gave him the incentive to realize his old hunch that the transition probability density can be cast in the form of the principle of least dissipation of energy. From a modern viewpoint, this may appear as almost "obvious" but it wasn't then. Onsager's intuition shows a deep appreciation between the stochastic nature of irreversible processes and the thermodynamic principles that govern them.

From a rigorous mathematical viewpoint, Cramér, in

- H. Cramér, "Sur un nouveau théorème limite de la théorie des probabilités," *Acta Sci. Ind.*, 736 (1938),

initiated a study of "large" deviations, or the probabilities of improbable events, using the method of conjugate distributions. In

- A.D. Wentzell and M.I. Freidlin, "On small random perturbations of dynamical systems," *Russ. Math. Surv.* **25**, No. 1, 1-55 (1970)

Wentzell and Freidlin showed that the same quadratic form could be used in the non-Gaussian cases as the small parameter, measuring the intensity of the noise, tends to zero.

More detailed references to the physical literature can be found in our monograph while those to the mathematical literature are given in

- M.I. Freidlin and A.D. Wentzell, *Random Perturbations of Dynamical Systems*, transl. by J. Szücs (Springer, New York, 1984).

A succinct account of large deviation theory, which includes both Cramér's theorem and the Wentzell and Friedlin formulation, is given in

- S.R.S. Varadhan, *Large Deviations and Applications* (Society for Industrial and Applied Mathematics, Philadelphia, PA, 1984).

Index